MESOPOTAMIAN PLANETARY ASTRONOMY-ASTROLOGY

CUNEIFORM MONOGRAPHS 18

Edited by

STYX
PUBLICATIONS
GRONINGEN
2000

CUNEIFORM MONOGRAPHS 18

MESOPOTAMIAN PLANETARY ASTRONOMY-ASTROLOGY

David Brown

STYX
PUBLICATIONS
GRONINGEN
2000

ISBN 90 5693 036 2
ISSN 0929-0052

STYX Publications
Postbus 2659
9704 CR GRONINGEN
THE NETHERLANDS
Tel. # 31 (0)50–5717502
Fax. # 31 (0)50–5733325
E-mail: styxnl@compuserve.com

Contents

Foreword vii

Abbreviations xi

Introduction 1
- I.1 Terminology 7
- I.2 Methodology 8
- I.3 Textual Sources Used in the Period c.750 – c.612 BC 11
- I.4 The Archaeological Context of the Source Material 17
- I.5 The Distribution of the Datable Source Material between c. 750 and 612 BC 23
- I.6 The Languages and Scripts Used in the Textual Sources 29

CHAPTER 1 The Astronomer-Astrologers – the Scholars 33
- 1.1 Familial and Ancestral Relationships 36
- 1.2 The Locations of the Neo-Assyrian and Neo-Babylonian Scholars 39
- 1.3 The Scholars' Relationship to the King 42
- 1.4 Scholarly Interrelationships 48

CHAPTER 2 The Planets and their Ominous Phenomena c. 750–612 BC – Names and Terms 53
- 2.1 The Planet-Names in Cuneiform, c. 750–612 BC 54
  - 2.1.1 Discussion of the Associations between the Planets and their Names 63
  - 2.1.2 "Learned" Associations and Interpreting Chart 2.1 75
- 2.2 The Ominous Phenomena and Configurations 81
  - 2.2.1 A Description of the Celestial Phenomena Afforded by Planets 81
  - 2.2.2 From c. 750 to 612 BC the following Planetary Phenomena were Ominous 85
  - 2.2.3 Comments and Comparisons with the Uninterpreted Observational Records 93

CHAPTER 3 Celestial Divination – The *Enūma Anu Ellil* (EAE) Paradigm 105
- 3.1 Defining the Paradigm 107
  - 3.1.1 Omens 108
  - 3.1.2 Period Schemes 113
  - 3.1.3 Observational Texts 122
  - 3.1.4 Other Related Material 124
- 3.2 Making the Heavens Interpretable with the EAE Paradigm 126
  - 3.2.1 The Rules of Omen Invention 126
  - 3.2.2 The EAE Paradigm Code 139
  - 3.2.3 Categorising the Universe – Variable Reducing and Anomaly Producing 153
- 3.3 Reflections on the EAE Paradigm – Canonisation 156

CHAPTER 4 The Prediction of Celestial Phenomena (PCP) Paradigm 161
4.1 Defining the Paradigm 162
4.1.1 What Phenomena were Predicted by the MAATs and the NMAATs? 163
4.1.2 How were the Predictions Made? 173
4.2 Evidence for the use of the PCP Paradigm between c. 750 and 612 BC 189
4.2.1 Accurate Records of Phenomena 190
4.2.2 Knowledge of Characteristic Planetary Periods 193
4.2.3 Calendar Regulation 195
4.2.4 Pre-612 BC Predictions of Celestial Phenomena 197
4.2.4.1 Prediction of the Movements of the Inferior and Superior Planets 197
4.2.4.2 Predicting Month Lengths and the Dates of Lunar "Opposition" 198
4.2.4.3 Predicting Eclipses 200

CHAPTER 5 A Revolution of Wisdom 209
5.1 From the EAE Paradigm to the PCP Paradigm 209
5.1.1 Internal Considerations 211
5.1.2 External Considerations 219
5.1.3 Philosophical Considerations 227
5.2 Conclusions 239

APPENDIX 1 A Chronological Bibliography of Cuneiform Astrological-Astronomical Texts 245

APPENDIX 2 Comments on the Dating of the Letters and Reports 265

APPENDIX 3 An Analysis of the Published EAE Planetary Omens 279

Bibliography 287

Indices 305
Index of Cuneiform Texts Quoted or Discussed (including only those texts in the SAA volumes quoted at length) 305
Index of Akkadian and Sumerian Words Discussed 312
Subject Index 313

Foreword

This is a study of Mesopotamian cuneiform texts concerned with the seven planets visible to the unaided eye. It incorporates an analysis of texts ranging from the divinatory to the mathematical. The studies are placed in an historical context as far as this is possible. The conclusions derived are, I believe, important both in terms of how cuneiform texts concerned with the sky were understood in their own period, and in terms of the wider issue of science in the ancient world. This work is aimed at both Assyriologists and Historians of Science, and by bridging two disciplines no doubt includes material some of which will appear elementary and other of which overly complex to one or other group.

I wish to be clear about my motivations for writing this book and not hide behind the idea that the study of the history of cuneiform astronomy-astrology is valuable for its own sake. My first degree was in physics and this background put me at a small but significant advantage when it came to studying those fantastically interesting, but often quite difficult mathematical texts that have occasionally been unearthed in Mesopotamia. In due course I undertook a Ph.D. in Assyriology, which Professor Nicholas Postgate kindly agreed to supervise, and decided to work on cuneiform astronomy and celestial divination. As I had had a long-term interest in the historical development of the so-called "exact sciences", I concerned myself with the evolution of the techniques used to predict celestial phenomena and was drawn to the wealth of material dating to the 8th and 7th centuries BC from both Assyria and Babylonia. My studies were funded firstly by a three-year internal graduate studentship and then by a one-year senior Rouse-Ball studentship awarded me by Trinity College, Cambridge. Poverty and marriage encouraged me to complete my thesis on time, though this was helped greatly by my supervisor and by the occasional but crucial suggestions about my work made by Christopher Walker of the British Museum. The result was "Neo-Assyrian and Neo-Babylonian Planetary Astronomy-Astrology (747–612 BC)" on which this book is based.

Over the last three years I have been a Research Fellow at Wolfson College, Oxford, funded by a Postdoctoral Fellowship of the British Academy and have worked on a number of topics including the water clock, celestial units, the eclipse ritual, Hellenistic legal texts, and records concerning the levels of the Euphrates. Each topic has cast new light on my doctoral work and this and my exposure to the world of Assyriology more generally has meant, I hope, that this book is a more mature endeavour than was my Ph.D.

It has long been clear that histories do not stand in isolation from the periods in which they were written, and the influence of the prevailing dominance of liberal bourgeois democracies in the West can be seen in all current studies of cuneiform science. For example, the play of economic forces are seen to explain many things in Assyriology, from the rise of writing to the demise of empires, and this mode of explanation is not excluded here. This is also the age of explicit self-awareness and throughout the following I have tried to be clear about my preconceptions and prejudices when it came to interpreting the texts. This, however, is not all that is needed, I maintain. It is currently fashionable to write a history purportedly for its own sake, but the suspicion remains that this may be only so as to prevent any challenge to prevailing attitudes. Modern histories are supposedly adjusted in order to avoid being centred in the present. While self-awareness of the motivations that lie behind an author's history is entirely laudable, this should mean more than that the author is up-to-speed with contemporary thinking. The present-centred motivations of the author should not be seemingly eliminated, or limited to a few disclaiming sentences in the introduction, but be recognised as permeating the entire work and indeed encouraged. If a work of history challenges the conventions concerning the way in which academia, politics or whatever are done today, so be it. My study challenges, elaborates, refines current thinking on Mesopotamian astrology-astronomy, much as any new study will attempt to do, but in a wider sense it offers a threat to conventional, widespread wisdom on the part that astrology and science are thought to play in our modern democratic society. The challenges to contemporary norms which emerge from my work are two fold. Firstly, celestial divination, which ultimately underpins contemporary astrology, was, I argue, not based on the results of observation, but on assignations of value and hermeneutic elaboration on the part of particular scholars. Secondly, the emergence of the world's first mathematical science of planetary prediction was almost entirely dependent on the structure and conventions of the preceding celestial divination and its fascination with ominous phenomena. The first threatens the claim to age-old empirical vindication of the astrological assignation of value to the planets and constellations, the second questions any claims that the exact sciences from their inception were free from the "constraints" of astrological thinking.

Aside from the desire to write a history that is not neutralised, the self-evident present-centred motivations of needing to complete a Ph.D. on time, to get a job, to continue in employment, and for the sake of pride to do the task well, even to entertain, account for this book. Much that is good in it can also be explained as being the result of the kind assistance of the following scholars (anything that is bad, falsified, or woefully written is my fault alone):

My greatest thanks are reserved for Nicholas Postgate who introduced me to Assyriology and demonstrated the need for insight gleaned from years of exposure to all aspects of the cultures in question. On innumerable occasions he steered me away from excesses and sought out profitable routes that I might explore. Christopher Walker, acknowledged expert in cuneiform divination and astronomy, provided me with much vital information that has greatly enhanced this work. In recent years I have had the privilege of collaborating on a number of projects with him, and he has consistently offered me the wisest of advice. Thanks also go to Joan Oates who taught me during my M.Phil. at Cambridge and who has been extremely helpful on many occasions since. I owe a particular debt to three scholars, whose works inspired much that is in this study. Their respective personal involvement in my Ph.D. at various stages in its development was also significant. Noel Swerdlow's work demonstrated how one further link between the extant observational material and the predictive texts could be made, and this informed my thinking in Chapter 4. He was particularly encouraging during a visit here to the UK while enjoying residency at All Souls, Oxford. Nicholas Denyer of Trinity College, Cambridge acted as my manager for the Rouse-Ball studentship and kindly brought my attention to his own work on the rationale behind divination. This enhanced my study of cuneiform celestial divination in Chapter 3. Sir Geoffrey Lloyd, Master of Darwin College, Cambridge, read my Ph.D. soon after its completion and offered me the benefit of his wide expertise on questions relating to the nature and practice of ancient science. This advice I incorporated particularly into Chapter 5.

I am also grateful to Professor Richard Stephenson and Dr Wilfred van Soldt, my Ph.D. examiners for their many suggestions on that particular day. Finally, many thanks go to Niek

Veldhuis who also read my Ph.D. and made some extremely useful suggestions concerning its layout and presentation as well as numerous comments of a technical nature.

Abbreviations

Abbreviations not listed below can be found in *The Assyrian Dictionary of the Oriental Institute of the University of Chicago = CAD*

Fs.Aaboe	*From Ancient Omens to Statistical Mechanics. Essays on the Exact Sciences Presented to Asgar Aaboe* eds. J.L. Bergren and B.R. Goldstein (1987)
ABABR	de Meis & Hunger (1998)
ABCD	Rochberg (1989a)
ACh.	Virolleaud (1905–12)
ACT	Neugebauer (1955)
AHES	*Archive for History of Exact Sciences*
ALCA	Pedersén (1986)
Fs.Böhl	Symbolae biblicae et Mesopotamicae Francisco Mario Theodoro de Liagre Böhl dedicatae (1973)
Fs.Borger	*Festschrift für Rykle Borger zu seinem 65. Geburtstag am 24.Mai 1994* ed. S.Maul
BPO1/2/3	Reiner & Pingree (1975b, 1981, 1998)
BSMS	*Bulletin of The Society for Mesopotamian Studies (Toronto)*
CAH	*The Cambridge Ancient History*
CAJ	*Cambridge Archaeological Journal*
Diary	Hunger & Sachs (1988, 1989, 1996)
EAE	Enūma Anu Ellil (see Appendix 1 §21)
ef	evening first (see Ch .2.2.1)
el	evening last (see Ch .2.2.1)
Fs.Finkelstein	*Connecticut Academy of Arts and Sciences, Memoir 19. Essays on the Ancient Near East in Memory of J.J. Finkelstein* ed. M.deJ. Ellis (1977)
Graz Vol.	*Die Rolle der Astronomie in den Kulturen Mesopotamiens* 3. Grazer Morgenländischen Symposium (23–27 Sept 1991) ed. Galter H.D. (1993)
GSL	Great Star List - see n170
Fs.Güterbock	*Anatolian Studies Presented to Hans Gustav Güterbock on the Occasion of his 65th Birthday* eds. Bittel K. et al. (1973)
Fs.Hartner	*Prismata (Festschrift für Willy Hartner)* eds. Y. Maeyama & W. Salzer (1977)
HOS	*Harvard Oriental Series*
Fs.Jacobsen	*Sumerological Studies in Honor of Thorkild Jacobsen on his Seventieth Birthday, June 7, 1974* (1976)
JHA	*Journal for the History of Astronomy*
Fs.Kramer	*AOAT 25 Kramer Anniversary Volume* (1976)
Fs.Kraus	*Zikir šumim. Assyriological Studies Presented to F.R. Kraus on the occasion of his Seventieth Birthday* eds. G.van Driel, Th.J.H. Krispijn, M. Stol & K.R. Veenhof (1982)
LAS I	Parpola (1970)

LAS II	Parpola (1983a)
LB	Late Babylonian (c.750–0 BC)
MA	Middle Assyrian
MAATs	Mathematical Astronomical-Astrological Texts including ephemerides and procedure texts
MB	Middle Babylonian
mf	morning first (see Ch .2.2.1)
ml	morning last (see Ch .2.2.1)
MMEW	Livingstone (1986)
Fs.Moran	*Lingering over Words* ed. Abusch T., Huehnergard J. and Steinkeller P. (1990).
Mul.Apin	Hunger & Pingree (1989)
NA	Neo-Assyrian (c.1000–612 BC)
Nachbarn	*Mesopotamien und seine Nachbarn* eds. Nissen and Renger (1982)
NB	Neo-Babylonian (c.1000–625 BC)
NMAAT	Non-Mathematical Astronomical-Astrological Text, including Eclipse Records, Diaries, GYTs, Horoscopes and Almanacs
OAkk	Old Akkadian (c.2350–2150 BC)
OB	Old Babylonian (c.1950–1530 BC)
Planetarium	Gössmann (1950)
Fs.Reiner	*Language, Literature, and History: Philological and Historical Studies Presented to Erica Reiner* ed. Rochberg-Halton F. (1987)
RMA	Campbell-Thompson (1900)
Fs.Römer	*dubsar anta-men Studien zur Altorientalistik* ed. T. Balke, M. Dietrich, O. Loretz (1998).
SAA	State Archives of Assyria
SAA 1	Parpola (1987b)
SAA 2	Parpola (1988)
SAA 3	Livingstone (1989)
SAA 4	Starr (1990)
SAA 5	Lanfranchi (1990)
SAA 6	Kwasman (1991)
SAA 7	Fales (1992)
SAA 8	Hunger (1992)
SAA 9	Parpola (1997)
SAA X	Parpola 1993c)
SAA XI	Fales (1995)
SAA XII	Kataja (1995)
Fs.Sachs	*A Scientific Humanist: Studies in memory of Abraham Sachs* eds. E. Leichty et al. (1988).
SB	Standard Babylonian
Fs.Sjöberg	*DUMU-E$_2$-DUB-BA-A: Studies in honor of Åke W. Sjöberg* eds. H. Behrens et al. (1989)
Fs.von Soden	*Vom Alten Orient zum Alten Testament: Festschrift für W.F. von Soden zum 85. Geburtstag am 19. Juni 1993* eds. O. Loretz & M. Dietrich (1995)
Ur III	Third dynasty of Ur (c. 2100–2000 BC)

Introduction

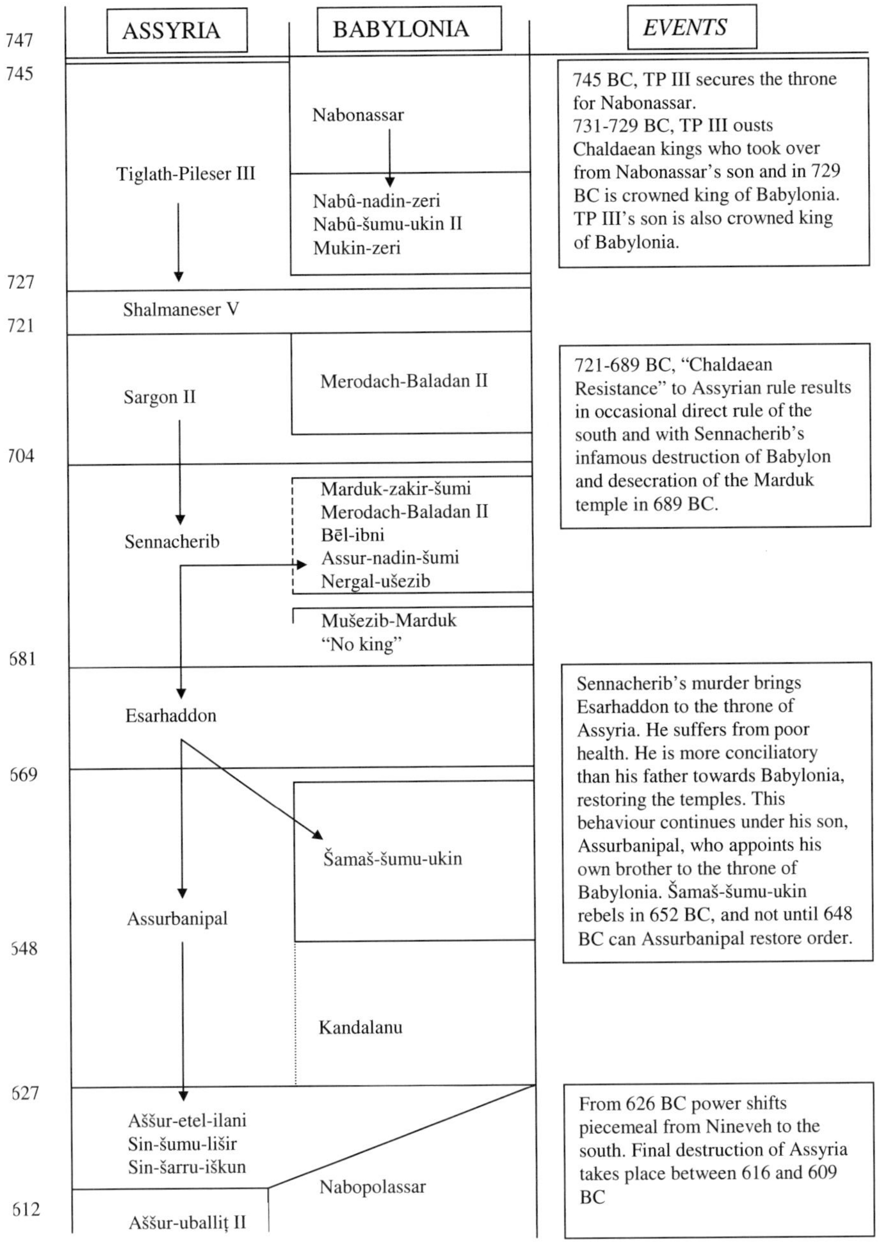

This book guides the reader towards the thesis espoused in the final chapter that the ability to predict certain planetary phenomena accurately first takes place in Mesopotamia during the 8th and 7th centuries BC. In this introduction the issues at stake in this bald statement are outlined, as is the background to the textual material studied. In Chapter 1 the "Scholars", the scribes who wrote the texts that concern us, are considered in detail. The changes that take place in the 8th and 7th centuries BC are brought about by these individuals, and the explanation for the revolution in scientific thinking at that time depends on a detailed understanding of their motivations. In Chapter 2 all the textual material know to me which concerns the planets and which is know to have been used in the period c. 750 BC to c. 612 BC is analysed. A detailed synchronic slice through the two millennia of cuneiform celestial textual material, as nearly comprehensive as possible, is thereby taken. There are particular methodological reasons for studying the material in this way. The texts considered are both divinatory, and concerned with predicting celestial phenomena. In Chapter 3 texts concerned with celestial divination are approached more generally, and the underpinning model is determined. In Chapter 4 the Paradigm which underlies the accurate prediction of celestial phenomena in Mesopotamia is characterised. Commonly regarded as commencing in the 5th or later centuries BC, this Paradigm is shown here to have begun in the 8th and 7th centuries BC, and in Chapter 5 the reasons for this development are considered. Chapters 1–5 raise many issues of direct interest to historians of science and thought, including the rôle of writing and individuals in the development of particular practices, the scientific status of celestial divination, the nature of cuneiform astronomical prediction and others. In Appendix 1 an outline of all the cuneiform texts that concern the sky is provided, and it is recommended that the reader totally unfamiliar with the textual material and Mesopotamian chronology begins there.

Texts[1] dating from as early as the first half of the second millennium BC include celestial omens, lists of stars and constellations, and stylised schemes of daylight length and other phenomena based on a simplified calendar. Celestial omens are of the form:

> "If the Moon eclipses and the north wind blows: the gods will have mercy on the land."[2]

Many celestial phenomena were considered ominous, including meteorological effects, stellar happenings, planet-planet interactions and meteorites, and most importantly the so-called heliacal phenomena of the planets. By this is meant the particular events brought about by certain spatial relationships that sometimes exist between the planets and the Sun, such as cosmical setting (when the Sun and the planet appear on opposite horizons), eclipses, first appearances, and so forth.

Star lists include a variety of tables wherein the celestial bodies are arranged into patterns which broadly correspond with their arrangements in the sky, and those wherein no such arrangements exist. Examples composed in the Sumerian language and dating to the end of the third millennium are known.

One Old Babylonian (OB) daylight scheme[3] is of the form:

[1] For references to this section consult Appendix 1.
[2] SAA Vol. 8, text 103, line 6 = 8103: 6.
[3] BM 17175+ published on p163 of Hunger & Pingree *Mul.Apin*.

"On the 15th of month XII, the watch of the day is 3 (units), the watch of the night is 3 (units). Day and night are equal. From the 15th of month XII to the 15th of month III is three months. On the 15th of month III the night transfers 1 (unit) of the watch to the day. The watch of the day is 4 (units). The watch of the night is 2 (units)."

As we shall see, celestial omens, star lists and schemes such as the one above epitomise cuneiform celestial divination from the 16th century BC on. A consistent scholarly enterprise was the norm throughout this period, and I have entitled it the "EAE (*Enūma Anu Ellil*) Paradigm[4]" after the incipit of the greatest of the celestial divination series. This Paradigm of celestial inspection and interpretation was one of the most important of all literate achievements of Mesopotamia, and was recognised as such at that time. The Paradigm has been found to be present in texts recovered from many of the ancient Near Eastern lands that came into contact with Mesopotamia. As we shall see it was of very great importance to the cuneiform scholars, and to the royalty who supported them. It influenced many other text genres. Many of the names of stars and their groupings into constellations are still used today, and many of the assignations of value to the planets, the stars, and the celestial phenomena heavily influenced Greek, Indian, Roman and thus contemporary astrology. The analysis of the Paradigm in Chapter 3 here is unlike any other previously attempted, and points to the cognitive background lying behind celestial divination, showing for the first time how small the rôle played by observation was in the creation of this important tradition.

Although some texts known to have been written before c.750 BC include the records of the observations of some celestial phenomena, and some round values for the periods after which certain phenomena recur, one key aspect of the texts which belong to the EAE Paradigm is that they never include or refer to any accurate predictions of celestial phenomena. I offer a new interpretation of the rôle played by those round values, and that rôle is divinatory.

In contrast, cuneiform texts composed in the late Persian, Hellenistic and Parthian periods predicted planetary movement and phenomena sometimes years ahead. They were found in Babylon and Uruk (and perhaps Sippar[5]) in the south of Mesopotamia. One group of such texts is here referred to as "mathematical astronomical-astrological texts" or MAATs, and these use linear methods to model planetary movement. For example the table *ACT* 600 begins:

---

[4] I am borrowing the term from Kuhn (1962) for whom (px) a Paradigm is "a universally recognised scientific achievement that for a time provides model problems and solutions to a community of practitioners." Masterman (1970) 65f pointed out that Kuhn had used the term *also* to mean a point of view associated with a set of beliefs, *and* to mean a construct which turned problems of understanding how something occurs into puzzles whose solution requires the use of standardised textbooks and methods. This more precise definition also suits my needs well. I use the term "Paradigm" for the sake of simplicity, without implying by its use that I agree with Kuhn's analysis of the manner in which science develops, though it is apparent that many intriguing parallels are found between the results he gleaned and the evidence from Mesopotamia.

[5] Pliny *Naturalis Historia* VI 121–23 and VII 193 speaks of three Babylonian astronomical schools at Uruk, Babylon and Sippar (Hipparenum – possibly Nippur). Sachs and Neugebauer in *ACT* I p5 showed that there was no direct evidence (thus far) of any mathematical-astronomical texts deriving from Sippar. However, the University of Baghdad 1984–7 season at Sippar discovered a small library (see *Iraq* 49 248–9) containing an EAE 14 day-night scheme and a circular star diagram noted in Al-Rawi & George (1990) 149 n.1. Both attest to the continued interest in celestial divination in Sippar in the late 6th century. Perhaps, subsequent discoveries will prove Pliny correct.

"(Year)113	48,05,10	month I 28,41,40 Jupiter	8,6 Capricorn, 1st station
114	48,05,10	month II 16,46,50 Jupiter	14,6 Aquarius, 1st station
115	48,05,10	month III 4,52	20,6 Pisces, 1st station"

The first column gives the year in the Seleucid calendar, the next column the difference between successive values in the third column, which itself gives the predicted time in terms of the month and thirtieths of month for when Jupiter is in "first station". The final column gives the calculated location of the stationary point in terms of degrees of the zodiac. Thus the first line states that in the 113th year of the Seleucid era (199 BC), Jupiter will be at first station 28;41,40[6] *tithis* (or thirtieths of a month) after the start of month I at 8;6°of Capricorn. The fixed temporal and spatial intervals between successive calculated stations step up or down after a certain point in the table to new values. The times and distances along the zodiac between these successive stations for Jupiter are being modelled in this text by what is termed a "step function". In other texts the intervals between planetary phenomena are approximated with linear zigzag functions. The phenomena predicted in these mathematical astronomical texts include oppositions, first and last appearances, stationary points, and eclipses.

Other texts found in Babylonia use no more than the average periods of time between identical planetary phenomena in order to predict them. I refer to these compositions as "non-mathematical astronomical-astrological texts" or NMAATs. They attest to a high degree of knowledge concerning the periods between celestial phenomena, but do not model the variations around the average intervals mathematically. They include "Eclipse Records", some of which are arranged into tables divided up into some characteristic periods between eclipses, dating back to the mid-8th century BC. One example reads:

> "Year 16 of Kandalanu, month III... day 15 (an eclipse of) 2 fingers (in magnitude). On the northeast side it was covered and it brightened to the north... from lamentation to brightening was 20 (UŠ)[7]"

This particular eclipse dating to –631, May 24 was recorded in a compilation that included some 912 lunar eclipse possibilities covering some 432 years and beginning somewhere between 750 and 740 BC. Records such as these, arranged in this way, indicate that the compilers (not necessarily the original recorders) were aware that eclipses recurred after certain characteristic intervals.

Another NMAAT group is now titled "Astronomical Diaries", the earliest of which dates back to 652 BC. These Diaries are made up mainly of daily records of celestial happenings, but some values therein were calculated by what we now believe to be methods using characteristic intervals between certain phenomena. Extracts from one such example read:

> "Year 37 of Nebukadnezar, king of Babylon (568 BC). Month I... Saturn was in front of the Swallow star... the 12th, Jupiter's acronychal rising. On the 14th Sunrise to Moonset lasted 4 UŠ (c. 12 minutes)... month XI (mid-month) Sunrise to Moonset 17 UŠ (c. 68 minutes), not observed."

6 This is sexagesimal script for 28 + 41/60 + 40/3600.

7 *LBAT* 1417 Obv. IV. 1 UŠ lasts approximately 4 minutes. The term "lamentation" (ér) refers to the start of the eclipse and alludes to the commencement of the ritual and associated laments that accompanied the obscuration of the heavenly body. See Brown & Linssen (1997) n14.

The second mid-month morning interval between Sunrise and Moonset was not observed but calculated. The methods by which this was done are not specified, but recently evidence has come to light of simple techniques by which such luni-solar intervals could have been calculated. These entailed the use of periods lasting 223 months or 229 months, or simple multiples thereof.[8] Some of the evidence for the use of these and other periods in the calculation of celestial phenomena is to found in a further group of NMAATs called "Goal Year Texts". One such, *LBAT* 1285, reads (obv.20f):

> "Year 148... month III... first part of the night of the 9th, Mercury's evening rising, 1 kùš 4 si (before) 'the front star of the twin's feet' (ν-Geminorum)... (r. 4f) Year 175, month VII the (lunar) šú + NA was 15 (UŠ), me + gi_6 was 8 (UŠ)"

Goal Year Texts provide information concerning the planets during spans of time that are a certain number of years prior to one particular year – the goal year. In the case of *LBAT* 1285 the goal year is the 194th of the Seleucid Era, or 135 BC. The information concerning Mercury is dated to a period 46 years before this, because after 46 years Mercury repeats many of its celestial phenomena on the same date and in the same part of the sky. In the case of the Moon data from some 18 years (223 months) earlier are presented – in this case the sums of certain characteristic intervals determined by the rising and setting of the Moon and Sun. These summed values repeat themselves after 18 years. *LBAT* 1285 presents data that permitted predictions to be made for the year 135 BC, and it and similar compositions indicate that a variety of planetary periods were known, some of which could only have been gleaned from the careful analysis of extensive records of observations, such as we find in the Astronomical Diaries.

Both MAATs and NMAATs use empirically determined facts about the planets, and it is very probable that the predictions determined in both cases were used in divination. In Chapter 4 I describe in more detail the methods used in both kinds of text, showing that both form part of what I describe there as the "Celestial Phenomena Predicting (PCP) Paradigm", which is to be differentiated from the EAE Paradigm by the intention on the part of its exponents to predict some celestial phenomena to a high degree of accuracy. In the words of Aaboe (1991) 278 "the mathematical astronomical texts...represent one of the last, as well as one of the finest contributions of Mesopotamian culture." He goes on to say (p292) "...we claim Babylonian mathematical astronomy as the common ancestor of modern efforts in the exact sciences." Pliny the Elder, who died in AD 79, wrote of Babylon:[9]

> "The temple of Jupiter Belus still remains – it was here the creator of the science of astronomy was – the rest has reverted to desert."

Clearly, it is important to establish the situation that brought about this late flowering of cuneiform scholarship.

Aaboe (1991) 285 believes that "at the moment it seems likely that Babylonian theoretical astronomy was created sometime in the fourth century BC." In Chapter 4 I indicate that the PCP Paradigm was fully established by the 7th century BC, and in Chapter 5 terms such as "theoretical astronomy" will be discussed. I conclude that it was during the 8th and 7th

[8] See now Brack-Bernsen (1997) Chapters 14 –15.

[9] *Naturalis Historia* VI: 121f.

centuries BC that a *revolution* took place from the EAE to the PCP Paradigm[10] – a scientific revolution or revolution of wisdom that can be compared profitably with other such revolutions. Specific evidence for this is found in the texts from this period, and the reasons for the transition, I argue, can be seen in the environment that surrounded the diviners.

The period of the 8th and 7th centuries BC has provided us with an abundance of astrological-astronomical material. The most informative textual sources are termed the "Letters" and "Reports"[11] written by certain elite Neo-Assyrian (NA) and Neo-Babylonian (NB) scribes to the Assyrian kings, most of which were found in Nineveh and are now in the British Museum. A typical Report reads as follows:

> "If in month X the Moon appears on the 30th day; the Ahlamû will devour the land of Subartu; a foreigner will rule the Westland. If the Moon is low at its appearance; the products of a distant land will come to the king of the world. From Bulluṭu." (8121 = SAA Vol 8 Text No. 121)

Frequently the Reports include only the omens culled from the divination series that pertained to the particular phenomena observed by the scholar. This alone tells us which phenomena were considered ominous in those centuries, and a great deal about the manner in which the divination series were used. Sometimes, though, the Scholars also included either in Reports or Letters statements that reveal to us their attempts to predict celestial phenomena, the extent of their abilities in this, and much else that tells us about the workings of the divination industry at that time:

> "If the Moon becomes late at an inappropriate time and is not seen; attack of a ruling city. It was seen on the 16th day... Within one month the Moon and Sun will make an eclipse... the king must not ignore these observations of the Moon; let the king perform either an (apotropaic) *namburbi* or some ritual which is pertinent to it. From Munnabitu." (8320)

A Report such as this one shows us that eclipses were predicted, albeit only a month in advance in this case, that the Scholars advised the king forcibly, and that a ritual industry concerned with averting the evil portended by certain celestial phenomena coexisted with divination.

Much other material from this period exists, and what has been used in this study is outlined in the following sections. In Chapter 2 I have analysed all of these data in order to provide a snapshot of the state of celestial divination under the auspices of the late Assyrian kings. In Chapter 4 I consider all of the data from these texts which indicate to us the state of the science of predicting celestial phenomena in this period. We will see that this science was in its infancy, but most assuredly ancestral to the science of prediction we find in the MAATs of the Hellenistic period.

---

[10] That the period under discussion was one of transformation in Mesopotamian celestial concerns is not wholly new in Assyriology. E.g. Neugebauer mentions the idea in passing in *HAMA* p2. I have also been heartened to discover that Koch-Westenholz (1995) 52 writes "what we have here (Scholarly work from Sargon to Assurbanipal) may well be the earliest documented instance of scientific revolution" and notes Kuhn. However, short of this observation, no attempt has previously been made to my knowledge to define the Paradigms, nor to elicit the details of the transformation or the reasons behind it.

[11] These will be defined more carefully shortly.

I.1 Terminology

"Mesopotamian" of the title of the book is used somewhat more broadly than the geographical designation of land between the Euphrates and the Tigris, for it applies to some texts found in Assyrian and Babylonian sites that are a little distant from the rivers. "Assyria" (in the north of Mesopotamia) is used to refer to both the Empire and to the Assyrian heartland proper.[12] "Babylonia" (in the south of Mesopotamia) is used to refer to the geographical region of ancient Akkad and Sumer, or to the land occupied under the empire of Hammurapi, even though the hegemony of Babylon itself over much of this area did not exist during many of the periods under consideration here.

In earlier times "planet" referred to one of the seven celestial bodies that moved against the background stars that were visible to the naked eye – the Sun, the Moon, Saturn, Jupiter, Mars, Venus, and Mercury – though today it is commonly used to refer only to those bodies which circle the Sun. Since the cuneiform terms ^{mul}udu.idim or *bibbu*[13] refer to the seven moving celestial bodies, "planet" will be used in that sense here.

The term "astrology-astronomy" will be used to refer to the particular branch of Mesopotamian scholarship herein considered. It is to be differentiated from cosmological or cosmogonical speculation – theories concerning the universe as a whole, or concerning the creation of the universe as a whole.[14] Astrology and astronomy mean different things today,[15] but the two words were used interchangeably at least until the 6th century AD.[16] This is not to imply that before this time no difference was ever appreciated between what we would term astrology and what we would term astronomy.[17] This difference is of course at the heart of many attempts to find "science" in the ancient world, and will be discussed in detail in Ch.5. Since there are no texts of the genre under consideration from Mesopotamia for which it cannot be argued that the inscribed "astronomy" had some potential "astrological" purpose, the combined term "astrology-astronomy" will be used in order not to prejudice the interpretation of the text under scrutiny.

The use of a capital first letter for Reports, Letters, Diaries, Eclipse Records, Goal Year Texts, and so forth implies that each refers to a specific text group to be identified in due course. The same applies to Scholars, who are a group of scribes defined in the next chapter. Abbreviations should be clear, but a list is provided with the bibliography.

---

[12] "The land" – *mātu* "a triangle with its apex at the city Assur on the Tigris, and its base stretching from Arbaʾil in the East to Nineveh in the West." (Grayson, 1991d 203). The fourth major city in the heartland is Kalḫu.

[13] Italic is reserved for Akkadian, normal font for Sumerian.

[14] There is, of course, some overlap between Mesopotamian cosmogonical speculation, and astrological-astronomical writings. For example in *Enūma Eliš* V, it is said to be Marduk who fashions the universe into the form described by such astrological-astronomical texts as Mul.Apin. For a comprehensive collection and study of this material from Mesopotamia see now Horowitz (1998).

[15] Astro-*logia* literally means the branch of knowledge concerned with the stars (and planets), but today means the art of judging the supposed effect on the world of the influence of the heavenly bodies and phenomena. Astro-*nomos* literally means "star arranging", but today is used to describe the measurement of position and movement of the heavenly bodies and their phenomena with the explicit intention of categorising, ordering in space and time, and of predicting, *without* any reference to the astrological implications of this.

[16] French (1994) xff no doubt referring to Isidore of Seville's distinction between "superstitious astrology" and "natural astrology".

[17] Ptolemy distinguishes them in *Tetrabiblos* I.1–2, for example, as two aspects of "astronomy". See Barton (1994) 60 and Lloyd (1992) 570.

I.2 Methodology

The main period of concern in this work is from c.750 BC to c. 612 BC. I decided that rather than begin by studying texts from all periods and imposing on them some overall characterisation, I would begin by studying all the available material from a narrower interval of time. Provided enough material existed, I could be confident that I was not studying and elaborating on the basis of texts that had survived precisely because they were exceptional. Having established the main features of the chosen interval, I could approach the earlier and later material with a greater sense of their historical contexts. The period c. 750–612 BC was chosen precisely because a large number of texts, many recently published, date from then. It is also a period relatively well understood historically.

The fundamental characteristic of the period is that Mesopotamia was then dominated by the Assyrians, and their rule undoubtedly influenced the development of predictive astronomy, as we shall see. 612 BC marks the date of the destruction by the combined forces of the Babylonians and others of their capital Nineveh, whence the largest number of texts from this period come. The other Assyrian sites were similarly abandoned around that time. 612 BC or thereabouts is clearly a useful date with which to end the main period of concern here. In 747 (or perhaps 748) BC[18] Nabonassar came to power in Babylonia, and this subsequently came to be seen as a turning point in the fortunes of Southern Mesopotamia, though there is some doubt as to whether or not this was really the case.[19] 746 BC marked the beginning of the reign of Tiglath-Pileser III and in the words of Grayson (1991a) 71: "the rebirth of the Assyrian empire after the dark days of 'the Interval'..." took place. His "reign was a brilliant beginning to a new and final era in the history of the Neo-Assyrian empire."[20] Tiglath-Pileser III immediately concerned himself with the Babylonian question[21]

[18] See Brinkman (1984) 39–40 n195. and idem (1991) 1 n1. The Mesopotamian year begins in spring, and the first regnal year of a Mesopotamian monarch was the first **full** year of reign. See now Walker (1997) n1.

[19] Hallo (1988) 187–90 lists ten pieces of evidence concerning the possible beginning of an epoch (a millennial epoch, he argues) with the coming to power of Nabonassar. In his opinion 747 did mark an actual transition in Babylonian fortunes, and not just a later retrospective historicising. This is not the view taken by Brinkman (1991) 24: "Although later ages were to view Nabonassar's accession as a turning point in Babylonian history, it is difficult to discern qualities in Nabonassar or his reign that were epoch making. Babylonia continued to suffer from weak central government...stabilisation seems to have taken place because (the Assyrian king) Tiglath-Pileser (III) was propping up the Babylonian throne against domination by the Chaldaeans." Nevertheless, Brinkman's 1984 monograph, and his 1991 chapter for the Cambridge Ancient History series effectively draw a line between the "dark age" (p6) which preceded 750 and the process of Babylonian "transformation and revitalisation...on many levels – demographic, political, socio-economic, and cultural..." (p68) that followed. He discusses (p3f) the results of the archaeological surface surveys in parts of Babylonia that have shown that a dramatic drop in population took place in the late second and early first millennium, and that this phenomenon is also reflected in the limited textual material that has survived from this period. Taking into consideration the limitations of the surface surveys, the skewing of the historical picture by the urban origins of most surviving documentation, and so forth, he concludes that "the period of worst decline ended in the second half of the eighth century" (p6). It is, therefore, not unreasonable to argue that c.750 BC did mark in reality, and not solely in retrospect, a point of significant importance in Babylonian history.

[20] ibid. 85.

[21] Babylonia was particularly significant to Assyria in terms of trade, language and heritage, and this is reflected in 9th century treaty arrangements between Assyria and Babylonia made to secure the throne of the king of Babylon (Grayson, 1991d 204 and Roaf, 1990 167). It is perhaps significant that Shalmaneser III, the last Assyrian king prior to the "dark Interval" which preceded Tiglath-Pileser III, is shown in a carving on a throne base in Kalḫu shaking hands with the then king of Babylonia, Marduk-zakir-šumi. The two kings are shown as equals, unlike in later representations where the Assyrian king is invariably shown superior to all (human) others. See also Brinkman (1973).

and before long ascended to the throne of Babylonia itself. From that point until c. 612 BC the socio-political circumstances that surrounded the implementation of the divinatory Paradigm and the development of predictive astronomy in both Assyria and Babylonia were interrelated. From the point of view of historical background, c. 750 BC is thus a convenient date with which to begin my main period of concern. In addition, Ptolemy[22] refers to records of eclipses known to him from Babylonia which date from 747 BC on, and this more or less corresponds, extraordinarily, with the earliest dating of the eclipses listed in the texts *LBAT* 1413ff from Babylon. These texts are the oldest known NMAATs, and although they do not prove that eclipse *predictions* were being made in the mid-8th century BC, they suggest that records were being taken for that purpose. Incidentally, Berossus[23] asserts that Nabonassar destroyed all records pertaining to earlier reigns, so even if regular accurate records of celestial phenomena had been made prior to 747 BC, we should perhaps not expect to find them. Walker (1997) 18 suggests that Berossus's story may in fact refer only to these Babylonian observational records rather than to more general historical records. Thus, while c. 750 BC is a somewhat arbitrary start date for the period of major concern here, I argue that it is convenient in that it corresponds with the earliest datable NMAATs, and justified from an historical point of view in the light of the influence I propose the Assyrians had on the scientific revolution here outlined.

It is assumed that this transition in Paradigms cannot be accounted for by a change in the "technology of communication", for similar scribal practices are known both prior to and following this period. However, increased scholarship, brought about by the socio-political situation, does appear to have taken place, particularly during the reigns of Esarhaddon and Assurbanipal. So while accepting the essential validity of Goody's (1977) argument,[24] a more traditionally functionalist approach will be considered in accounting for the development of the PCP Paradigm. That is, the (often hidden) functionalist assumptions of accounting for developments on the basis of economic need and survival will be made.

Kuhn's (1962, 70, and 72) model of the development of science through Paradigms and revolutions has been considered to see if it usefully describes the Mesopotamian evidence. By borrowing Kuhn's perhaps somewhat dated terminology, I am not assuming that the reader need know the details of his model, nor am I presuming its general validity. I have found that his model corresponds well with my discoveries as to the development of cuneiform astronomy-astrology. I in effect posited an hypothesis, essentially provisional and revisable (Lloyd, 1992, 575), concerning the development of cuneiform astrology-astronomy along Kuhnian lines. This hypothesis allowed me thereby to *deduce* aspects of the intellectual endeavour that might be found in Mesopotamia, and the results of my conclusions on this can now be readily compared with those endeavours that characterise other sciences ancient and modern. Part of the intention in approaching the material with

[22] E.g. *Almagest* III 7, and the "Ptolemaic Canon" or "Chronological Table of the Kings" published as an appendix to the *Almagest*, refer to the Nabonassar epoch commencing on Feb. 26th midday, -746.

[23] Burstein (1978) 164 *babyloniaca* Book 2 §5.1 "*Nabonasoros* collected together and destroyed the records of the kings before him in order that the list of Chaldaean kings might begin with him."

[24] Goody writes for example on pages 36–7: "...differences in intellectual processes...can be related not so much to differences in 'mind' but to differences in systems of communication...especially alphabetic literacy." The alleged significance of the technology of writing (though clearly not of alphabetic literacy in Mesopotamia) on the development of science is accepted in its broadest terms. Ditto Larsen (1987) 223, who instead grounds Goody's abstraction as to the effects of writing in terms of the "social and political structures". Also, for discussions of the rôle of writing in the formulation of celestial divination see Chs.2 and 3.

this methodology was to ensure that cuneiform celestial writing could begin to take its place alongside Greek, Islamic, Chinese and European works as material worthy of investigation when it comes to the question of what constitutes a science. I therefore approach the question of the scientific nature of these texts and deliberate distinguish this analysis from the more relativistic definitions of ancient science offered recently by Pingree, for example. He argues (1992c) 554 that "astral omens (and others)...were or are sciences within the contexts of the culture in which they once flourished." While, I agree that cuneiform astrology-astronomy was undoubtedly significant in *its* world, it also happens to have had many features in common with modern science.

I would certainly not advocate a return to Eurocentric nomenclature such as "une science bloquée",[25] or "proto-science"[26] for defining early achievements, but equally there is so much in cuneiform astronomy-astrology that is challenging and intriguing to the contemporary mind, that the parallels and contrasts with modern science ought to be highlighted. Parallels include the mathematical modelling of complex periodic functions with a summation of a series of simple linear functions of various periods. Contrasts include the supposed absence of any underpinning models of the universe (a supposition challenged in Ch.5), and the assumption that predicted phenomena were still ominous. Small importance was attached by the Scholars to mechanistic explanations, and great importance was attached by them to teleological ones. The debt owed by predictive astronomy to the celestial divination industry will be stressed, for this is a very different basis for a science than those that characterise the sciences of today. This comparative approach will be more rewarding than an analysis of the subject only in terms of its own context, even if this should result in the raising of some questions as to the nature of science today. The evidence of cuneiform astronomy-astrology can contribute to a fuller understanding of the modern idea of science, I suggest.

Nevertheless, in awareness of contemporary thought on matters of cultural relativism, on the impossibility of establishing cultural universals and pan-historical human mind-sets, and on the location of meaning not with things "in the world", but in the interrelationships of things which signify them, an attempt has also been made to determine Babylonian and Assyrian attitudes towards their own work in astrology-astronomy. Specifically, the results concerning changes in cognition brought about by the scientific revolution of the 8th and 7th centuries BC pertain not to Mesopotamia as whole, but to the Scholars who authored these compositions. The apparent lack of reflexive texts in cuneiform is well known,[27] but astrology-astronomy does lend itself to some measure of being able to determine the scribes' underlying assumptions and the significance to *them* of their own work.

---

[25] Limet (1982) E.g. 28.

[26] Barton (1994) 5. Rochberg (1992) 549 describes the "middle ground" between a relativistic view that ancient science can only be understood in its own context (incommensurability), that it is a social construct first and last, and the notion that modern science is what ancient practice inevitably leads towards, that only those things which did anticipate today's achievements are to be valued. It is precisely this middle ground that I am taking here. See Ch.5.1.3.

[27] Even to students of Greek science: Lloyd (1979) 232–3. In fact, it is not true that "investigations... into...questions concerning the nature of the inquiry itself (into astronomy, medicine, or mathematics)..." did not take place outside Greece. A need for a careful assessment of abstraction in Mesopotamia has been recognised by Bottéro (1974) 190 n1 and Larsen op.cit. 216 n40.

I.3 Textual Sources Used in the Period c.750 – c.612 BC

Many of the texts known to have been written in this period contain, copy, or allude to much older material. This is believed to be the case when older exemplars are attested from other sites, when the orthography or grammar appears to be in an old style, and sometimes on the basis of content alone. Texts that record planetary and stellar positions lend themselves to this particular endeavour better than most.[28] It is, however, the fact of this older material's use between c. 750 and 612 BC that is of importance in this analysis. Texts that are copies of older material, even if updated, but for which there is no direct evidence that they were being used by the Scholars and scribes, will be ignored in the first instance.[29] Some of the omens in NA and NB copies of the canonical[30] omen series, whose content was mostly

---

[28] E.g. the attempted dating of the composition of Mul.Apin to the second (or even third) millennium (described in Hunger & Pingree, 1989, 11), and the attempted datings of K.250, the astrolabes, and HS 245. For details see App.1. Often the datings reveal little more than the wish on the part of the modern student to have the text in question fit his or her view of the state of development of Mesopotamian astrology-astronomy.

[29] Antiquarianism was practised by many Mesopotamian monarchs. Wiseman (1991) 242, for example, makes this point. See also Porter (1993) n207. The existence of copies of older texts, even recent copies, does not indicate that the content of the said texts represented the state of current thinking.

[30] "Canonical" is used to refer to the broadly standardised form of a text, such as an omen series. The canonical form of a text may have come about through the bringing together of various different forms of the text, and reconciling their differences. It was perhaps motivated by a desire to preserve texts in dialects of the Akkadian language which were dying out, and this perhaps explains the largely "logographic" nature of the canonical omen series (Leichty, 1970, 30). The precise definition of "canonical" in this context is elusive, but the term is endemic to Assyriological secondary literature. Useful recent contributions have been provided by Elman (1975), Rochberg-Halton (1984b), and Lieberman (1990). For Rochberg-Halton, the term applies to those cuneiform series which show "text stability and (a) fixed sequence of tablets" (op.cit. 129). Koch-Westenholz (1995) 75 identifies "canonical" with Oppenheim's *AM* 13 phrase – "the stream of tradition". The extent to which "canonical" cuneiform texts are divinely sanctioned is less clear, as Lieberman (op.cit. 306) discusses. Rochberg-Halton's analysis (after Elman) showed that three main streams of textual transmission are identified by the Assyrians, the *iškaru* "official series", *aḫû* "extraneous", and *ša pî ummâni* "from the Scholar's mouth" forms. The *aḫû* texts do not appear to be "non-authoritative", and are therefore not "non-canonical", as frequently translated (e.g. Parpola in SAAX, but not Hunger in SAA8). They are "external"; perhaps an appendix or excursus, as Lieberman (op.cit. 308) suggests. Lieberman extends the discussion by looking at the term "official" when applied to texts. He argues that the implication of this term is a form of government sanction applied to the texts and that, in fact, none of the texts from Assurbanipal's library should be designated in this way. He characterises the collection as "personal" rather than "official" (op.cit. 319). However, it seems to me absurd to argue that texts such as EAE did not have some form of royal sanction. Assurbanipal's collection may well have been personal, and designed to give him some measure of control over the Scholars (loc. cit. 320), but he and his royal predecessors *funded* the Scholars, thereby sanctioning their work – work which was dependent on such compositions as EAE. If it was not the king who gave the Scholars the go-ahead to practise their art, then who was it? If the sanctioning were theological, the king was also the high priest of Aššur. I strongly favour the idea of a royal sanctioning of the Scholars, of the giving of official status to the divinatory series, because power and astrology traditionally justify each other (Barton, 1994, 211: "Which ruler could deny himself the inscription of his imperial destiny in the cosmos.")

Contemporary students tend to view the period just prior to the end of the second millennium BC as the time when the great omen series were more or less established in their so-called "canonical" forms (for the reasons see Koch-Westenholz, 1995, 42f). E.g. Hunger & Pingree's (1989) 10–12 dating of Mul.Apin, and Livingstone's (1986) dating of i.NAM.giš.ḫur.an.ki.a. correspond with the views expressed by Weidner (1941/4a, 176: "...nach unserem heutigen Wissen...die Serie Enûma Anu Enlil ein Werk babylonischer Kompilatoren aus der zweiten Hälfte des 2. oder dem Anfang des 1. Jahrtausends ist") and Leichty (1970) 21 on the period during which EAE and Šumma izbu were put into their canonical forms. The effect is of mutual corroboration. Note Lambert's (1967) 9 warning. See also the comments in Reiner (1991) 304 who pinpoints the reigns of Nebuchadrezzar 1 (1125–1104 BC) and of Nabonassar as "canonising" periods, and yet on page 320 she comments on the continuous expansion of EAE throughout the first millennium BC. Similarly Jeyes (1991–2) 27f argues that extispicy was from the OB to NA periods a "developing science". See Ch.3.3, here. The extent to which Assyrians were involved in "canon-

fixed long before the 8th century, show subtle variations from those used by the Scholars. These differences attest to the continued development of EAE, and are discussed in Chapter 2.1.2 & Ch.3.3. The intention is not to delineate a corpus, but for methodological reasons to restrict the material used to that which will indicate to us the *state* of cuneiform astronomy-astrology and its socio-political background in the 7th and 8th centuries BC.

Most important amongst the text-types studied here is the *correspondence* of the Assyrian and Babylonian Scholars and scribes to the NA kings.[31] Two main types are identified, though the difference between a *Report* and a *Letter* is not always hard and fast.[32] Reports and Letters differ in a number of particulars. Oblong tablets with the text written parallel to the long sides are entitled *u'iltu* in Akkadian. It means a "binding", and the documents we call Reports were generally written in this way, as were loan documents, excerpt texts and court proceedings. *Egirtu* tablets have their text written parallel to the short sides. Scholars' Letters and legal transactions were usually written in this form.[33] Apart from these designations, Reports and Letters can broadly be differentiated on the basis of content. Reports usually contain just the results of observation, either in the form of raw data on position and time, and so forth, or in the form of omens whose protases describe the phenomena observed. Letters from the Scholars may also contain observational data and omens, but they usually contain an introductory greeting to the recipient and other information.

Many of the Reports and Letters sent by the Scholars contain celestial omens, observational data, planetary predictions, and other comments obviously relevant to this work. However, some Scholars sent Reports with omens relating to non-celestial phenomena,[34] or Letters that do not concern celestial matters at all. The guiding principle in this book is that all the works known to have been composed by Scholars and scribes who are known to have written texts concerned with astrology-astronomy will be considered. Much that they wrote that is not strictly concerned with planets tells of their relationships to the kings and other Scholars, of the breadth of their expertise, and of further details significant to an understanding of their rôle and function in society. This context, in which the planetary astrology-astronomy is embedded, is vital to a fuller understanding of the subject.

ising" omen series and the like is also virtually impossible to assess at the moment. For a few comments see App.1 §21 and Koch-Westenholz (1995) 43f.

31 The letters uncovered from the NA courts and dating to between 750 and 612 BC have been estimated by Parpola (1981) 118 and Brinkman (1984) 113 to number about 3200, of which about 2300 are in Assyrian (script). The vast majority were found in Nineveh and Nimrūd and almost all of them date to the century 745–646 BC. Of these, those dating to the reigns of Tiglath-Pileser III, Shalmaneser V (see Parpola ibid.119 n1), Sargon II and to the middle of Assurbanipal's reign mainly concern politics, administration and warfare. They are letters sent by officials, mainly written by unnamed scribes. Some 390 Letters are written by Assyrian and Babylonian Scholars. Most of them can be assigned to named individuals. To this number must be added some 570 astrological-astronomical Reports and related fragments (which Parpola reasonably does not include in his figures for letters, but which certainly form part of the NA "correspondence" – see Brinkman ibid.) Most of these Reports can be assigned to named Scholars, some of whom are also attested from the Letter corpus. Their temporal distribution reflects that of the Scholars' Letters – see below. The Ninevite letters (please note "Letters" with a capital "L" refers to those of the Scholars) have been published principally in Harper *ABL* (transcribed and translated in Waterman *RCAE* and Pfeiffer *SLA*), in Parpola (1970, 1972, 1979, 1983a, 1987a, 1988, 1990, 1993c), and Dietrich (*WO* 1967–71, 1979). Some of the Nimrūd letters have been published by Saggs (1955–74). Most of the Reports were first published by Campbell-Thompson *RMA*. For the most recent publications concerning the Scholars' Letters and Reports see below.

32 Shown, for example, by the presence of some SAA8 texts in *LAS* I and II.

33 See Parpola (1983b) 2 n.5, where he argues that *egirtu* does not specifically mean "letter", but rather the form of tablet in which letters are often composed, as in *CAD* E 46,2) b).

34 Hunger SAA8 xviii.

Too often, studies in Mesopotamian astrology-astronomy have treated this subject (or worse still, astronomy and astrology as separate disciplines) as an achievement of the abstract "Babylonians" and "Assyrians" without attempting to find to which group of people, or to which individuals, and in which contexts, these achievements can be attributed. Goody (1977) Ch.2, attempts to re-establish the significance of creative, imaginative "intellectuals" in early societies, and move away from the abstract notions of socially determined "thinking". To this end, further texts which reveal aspects of the Scholars' lives and ancestry will be considered. These include *chronographic texts*[35] which list Scholars, *administrative records*[36] from the palaces, *extispicy reports*[37] (which demonstrate some of the activities of the *bārû* class of diviner) and a number of *colophons*[38] to tablets which contain biographical and prosopographical information. This work on the Scholars and scribes will be covered in Chapter 1.

A few, but important, details on planetary astronomy-astrology between c. 750 and 612 BC are to be found amongst the following text-types composed during this period:

"Literary" texts

Some of the few attested original Neo-Assyrian compositions mention the planets and related themes. (Livingstone SAA3 texts 1: 21; 2: 41-r.9; 25: ii20'f; 32: r19; 37: 20; 38: 24f; 39: 33&r4.) Similarly, the Babylonian text *Erra and Išum*, the composition of which is perhaps to be dated to the period soon after the "dark age", contains several astrological-astronomical themes.[39] Texts such as these indicate how widespread the influence of astronomy-astrology was in the late NA period. They also allude to themes which were thought to underpin the discipline at that time, themes which show a great deal of similarity with those in some older "literary" texts. This will be discussed further in Chapters 3 and 5.

NA treaties and loyalty oaths

Some of these were concluded in front of the planets, e.g. Parpola and Watanabe SAA2 texts 6: i13f; 8e26f; 11: 6; 14: 4&ii1. SAA2 text 6 §10 imposes an obligation on the people to report to Esarhaddon the evil words of prophets, ecstatics, and diviners. It is interesting, though perhaps not very significant, that none of the treaties prior to Esarhaddon's rule mention the planets. This may possibly indicate an elevation in the status of the planets in the minds of the compilers after Esarhaddon's time and continuing into Assurbanipal's reign, though it must be noted that Esarhaddon's accession treaty (SAA2 text 4), and his treaty with Baal, king of Tyre (SAA2 text 5) do not mention the planets either. Note also the reference to a treaty (*adê*) concluded in c.670 BC before Jupiter and Sirius in the text edited by Parpola (1972). Were there to have been any rise in royal concern with celestial divina-

---

35 That is, chronicles and king lists, for which now see Grayson *ABC* and for a summary Grayson (1980–3), in particular §3.12 III 15 – IV 16 and §3.14 10 which lists the kings and their *ummânus*. The understanding of the relationship between the chronographic texts and the omen series is important for an understanding of Mesopotamian historiography. The matter is discussed briefly by Grayson (1966), Starr (1986), and Koch-Westenholz (1995) 15f. See also Leichty (1970) 4 and Cooper (1980). My analyses in Ch.3.1.1 and 3.2 reflect on this debate.

36 See Postgate and Fales SAA7 and SAAXI, and Parpola (1983b).

37 Starr SAA4.

38 Hunger *Kolophone* and Streck (1916) 354–75.

39 For editions see Cagni (1969 and 1977) and Al-Rawi and Black (1989). See also Reiner (1960b) and App.1 §24, here.

tion in this period, it would likely have had a significant influence on the development of the PCP Paradigm.

Royal inscriptions

These are attested for all the NA kings relevant to the period under discussion, and for the Babylonian kings Merodach-Baladan II and Šamaš-šumu-ukīn. Inscriptions of local dignitaries are also attested.[40] To my knowledge the earliest attestation of planets in the inscriptions is from 714 BC in *Sargon's Eighth Campaign*.[41] The planets are mentioned several times in Esarhaddon's and later inscriptions,[42] but apart from in this unusually "literary" Sargon inscription, are not found in any inscriptions of the other kings. As in the treaties and loyalty oaths, it is only from Esarhaddon's time on that the unambiguous names (see Chapter 2) for the planets are used in royal inscriptions.

Prophecies and Oracles

For Neo-Assyrian examples of this genre see now Parpola SAA9. The planets are not mentioned explicitly in these examples, but the genre is of interest because of its relationship to the celestial omen series. This is particularly the case in the so-called Prophecy Texts A and B, which are much older, and has been discussed by Grayson *BHL* 17 and Biggs (1985 and 1987).[43] There is some overlap between the historical information contained in the prophe-

[40] The principal publications of these inscriptions are conveniently listed by Brinkman (1984) nn. 560–566. See also Frame (1992) 9–10 and Porter (1993) Apps. 1–4.

[41] Thureau-Dangin (1912) line 317: *i-na qí-bi-it ṣir-te šá* ^{d}*Nabû* ^{d}*Marduk šá i-na man-za-az* mul.meš *šá... iṣ-ba-tu ta-lu-ku*. "At the august command of Nabû (and) Marduk, who had taken a course in a 'station' of stars, which..." Given the eclipse of the Moon which is described in the lines following (see App.1 §23), it is quite clear that planets pertaining to the gods were meant. The following options are possible: (I) Two planets were meant; namely Mercury (Nabû) and Jupiter (Marduk). (II) Only Marduk has taken a course in the sky (the subjunctive *u* hides any plural *u*). This ambiguity is also to be found in my English translation. In this case Marduk is more frequently attested as a name for Jupiter (see Ch.2), but is also attested as a name for Mercury. (III) Only one planet is meant, with the name Nabû-Marduk, which would be Mercury. This would account for the missing "and" in line 317. From the date of the eclipse it is possible to calculate that Jupiter was not visible during the eclipse (Dvorak & Hunger, 1981). Thus option (III) is the likeliest. It is worth noting that the unambiguous planetary names were not used here, as they are in Esarhaddon's annals and treaties, though this is perhaps to be understood in the context of the "*ina qibīt ṣīrte*". Gods, not their planetary manifestations, issue commands. Reiner (1995) 12 suggests that Sargon II, by describing how an eclipse portended the downfall of his enemy Urartu, was deliberately referring to the example set by the Old Akkadian king Sargon I for whom it is related in the OB *King of Battle* poem "the Sun became obscured, the stars came forth for the enemy". For *manzāzu* see also Horowitz (1998) 116.

[42] *Esarhaddon*: Borger (1956) Ass.A I: 31-II: 26; Ass.C II: 1–12; Bab. Ep.13: A: 34–41; §102^{a},a,12. *Assurbanipal*: Streck (1916) p189 K2652: 5; p217 K3087: 1; p223 K3405: 1 (all Venus) and Lambert (1957/8b) K4449: 21f (Jupiter, Mars). *Šamaš-šumu-ukīn*: Lambert ibid. CBS 733+1757: 6f (Mercury).

The statement by Oppenheim (1960) 137 and n10, that the only example of a celestial omen in the Assyrian royal inscriptions is the Sargon II one above, is wrong. The examples, Borger (1956) Nin.A II: 5 *uk-ki-ba-nim-ma i-da-at dum-qí ina šá-ma-me u qaq-qa-ri* "propitious signs in the sky and the earth followed each other for me", or op.cit. Bab. A I: 34f, II: 24f and Ass A I: 39f, in combination with those cited above which describe the positions of the planets, are very similar to the account found in the *Eighth Campaign*. See also Cogan (1983), who attempts to identify the audience to which inscriptions were directed, based on whether or not they mentioned omens (*contra* Porter, 1993 n225), and Koch-Westenholz (1995) 155–8, who suggests that the omens in the inscriptions may even correspond to those found in the Letters, Reports and EAE.

[43] Brief Bibliography: Prophecy texts A,B,C & D were published in Grayson & Lambert (1964).

A MA = *KAR* 421.

B OB. Kuyunjik sources are attested. See also Biggs *Iraq* 29 117–32. An additional NB source, PBS 13 84 was published in Biggs (1987). This shows strongly the influence of celestial divination, with several references to the planets and to omens.

cies and that in the celestial omens, indicating the probable use of similar sources.[44] This suggests one way in which omens were generated that did not derive from the simultaneous observation of celestial and terrestrial events. The NA examples demonstrate clearly the way in which the *ragimmu* "prophet" or *raggintu* "prophetess" endeavoured to look after their kings,[45] a rôle also played by the celestial diviners. They also indicate the manner in which the king was perceived by the literate experts who composed the prophecies and by those who wrote the Letters and Reports,[46] and to this extent tell us a great deal about the courtly background in which they worked. This in turn helps explain the emergence of the accurate prediction of celestial phenomena at this time.

Prayers, Incantations, Hymns and Rituals

Many prayers and hymns and some incantations are addressed to the planets and stars.[47] They indicate in part the manner in which the celestial bodies were thought to *affect* the terrestrial plain. In general this will not concern us here, for as I argue in Chapter 3, celestial signs were thought by the Scholars to *herald* good or ill fortune without *causing* affliction or good health directly. For this latter aspect of the Scholars' concerns with the heavenly bodies, particularly the stars, see Reiner (1995). Some rituals aimed to avert the consequences portended by celestial phenomena.[48] Those in which the suppliant's name is known to date to the period under consideration are of interest

---

C From Tukulti-Ninurta I's or Sennacherib's reign. Copies from Nineveh and Assur.

D Attested in a copy from Assur.

Uruk Prophecy LB. Published in Hunger (1976) No.3 and Hunger and Kaufman *JAOS* 95 371–5. Brinkman (1984) n582 suggests that this text might describe the period under consideration.

Dynastic Prophecy LB. Grayson *BHL* 24–37.

Marduk Prophecy Borger (1971). Kuyunjik version attested. MB see ibid. 21f.

Šulgi Prophecy Borger (1971). Kuyunjik version attested. MB see ibid. 22f.

The Birth Legend of Sargon (Lewis 1980, Westenholz 1997, 36f) is of the same genre (Reiner 1991, 305). See Grayson (1974–7), *ABC* 43–9 & 57 n60 and *BHL* 8 n11 for other Sargon revival texts. See also Ellis (1989).

44 Biggs (1987) 6.

45 SAA9 Text 2: 15'f "[Have no fe]ar, Esarhaddon... [I will go] around you and protect you (*a-na-ṣar*)"

46 Parpola SAA9 lxi suggests that prophecy text 8 was perhaps composed by the chief scribe Issar-šumu-ereš, author of many Reports and Letters.

47 Late copies of older prayers, incantations or hymns, such as the famous *The Prayer to the Gods of the Night* (see App.1 §11), do not prove that they were in use between 750 and 612 BC, however likely. The hymn to Ištar mentioning Ellil-bāni, the governor of Nippur under Šamaš-šumu-ukīn, does demonstrate that this text (BM 78903) was in use, however. See Frame (1992) 17. For a diachronic survey of the extant "lifting of the hand" (šu.íl.la) prayers concerned with the stars and planets see Mayer's (1976) section entitled "Gebete an Gestirne". Note that some of the prayers to the Moon and Sun gods allude to these deities' celestial manifestations. See Koch-Westenholz (1995) 113 n3 for those šu.íl.la prayers forming part of eclipse rituals at the time of Sargon II and Šamaš-šumu-ukīn and add Mayer loc.cit. Nergal 1 which concerns Mars on the occasion of a plague epidemic again under Šamaš-šumu-ukīn. Some so-called *lipšur*-litany prayers and incantations enumerate stars and planets, for which see Reiner (1995) 19–20.

48 See now Brown and Linssen (1997), who published BM 134701, which continues the text Clay *BRM* IV,6, for a discussion of the elaborate Hellenistic period ritual whose purpose was to avert the consequences of a lunar eclipse, and the NA examples which preceded it. The kettledrum referred to in line 9 of the NA letter SAA X 347 from Mar-Issar to Esarhaddon is one example indicating that this ritual was being practised (at least in Babylonia and probably in Assyria) during the period under consideration. Frame (1992) pp116–7 n77 provides a list of prayers and rituals that refer explicitly to Šamaš-šumu-ukīn. They are mainly from the series *bīt rimki* and designed to ward off the evil portended by lunar eclipses. See also Reiner 1995 n47.

Commentary texts
Some of these are known to have been composed between c. 750 and 612 BC. For example, the Assyrian "Marduk Ordeal" texts[49] are described by Grayson (1991b) 119 as learned commentaries, and are probably propaganda writings accounting for Sennacherib's destruction of Babylon's Marduk temple. Many specifically astronomical-astrological commentaries and other explanatory works[50] are attested from Assyrian sites dating to this period. They are characterised by instances of word play and learned allusions, and some examples are studied in Chapters 2 and 3. It is often unclear, however, when these texts were first composed. Some of them serve to explain the omens series, and may have been composed when the omen series were being put into their "canonical" forms. They were then perhaps copied along with the omen series. It is also possible that some of the commentaries and explanatory works accounted for omens that had otherwise become inexplicable since the omen series were assembled into their near-final forms. Their composition should perhaps then be dated to the period under consideration here.[51] Often the commentary texts, as with the commentary parts of the Letters and Reports, appear to permit an *elaboration* of an omen to take place, or justify its taking place. These elaborations were, I argue, a means by which the omens and schemes of celestial divination were adapted to suit contemporary circumstances from the earliest times. In fact it was the normal manner in which cuneiform celestial divination developed over the centuries.

Finally, the NMAATs that date to the period prior to c. 612 BC will be considered in detail in Chapter 4. One further work there studied is a short five tablet cryptic series found on the tablets DT 78, DT 72 and 81–6-25,136 which was composed prior to 612 BC and probably during the period of concern.[52] None of these texts, which exemplify the PCP Paradigm, are attested prior to c. 750 BC and those dating to our main period of concern manifest methods which anticipate those of the Hellenistic period, but which are also in their infancy.

All other texts employed in this study are listed Appendix 1. They have been used in a manner distinct from the material which can safely be dated to the period c. 750 – 612 BC. Little or no attempt has been made to place these texts in their social contexts. In particular, since only a small number date from the OB and MB or MA periods, it is unlikely that one could build up any picture of the state of cuneiform astrology-astronomy prior to 750 BC without recourse to the later material. Nevertheless, by working back and forward from a clearly defined *state* in the 8th and 7th centuries it has proven possible to argue when change has manifestly taken place and when it has not. As we will see in Chapter 3, celestial divination evolves gradually between the OB and the late NA period without any substantial changes in the underlying premises. In Chapter 4 I show that that the PCP Paradigm also evolves

---

49 See now Livingstone SAA3 texts 34 and 35 and idem *MMEW*.

50 For details see App.1 §§ 28 and 29, and for general descriptions Reiner (1991) 319, Koch-Westenholz (1995) 82f. See also Labat (1933) VII, XIII, XV & XX. Work on this important genre is currently being undertaken by E. Frahm.

51 This is the view adopted by Reiner & Pingree in *BPO2* and in *BPO3*. See also my discussion of the *mukallimtu "Šumma Sîn ina tāmartīšu" ACh.* Sîn3 in Ch.4.2.4.3.

52 DT 78 (BM92685), upper edge, line 2 reads A$^{!m}$an.šár-dù-a šàr ku[r...] where Gadd (1967) 61 reads the first sign as *apil* "son of". Hunger *Kolophone* No. 496 reads it as *šá* "of", which is more likely. The remainder reads: "Assurbanipal, king of the lan[d of Assyria]". See also App.1 §29.

gradually, from c. 750 BC until the Hellenistic period, without any substantial changes in its underlying premises.

I.4 The Archaeological Context of the Source Material

Much can be gleaned about the purpose to which the texts considered were put from the contexts in which they were found. Most of the textual material dating to the NA period is believed to have come from archives and libraries associated with the institutions of the temple and royalty. Very little is known to have come from private collections. This is partly due to the nature of what has been considered worthy of excavating, but it is also clear that celestial divination in particular was mainly for use by the king, a matter to which I shall return in the next chapter. The archives and libraries differ somewhat in their contents. Some attest to an extensive collection of literary texts and to those materials necessary for learning the scribal art, others to an emphasis on divination. This tells us something about the function of the scribes associated with each, and the extent of royal patronage. This in turn provides information on the conditions under which the particular experts I am considering here were working. It is apparent that the great celestial divination series EAE was found in most good libraries at this time, both north and south. Despite this, it is only in the Letters, Reports and a few texts from Nineveh and Babylon that evidence for the emergence of works connected to the accurate prediction of celestial phenomena can be found. This suggests that these developments were not merely associated with EAE-type divination, but with the particular circumstances under which the Scholars who wrote to the late Assyrian kings worked – in other words they can be ascribed to particular demands of royal divination *at that time*.

Most of the texts herein studied were discovered in what are loosely referred to as "the Royal Archives and Libraries[53]" of **Nineveh**, (*Ninua*) capital to the Assyrian kings Sennacherib, Esarhaddon and Assurbanipal. This vast assemblage of material included literary and divinatory texts, reference works and those associated with scribal education as well as a great deal of royal correspondence. One major collection was associated with the so-called North Palace completed in 645 BC,[54] another with the earlier South West Palace.[55] The former in-

[53] There is some overlap between "library" and "archival" texts, but broadly the former includes those texts which are "finished products" (Reade, 1986a, 219). They often have colophons. They are the texts we believe would have been regularly consulted by Scholars and scribes, often the "canonical" material. One word translatable as library is *gerginakku* written im.gú.lá, im.gú, or im.lá. It appears to be a Sumerian loan word from gìr.gin.na = "sequence of tablets" and perhaps refers to a series (*CAD* G 87 and Hunger Kolphone 162). In Nineveh it is attested only in reference to the Nabû temple library and otherwise attested to refer to Uruk's and Huzirīna's Ištar temples (Hunger *Kolophone* no.106), so perhaps *gerginnaku* described temple libraries. The *bīt ṭuppāti* "tablet house" (Hunger *Kolophone* 314: 6, *AHw* ṭ/tuppu(m) 2) a) can refer to a library, archival room, scriptorium, or school, the distinctions between which are probably clearer to us than to them. "Archives" normally refer to collections or repositories of texts deemed to be significant, but no longer in use. Veenhof (1986) 7 notes, however, that in Assyriological contexts "archives" describe "the total records accumulated during the time a particular task was performed by a particular institution." This usually means a collection of correspondence or administrative texts. In Nineveh, archival texts were usually made of an inferior clay, were also unbaked, and lacked the colophons and "finished" quality of the library texts. On the nature of archives in the Ancient Near East we await the publication of the volume based on the 17th-19th September 1998 symposium in Christchurch, Oxford.

[54] Assurbanipal's Palace, built on the site of the *bīt redûti*, the "house of succession" or crown prince's residence.

[55] Sennacherib's "Palace without a rival" which was constructed between 703 and 694 BC and used by him, Esarhaddon and Assurbanipal until 645 BC.

cluded the famous "Assurbanipal library", the latter a series of libraries and archives formed in a variety of ways. Parpola (1983b) 10f discusses the production of tablets on site, the use of private collections as the core of the palace or temple libraries,[56] and the formal sequestration of the collections of conquered lands or dominated peoples.[57] Some library records from 647 BC[58] list the "contributions" from private individuals (and in one case from the "house of Ibâ") mainly based in Babylonia which were, Parpola suggests, the result of the 652–48 rebellion in that country and the direct control over its southern neighbour that Assyria had then once more assumed. No doubt the acquisition and re-use of such "spoils of war" as these texts was partly due to Assurbanipal's particular literary bent (see below), but I suggest that it may also have represented the assertion of new Assyrian confidence in matters "scientific". Such is its scale that it no doubt enhanced Assyrian scholarship and impeded Babylonian.[59] It is possible to assume that the North Palace library was built to house a new comprehensive and definitive collection for the Assyrian Scholars, now that Babylonia had been crushed once more. It seems to me plausible that the absence of any Letters or Reports datable to the period after 647 BC (see below) is connected to the creation of this library, itself partially inspired by a post-Great Rebellion Assyrian desire for self-sufficiency in divination and other scholarly crafts.

Often the precise find-spots were not recorded and some collections were mixed, so that even a K. number does not absolutely guarantee a Kuyunjik[60] origin. Of the texts herein discussed, the following is known about their original locations:

- Texts with K. numbers 1–278 were almost all found by Layard in the SW Palace in rooms XL-XLI. Assyrian and Babylonian Scholars' Letters and Reports, and commentary texts form part of this group.
- S/Sm texts are mostly from the SW Palace. Some Assyrian Letters, and Assyrian and Babylonian Reports have Sm numbers.
- The collection 83–1–18 of 900 tablets and fragments, including many Letters and Reports, seems likely to be entirely from the SW Palace, as Parpola (1986) 228–30 has shown. This collection includes mainly archival texts, but some library texts as well.
- 90% of the Rm2 collection[61] of 606 tablets and fragments, most of which are library type, comes from the North Palace. Only two Letters and two Reports have Rm2 designations.[62]

[56] Most famously Nabû-zuqup-kēna, a Scholar of Sargon II and Sennacherib is attested as the owner, scribe (*šaṭāru*) and collationer (*barû*) of tablets originally from Kalḫu, many of which have been found at Nineveh in library context.

[57] The well known text *CT* 22 1 (Waterman *RCAE* IV 212–5) from an Assyrian king to the governor of Borsippa explicitly demands the confiscation of texts from private and temple libraries.

[58] Parpola, 1983b 6 and SAA7 49–56.

[59] I. Finkel (lecture at Inst.Arch., London, 30/1/95) suggests that the individuals named in these library Records may have been authorised to collect the texts listed from their local vicinity, and that therefore the texts represent the collections of many individuals. This would also partly explain the observation made by Parpola (1983b) 8–9 that the tablets listed against the individuals do not include works related to the specialisations of those individuals. They represent instead what they had collected from their local region. Perhaps, by being authorised to do the collecting they were exempt from submitting their own collections. Finkel's suggestion also helps explain the variation in the number of texts given by scholars of equal repute, and indeed the large number of copies of works submitted by the individuals.

[60] The name of the larger of the two mounds near Mosul on which the remains of Nineveh lie.

[61] Parpola (1986) 230–1.

[62] Rm2,6 = x033 Assyrian Letter probably dating to the end of Sennacherib's reign at the earliest.

- DT texts were mainly from the North Palace. Only three Assyrian Letters have DT designations, two of which (x137 & x347) seem to date to Esarhaddon's reign. Four undated Reports, three of which are Babylonian, have DT designations. For DT 72+78, and 81–6-25,136 see n52. Despite their DT numbers these few DT texts probably belong to the SW Palace archive.
- Report BM 123358, and Letters BM 123359, 134556 were found in the Ištar temple, probably in secondary context.

In summary, it is known that the North Palace housed a library collection from Assurbanipal's time and archives from the *bīt rēdûti* (n54) dating to the period before the construction of the SW Palace (before c.703 BC).[63] The SW Palace housed the great mass of the Scholars' Letters and Reports. Not one of these can be dated to Sennacherib's reign (c. 704 – 681 BC), or indeed to between 708 and 681 (see below). It is probable that the Scholars' archive in the SW Palace *deliberately* did not contain any Letters or Reports dating to Sennacherib's reign, and yet the many legal documents found with them have been dated to almost every year between 710 and 680.[64] Apparently, the correspondence sent by Scholars to Sennacherib (which is certain given what is stated in x109 r.1ff & x076: 11f) were deliberately isolated from the legal records (and perhaps destroyed) in antiquity. This was, I tentatively propose, because of the sensitive nature of the material possibly contained therein.

Similarly, the rapid diminution in the number of Scholars' Letters and Reports datable to the years after 650 BC suggests that those written after Assurbanipal had built the North Palace in 645 BC were stored elsewhere.[65] The legal texts show that the archive in which the Letters and Reports were found in the SW Palace, known as the "Chamber of Records", was used until the destruction of Nineveh in 612 BC. Thus, the attested collection of Scholars' Letters and Reports include *only* those that were written during Esarhaddon's and Assurbanipal's residence in that Palace, which is strongly suggestive of the *royal* nature of

Rm2,409 = x250 Assyrian Letter probably dating to Esarhaddon's reign at the earliest.

Rm2,345 = 8501 unassigned Report in Babylonian, but convincingly dated to Nov. 27 -708. Perhaps this text was originally housed in the *bīt rēdûti*, and returned to the same site after 645 BC.

Rm2,254 = 8546 unassigned Report in Babylonian and undated.

63 See Parrot (1955), Parpola (1986), and Reade (1986a).

64 See SAA 6 xviiif for the distribution of the NA legal archives. This point has already been made by Parpola (1981) 120 n.3, and repeated idem (1986) 235. To the best of our knowledge it applies to all the correspondence (not just the Scholarly) found in the SW Palace. The further c.400 letters from priests and various officials involved in temple and palace administration appear to follow the same pattern of temporal distribution as the Scholars' correspondence, though only very few can be dated accurately.

65 Only one Letter, unassigned, has been dated by Parpola to the period after 645 BC. This is K.1216 = *ABL* 1444 = *LAS* 105 = SAAX 149 which he dates to 22/4/621 BC. However, on close inspection of the data (presented by Parpola *LAS* II p90–1) it appears to me that the most likely date for the eclipse this Letter records is June 2nd, 679 BC. This date corresponds to the Simānu 14th derived from new-Moon computations in *LAS* App.A, whereas the April 22nd, 621 BC date implies that the Mesopotamian year 621 BC instead of starting within a month of the spring equinox, started more than one whole month too early (*LAS* p90). This seems to me to have been extremely unlikely, much more unlikely than that which argues against the 679 BC date. This is that in rev.1 of the text it states *ina* ki mul.gír.tab *a-dir*, "It was eclipsed in the region of the Scorpion constellation", where ki is read *qaqqaru*. The calculations show that in 679 BC the partial (0.13 – see now Steele & Stephenson, 1997/8) lunar eclipse took place in the morning, with the southern quadrant eclipsing, *only* some 12° beyond the boundary seen by Parpola to describe the limit of Scorpius (*LAS* App.C). Not to see the 679 BC date as the most likely, seems to me to overdefine *qaqqaru* (*CAD* Q 121 5 b) in order to date one text to a period almost thirty years later than any other Report or Letter, and to suggest that this was also a time of severe calendrical neglect.

these texts, by which I mean they were intimately connected to the person of the king and not merely to the ruling institution.

In **Assur**, modern Qal'at Širqaṭ, destroyed in 614 BC, a number of institutional and private libraries and archives dating to the late Assyrian period were discovered. Assur was not the capital of Assyria at this time and the texts discovered there reflect this. No celestial Reports or Letters to the kings have been found, for example. Nevertheless some astronomical-astrological texts have survived and the contexts in which they were found are of interest here:[66]

- The official library and archives of the Aššur temple housed texts dating from Old Akkadian times on, including celestial omens from both the MA and NA periods.[67] A private library and archive of a family of Assyrian scribes (*ṭupšarru aššurû*), dating from at least 687 BC to the time of Sîn-šarru-iškun, contained celestial and other omen texts and a copy of Mul.Apin, suggesting that the trade of astronomy-astrology was quite prevalent in this family during Sennacherib's reign.
- An archive and private library of the *nargallu* chief singers/musicians included texts from c.750–614 BC with at least one astronomical-astrological school tablet.
- A large library of the exorcists' guild included texts from as least as early as the MA period. Nabû-šallim-šunu, the *ummânu* of Sargon II, scribe of *Sargon's Eighth Campaign* is attested in a colophon. Most of the texts recovered concern exorcism (incantations, prayers, and rituals), and a very few celestial divination texts are known. Also of interest from this library is a text which concerns divination based on shooting stars and a ritual and prayer to Ursa Major on the obverse, and divination (broken) concerning the flights of birds with an associated ritual (broken) and prayer (fragmentary) on the reverse.[68] As another text[69] from this collection indicates, part of the repertoire of exorcistic knowledge included "oracles" of stars, birds, oxen, and wild animals, and (r.16) ud.an $^{d+}$en.lîl.lá = EAE. It appears as if these exorcists in Assur were competent to interpret celestial and other omens, but only occasionally did so, as the absence of omen texts would indicate. Reiner (1960a) 30 suggests that the type of divination performed by exorcists may often have been of the "yes-no" variety, and designed for private individuals, as against the EAE-type divination which was more or less exclusively designed for the royal family and the state. That kind of divination was perhaps more commonly undertaken by the chief singers or the Assyrian scribes.

The many archives located in private houses in Assur contain, with very few exceptions, only documents concerned with legal or financial matters. They probably belonged to people who could at most read, but for whom writing was not a necessary skill, as the absence of lexical material suggests. No doubt they employed scribes when needing to record significant events. If celestial divination was a craft they employed, no doubt the prognostications were delivered orally. Celestial divination *was* undoubtedly practised in Assur, perhaps on

66 Work on this material is currently underway in Heidelberg, but for the time being Pedersén (1985&6) has provided the relevant information.
67 See Weidner (1952/3).
68 Reiner (1960a) 28–9.
69 Ebeling *KAR* 44 – the "exorcist's manual". For an edition see Bottéro (1985).

behalf of the local dignitaries, perhaps only for members of the royal family. If an archive of Letters and Reports sent to such dignitaries was kept, it has evaded discovery thus far. Written celestial divination for the king was probably undertaken by Assur's Scholars, but then sent to him in the capitals – Kalḫu, Dūr-Šarkēn or Nineveh.

Kalḫu (modern Tell Nimrūd) was the capital of Assyria from the time of Assurnaṣirpal II (883–859 BC) until that of Sargon II.[70] It was destroyed between 614 and 612 BC. Celestial divination, amongst many other types of divination, was practised there before 750 BC.[71] This is important, for it demonstrates that prognosticating from heavenly phenomena was not a *new* royal activity in the late 8th and 7th centuries BC. Far from it, some royal connection with EAE Paradigm divination is perhaps as old as the OB period (see Chapter 3), particularly in Babylonia. Of the hundred or so letters from Kalḫu that have thus far been published (Saggs, 1955–72), most of which date to Tiglath-Pileser's reign, *none* concerns astrology-astronomy. Given the type of letter, this is not significant in itself, but I shall adduce later that although royal celestial divination is present in Assyria prior to c. 750 BC, it becomes more important after the reign of Sargon II. It is also interesting to note, as Kinnier-Wilson (1972) 75 argues, that the *bārû* or haruspex/diviners residing in Kalḫu in the first quarter of the 8th century BC were mostly Babylonians.[72] This was a period prior to direct Assyrian involvement in Babylonia, and suggests that at that time only a few Assyrian Scholars were considered worthy of royal patronage. In later times, perhaps as a result of Assyrian oppression of the south and as result of growing Assyrian confidence in intellectual matters, the relative number of elite Assyrian scribes increased.

The Esarhaddon Vassal Treaties (Wiseman, 1958 & SAA2: 6) were found in Kalḫu, and yet by the time these were written Nineveh was the Assyrian capital. It is apparent that Kalḫu remained an important scribal centre after the royal court had moved on. It is undoubtedly significant to the development of cuneiform astronomy-astrology that many scribal centres were engaged in producing works for the Ninevite kings, amongst which were the Letters and Reports containing celestial predictions. As we shall see the influence of one Nabû-zuqup-kēna of Kalḫu was particularly important in Ninevite scribal circles. If the developments which took place at this time can be ascribed to particular individuals, Nabû-zuqup-kēna has perhaps the greatest claim to be one such amongst the Assyrians.

Other relevant texts come from **Huzirīna**, modern Sultantepe,[73] situated in Southern Turkey, where in 1951–2 a small "library" of several hundred texts was uncovered. Attested dates range from 718–612 BC, and the collection appears to have been the product of a temple (Ištar?) school. *STT* 73, discussed in Reiner 1960a, describes the type of

---

70 Details from Mallowan (1966), Postgate and Reade (1977–80), Wiseman (1968), Kinnier-Wilson (1972) and Black & Wiseman (1997).

71 Several tablets of EAE were found amongst the literary tablets of the Nabû temple (*CTN* 4 texts 1–30). *CTN* 4 8 = ND 4367 dates to 787 BC. The Kalḫu versions of EAE differ slightly from the Ninevite versions. Compare for example *CTN* 4 10 (now in Hunger, 1998) incorporating part of EAE Tablet 14, with the manuscripts presented in Al-Rawi & George (1991/2).

72 In the Letters from Nineveh the *bārûs never* report on celestial matters, though this does not indicate that they did not do so in the 780s in Kalḫu. Perhaps this function came to be performed by others, specialists, scribes of EAE etc., as the discipline grew in popularity under Sargon II and the later kings.

73 Details from Finkelstein and Gurney (1957) = *STT* I, Gurney and Hulin (1964) = *STT* II, and Postgate (1972–75).

divination seemingly performed for private individuals by exorcists, some of which concerns shooting stars. Texts *STT* 329–339 include typical astrological-astronomical library material, including omens from EAE and fragments of Mul.Apin. Most interesting is *STT* 300, which dates from 619 BC, and which parallels the Persian-Seleucid period Uruk text *BRM* IV 19 (MLC 1886),[74] though without the "dodekatamoria" section. The existence of these texts in a site so distant from Nineveh attests to the widespread importance of celestial divination at this time, both of the royal sort and of the exorcistic sort. I am unaware of the existence of any relevant material in the largely unpublished collections from Dūr-Šarkēn and Tarbīṣu.

The shortage of textual material found in Babylonia, by comparison with that found in Assyrian cities dating to the period c.750–612 BC, is mainly explained by the absence of the c. 612 BC destruction wrought on the northern nation. To my knowledge, only **Babylon** (*Bābili*) has produced material relevant to this study, aside from one letter from *Sippar* (modern Abu Habba)[75] and some royal inscriptions from *Nippur* (modern Nuffar).[76] Most of the textual material from Babylon was retrieved from uncontrolled excavations, and the find-spots are generally unknown. The astronomical-astrological material from c. 750–612 BC was probably located with the later astronomical-astrological texts in what has become known as the "astronomical archive", though this term disguises more than it reveals. It was probably associated with the Marduk temple, but short of this little more archaeological information is known (see n377, below). Those texts of interest here are generally later copies of material which has been dated on astronomical grounds. They are discussed in full in Chapter 4. One tablet containing part of EAE, and dated in its colophon to Sargon II's reign, has been found.[77] Babylon was Sargon's residence prior to Khorsabad, and this may have come from a collection of astrological-astronomical texts that were being used there to guide his fortunes.

Many texts of interest here that were found in Assyria came originally from Babylonia. Scholars from Babylon, *Borsippa* (*Barsipa*, modern Birs Nimrūd), *Dilbat* (modern Tell Dulaim), *Cutha* (*Kutû*, modern Imam Ibrahim), *Ur* (*Uru,* modern Tell al-Muqayyar), and *Uruk* (modern Warka) wrote to the Assyrian kings. Nippur must have been a city in which Scholars made observations, as x114: 7 and x347: 7 make clear, though none can, as yet, be shown to have resided there. Mar-Issar made celestial observations in *Akkad* (location unknown) after its temporary resettlement, as x347 shows, though perhaps no Scholars resided there permanently (*LAS* II p269 6ff). *Bīt Ibâ* (unknown location) and Nippur are cited in the Assyrian Library records as the home towns of various Scholars required to (collect and) hand over library texts. *Der*, *Kiš*, *Eridu*, and *Larsa* were the source of library texts as, for example, the colophons on texts from Assur reveal (Hunger *Kolophone* Nos.292, 185–190.) They

74 Clay (1923), discussed in Ungnad (1944), and Neugebauer and Sachs (1952–3).

75 CBS 1471 – it is the sole example of a Letter or Report attested from Babylonia. This is not surprising since, in addition to the reason cited above, most correspondence between the Scholars and the Assyrian kings was from the former and to the latter. Indeed CBS 1471 (Parpola x295 = *LAS* 226) was from Assurbanipal to Urad-Gula. It probably dates to the period covered by Urad-Gula's other correspondence (672–668 BC), though Parpola (*LAS* II p218) suggests that it might be connected with Assurbanipal's compiling of his library at the time of the construction of the North Palace at Nineveh (c. 647 BC). Recent excavations in Sippar have uncovered astronomical-astrological material from the period soon after 612 BC.

76 Brinkman (1984) 116 n561.

77 Hunger *Kolophone* No.150. Loc. cit. No.154 is not unrelated.

were probably also major scribal centres during the period under discussion. Other administrative centres, and major towns in Babylonia would have employed scribes, but perhaps no Scholars.

To summarise, temple library collections in Assyria and Babylonia endeavoured to maintain a collection of texts that more or less reflected the canonical repertoire of scribal material, a repertoire that was probably established by the end of the second millennium BC. The Aššur temple collection exemplifies this, and was no doubt maintained throughout the period of Assur's status as the capital city. The collection in the Nabû temple in Kalḫu was probably only brought together around 800 BC[78] and the relative paucity of literary as opposed to divinatory material no doubt reflects the more practical, royal nature of the collection. In Nineveh the collections were mainly royal, and in the case of Assurbanipal's library specifically belonged to the person of the king. While some older material was associated with temples in Nineveh, the large collections in both palaces were rapidly assimilated only while the kings were in residence in that city. To some extent this collecting was a matter of prestige, but it also attests to a particular interest on the part of the last NA kings with divination and cultural literary heritage. These kings employed a large number of scribes and Scholars and took great pains to establish and maintain large library collections, an activity that was formerly performed largely by the temples and their associated personnel. Conceivably, this was a deliberate act of *secularisation* of the literary heritage, and was designed to bring the scribes under more direct control. The religious rôle of the late Assyrian kings must not be understated, however. Either way this act brought together, both in Nineveh and through extensive correspondence, the best Scholars in Mesopotamia, and it was this fundamental change in scribal practice which led, at least in part, to developments in astronomy-astrology of far-reaching significance.

I.5 The Distribution of the Datable Source Material between c. 750 and 612 BC

The establishment of the date of composition is critical both if a synchronic analysis of the "state" of Mesopotamian planetary considerations is to be attempted, and if the context in which they were written is to be determined. The few royal inscriptions which mention the planets can be dated: Sargon; *Eighth Campaign* (714 BC), Esarhaddon; Ass.A I: 30-II: 26, Ass C II: 1–12, and Bab.Ep.13: A: 34–41 all describe celestial events in 680 BC, Assurbanipal; K2652 (post 653), K3087 & 3405 (post defeat of Hazâ'ilu, king of the Arabs), K4449 (early Ass.), Šamaš-šumu-ukīn; CBS 733+ (pre 652). Similarly, some treaties and loyalty oaths are datable. SAA2 texts 6 (672 BC), 8 (after Nov.669 BC), 11 (627–612 BC), and 14 (680–669 BC). All the relevant Assyrian "literary" creations can be dated to Assurbanipal's reign, except SAA3 37 & 38, which are undated, and 39 which is older than the copy written by Kiṣir-Aššur, who is attested in 658 BC in Pedersén *ALCA* N4 No.69. *Erra* perhaps dates to Nabonassar's reign.

Some of the Letters and Reports have been dated astronomically by successive students, principally Schaumberger (1938), Schott & Schaumberger (1941/2), Hartner (1962), Parpola *LAS*, SAA8, and SAAX, and very recently de Meis & Hunger (1998). Parpola was also able to date many of the texts through a comparative analysis. His results can be found

[78] Wiseman & Black (1996) 4.

in *LAS* II Apps. I and J and SAAX xxix. Parpola has dated to within certain bands of years some 247 Assyrian and Babylonian Scholars' Letters, that is 247/389 or 63% of them:

Date (BC)	Number of Letters
680–675	13
674–672	18
671–669	170
668–665	25
664–658	0
657–655	5
654–652	0
651–648	15
621	1

Hunger in SAA8 xxii assigns dates to 120 Reports, some 120/567 = 21%:

Date (B.C)	Number of Reports
709	1
680	2
679	3
678	4
677	4
676	3
675	14
674	7
673	9
672	11
671	4
670	10
669	19
668	5
667	14
666	6
665	0
664	1
657	3
649	1

I have reconsidered the dating of all the Reports and all the Letters. Those texts which I now feel can be securely assigned a date accurate *to within a year* have been listed in Table 1. They number many fewer than those considered datable by Parpola and Hunger in SAAX and SAA8 respectively. My reasons for considering a date to be unreliable, or for dating, or redating a Letter or Report are outlined in Appendix 2. For completeness, I have added to the table those other texts being considered in this book whose date of composition is known. Note that I have included the few texts that were copied much later, but which contain material that records celestial events from known years between 747 and 612 BC. The assumption is that these late copies must have derived *originally* from written material composed very shortly after the celestial events recorded. I have plotted them according to the *earliest* datable record which they now contain or contained when complete. Texts written in Babylonian[79] are italicised.

Table 1. The Distribution of Cuneiform Texts Concerning Planets and Datable to the Year BC

Reports (8iii = SAA8); Letters (xiii = SAAX); Treaty Oaths (2iii = SAA2); Diary (Diar); Eclipse Recs (*LBAT* No. 141n, see App.1 §32); Saturn/Mars Records (Sat.R/Mar.R, see App.1 §§32 & 41); *Sargon's 8th Campaign* (8thC); BM36731 (App.1 §38); Inscriptions (AssA/AssC/BaEp, see n42).

747	*1413, 1414+*	TIGLATH-PILESER III
746		
745		
744		
743		
742		
741		
740		
739		
738		
737		
736		
735		
734		
733		
732		
731		
730		
729		
728		
727		
726		SHALMANESER V
725		
724		
723		
722		
721		
720		SARGON II
719		
718		
717		
716		
715		
714	8thC	
713		
712		
711		
710		
709	*8501*	
708		
707		
706		

[79] And thus we assume by Babylonians – see below.

705		
704		
703		SENNACHERIB
702		
701		
700		
699		
698		
697		
696		
695		
694		
693		
692		
691		
690		
689		
688		
687		
686		
685		
684		
683		
682		
681		
680	AssC, AssB, *BaEp, x109, Mar.R*	ESARHADDON
679	x149, *8502*	
678	*8300, 8289, 8316, 8336, 8500, 8535*	
677	*x113*	
676	x084, 8100, *8301, 8438*	
675	*x111, x112, 8317, 8324, 8339, 8356, 8369, 8456*	
674	*8247, 8248*	
673		
672	2006, x185, x238, *8253*	
671	x011, x012, x040, x041, x067, x189, x240, x314, x347, x348, x349, x350, x359, x377, *x168, 8244, 8340*	
670	x043, x044, x194, x195, x196, x197, x198, x199, x200, x201, x241, x242, x243, x244, x245, x246, x247, x248, x252, x253, x254, x255, x256, x257, x258, x259, x260, x261, x274, x297, x301, x302, x305, x306, x315, x316, x351, x352, x353, x354, x356, x357, x358, 8114, *8341*	
669	x023, x024, x025, x026, x027, x047, x048, x050, x051, x052, x055, x072, x074, x128, x148, x152, x362, x363, x364, 8004, 8049, 8050, 8082, 8083, 8102, 8115, 8168, 8169, 8170, *x371, 8327, 8381, 8383, 8416, 8491, 8505*	
668	8051, *x172*	ASSURBANIPAL
667	x057, x075, x076, x077, x174, x224, 8052, 8053, 8055, 8085, 8103, *8387, 8418*	
666	x090, x226, x227, x228	
665		
664		
663		
662		
661		
660		
659		
658		
657	x100, x159, x381, 8008, 8104, 8186, *8384*	
656		

655	
654	
653	
652	*Diar*
651	
650	x104, x138
649	x139, *8487*
648	x141
647	*Sat.R*
646	
645	
644	
643	
642	
641	
640	
639	
638	
637	
636	
635	
634	
633	
632	
631	
630	
629	
628	
627	
626	
625	
624	
623	
622	
621	
620	
619	
618	
617	
616	*BM36731*
615	
614	
613	
612	

From Table 1 it can be seen that I consider 163 Letters and Reports to be dated securely to within a year. This represents only 158/(389+567) = 17% of the total. This small percentage makes most statements concerning the distribution of the texts statistically unprovable. For example, it is tempting to note that early in Esarhaddon's reign there appear to be more Babylonian than Assyrian Letters and Reports (particularly considering Bēl-ušezib's Letters which date to this period, but which cannot all be assigned to specific years), a situation which is then seemingly reversed later in his reign. However, the number of texts datable to the early period of Esarhaddon's reign is so small as to make this observation unreliable. All that can be said is that at least one Assyrian Scholar (x149) was writing to the king in 679 BC. Combined with what is known about the Assyrian Scholars working in Kalḫu, from the date of 8501, and from the content of x109: 8 there is no reason to suppose that Assyrian Scholars did not write Letters and Reports to Sennacherib, Sargon II, and perhaps to the earlier kings.

The very large numbers of texts that can be dated to 670 and 669 BC are statistically interesting, however. Many can be dated because of their references in the greeting formulae to the crown prince and so to the joint reign of Esarhaddon and Assurbanipal (672–669). Nevertheless, the number of Reports and Letters datable to these two years is more than would be expected from an even distribution over the years of interest. As argued above there is good reason to suppose that the SW Palace collection of Scholars' Letters and Reports found at Nineveh cover the years 680–647, from the commencement of Esarhaddon's reign to the building of the North Palace, a span of some 34 years. This is fully confirmed by the distribution of the Letters and Reports datable to the year. The one exception is 8501 = Rm2,345 from 709 BC, which was perhaps found in the North Palace (n62) or in secondary context. 389+567 = 956 Letters and Reports are attested. Distributed evenly over 34 years this would imply some 28 texts per year. Clearly, the 45 texts datable to 670, and the 36 datable to 669, already exceed this average despite the fact that only 17% of the total archive can be assigned a year date. This uneven distribution cannot easily be explained by the vagaries of recovery, for the sample is statistically large. It is very unlikely that the excavators pulled out 950 odd texts, many of which can be dated to years other than 670 and 669, but which still constituted only a section of an evenly distributed archive. The reasons for the large number of texts dating to the end of Esarhaddon's reign must be sought elsewhere.

The increased activity of the Scholars in 669, 670 (and 671) BC might be partially explained by an increasing prosperity, particularly in the South prompted by Esarhaddon's conciliatory attitude towards Babylonia.[80] More importantly, the increased correspondence may well have been due to the king's increasing ill health. We know Esarhaddon fell ill repeatedly from 672 to 670 and died on November 1st, 669.[81] Many of the Letters from these years are from exorcists and from Urad-Nanaya, the chief physician. Similarly, the Letters and Reports concerning celestial divination from these years probably reflect a heightened interest in the celestial manifestations of the gods' decisions concerning the health and fate of the king. Significantly, most of the texts datable to these three years are from Assyrian Scholars whom one might expect to be more concerned with their king's well-being. The part played by the Scholars in "guarding" the king against misfortune will be pursued in the next chapter, and appears to be borne out by the evidence from Table 1.

There is clearly a diminution in the number of texts sent by the Scholars to the king once Assurbanipal ascended to the throne. This must partly be due to the other court of Šamaš-šumu-ukīn to whom the apodoses concerning Akkad then applied. He, no doubt, employed an entourage of Scholars whose Letters and Reports have probably been lost to the destructions wrought on, or to the rising water table beneath, Babylon. Whether or not some Scholars worked both for Assurbanipal and Šamaš-šumu-ukīn is unclear, for Table 1 shows that Babylonian scribes did send Reports to Assurbanipal (*8387*, *8418*, *8384* and *8487*). It would seem reasonable to assume that the Scholarly entourages of Assurbanipal and his brother the king of Babylonia were separate and home-based. However, this perhaps ignores the requirement of the celestial divinatory industry to have observation stations sufficiently dis-

[80] Porter (1993).

[81] Parpola *LAS* II App.K.

tant from each other to experience differing weather conditions such that one location might not be cloudy at the critical time of a celestially significant phenomenon. I suggest that, if part of the purpose of predicting celestial phenomena were to enable the celestial diviners to prognosticate in spite of inclement weather,[82] then this was further motivated by the increasingly restricted nature of the enterprise in the later periods. Under the Iranians and Greeks there were probably no networks of observation posts dotted around the empires as there had been under the rule of the NA kings. Isolated Babylonian astronomer-astrologers could not then expect reports from elsewhere in the event of cloudy conditions, but were forced to calculate the celestial phenomena instead.

Table 1 indicates that the distribution of Reports closely follows that of the Letters. This indicates, in combination with the argument made in the next chapter, that one can talk of an "archive", perhaps "dossier" of Letters *and* Reports. Finally, despite the fact that the table suggests that all the *innovative* texts datable to this period – the Diary, the Eclipse and Saturn Records and so forth – were Babylonian, it will be shown that there is no evidence that the Assyrian Scholars were any less advanced in the technologies of prediction than were their southern counterparts.

I.6 The Languages and Scripts Used in the Textual Sources

Three main dialects of Akkadian, the Sumerian language, and two main scripts are used in the texts herein studied.[83] The scripts are sometimes coarse, with large signs and a few lines, and sometimes condensed in order to accommodate a large amount of information. The dialects are Standard Babylonian (SB), derived from the Old Babylonian dialect, Neo-Babylonian and Neo-Assyrian. The scripts are Assyrian and Babylonian. A text whose script is NA is considered to have been written by a native Assyrian, even though the dialect may be one of all three. Similarly, Babylonians are considered to be the authors of those texts in NB script.[84]

Texts which are believed to have been composed and broadly standardised before the NA period are written in the SB dialect. EAE and its associated commentary texts are written in SB.[85] Even the quotations from the omen series found in the Letters and Reports are in SB, regardless of the nationality of the author. When not quoting, those who wrote the NA script generally used the NA dialect,[86] and vice versa for NB. There are some interesting exceptions, however. For example in 8316: r.2 Munnabitu, a Babylonian writing in Babylonian

[82] Swerdlow (1998) 18 "...inclement weather may have been of unexpected benefit as the principal motivation, perhaps the entire motivation, for the development of mathematical astronomy, in order to determine by calculation the dates of ominous phenomena concealed by clouds..." This is only one of many motivations lying behind the Scholars' development of predictive astronomy which I outline in Chapter 5.

[83] For a brief summary of the development and interrelationships of the Akkadian dialects and Sumerian see Livingstone SAA3 xvf and n1 for a short bibliography.

[84] Some texts in Babylonian script have colophons indicating the Assyrian Nabû-zuqup-kēna as author. He was perhaps merely the owner (Livingstone, 1997 171) though the ability to write in two scripts is not impossible.

[85] Unusually in line 16 of the EAE *mukallimtu* published in Borger (1973) text 1, is found the Assyrianism "*da-'u-ú-mat*".

[86] E.g. Issar-šumu-ereš in SAA8 1: 7 uses *lā* instead of *ul* before the verb to indicate negation, which is typically Neo-Assyrian. *Epāšu* is used by the Assyrians, where *epēšu* is used by the Babylonians – op.cit. p318, and so forth. For a rendition into American (the NA) and British English (the SB) of Report 8232 see Livingstone (1997) 169–70.

script, writes: 1-*en* nun lugal *lu-še-en-ni la qip-ti-šú lu-pe-et-tu-šú* "let the king change one prince (from amongst various nobles) and sack him from his office" which shows a Babylonian slant in the use of the *e* in the verb *petûm*, and an Assyrianism in the use of the precative prefixes *lu*. Ordinarily in the Babylonian dialect *li* is used for the 3rd person precative prefix. Munnabitu is perhaps trying to write in Assyrian because he is writing to the Assyrian king, though in general the Babylonian Scholars wrote to the Assyrian kings without attempting to modify their dialect. Babylonian Bēl-ušezib writes in x109: 11 *ú-še-zi-ba-am-ma*, where Assyrian Issar-šumu-ereš writes in x020: 5 *ú-še-za-ab*.

Compositions in the 8th and 7th centuries BC are attested in SB, NA, and in NB. This makes even the approximate dating of SB compositions difficult without the presence of clear Neo-Assyrianisms or Neo-Babylonianisms. Many of the texts edited in SAA3 are written in the NA dialect, demonstrating NA composition. The treaty oaths and royal inscriptions are essentially written in SB, although composed by Assyrians. This is often demonstrated by the presence of Assyrianisms, and by the absence of Babylonianisms (e.g. the Neo-Assyrian use of *u-ni* as the subjunctive marker in SAA2 text 6: 8). This seems to imply that the Assyrians emulated Babylonian scribal forms, but that their Babylonian teachers (from whatever generation), whilst using NB themselves, did not pass on anything except the *hoch*-Akkadian of SB to their NA counterparts. This would suggest that the transfer of literate knowledge from Babylonia to Assyria did not particularly take place during the late NA period. This, again, hints that Assyrian Scholars were not the cultural parochials they are sometimes thought to have been in the 8th and 7th centuries BC. They played a full part in the intellectual life of the region and contributed to the emergence of the PCP Paradigm, I suggest.

Other than clay and stone, which survived, it is known that wax-covered boards, some polyptychs of many leaves, papyrus, parchment, and leather were used for writing.[87] The *le'u* "board" was by the MA period, at least, covered with wax. They were employed in administration where their reusable nature was probably helpful, but were also used for the inscription of library texts (e.g. *CAD* L 159 b') including EAE (cf. SAA8 19), Mul.Apin (cf. SAAX 62) and others. From their frequent mention in the Letters and Reports (Assyrian and Babylonian), it is clear they formed a significant portion of the literary material the Scholars came into contact with.

On the basis of library Records from Nineveh dating to 647 BC in which *le'u* is used to refer to writing-boards of more than one leaf, namely polyptychs, Parpola (1983b) argues that in this one particular instance the Assurbanipal Library acquired around 2000 tablets and 300 writing board leaves. This compares with the estimated 10,000 complete tablets excavated at Nineveh (18% of which are *not* library texts) and gives an idea of the size of this particular acquisition and of the amount of writing material that has perished. It is, however, unlikely that any library texts written in Akkadian were recorded *only* on perishable materials. This cannot be said for texts written in languages using cursive scripts, for which clay is unsuited.[88]

In 8316: r.2, mentioned above, the word *la* is of interest. It is an Aramaic loan word for "from" which replaces perfectly acceptable Akkadian equivalents. It shows how pervasive the influence of this language was at the time, infiltrating the works of even the most schol-

87 Parpola (1986) nn 16–19 provides some references showing their usage in Nineveh.

88 But not unattested, e.g. Pédersen (1986) 11.

arly. Aramaic was indeed soon to become the *lingua franca* of the Persian empire. This West Semitic language had a profound effect on NA grammar, and clearly influenced Babylonians as well.[89] The Aramaeans are first attested in Mesopotamia at the end of the second millennium, and by the 8th century BC Tiglath-Pileser III is able to list in his annals some 36 Aramaean tribal houses in Babylonia. Aramaean scribes are attested at Kalḫu in the early 8th century BC.[90] Its popularity as a written language was no doubt enhanced by its cursive, alphabetic script, and it may have been used extensively for mercantile and administrative purposes (though not to the exclusion of cuneiform[91]). However, extremely little has survived and this puts an inevitable limit on the extent to which the cultural landscape of the ordinary Assyrians and Babylonians can be reconstructed. It is quite possible that much royal correspondence during the NA empire was conducted in Aramaic. When reconstructing the background to the emergence of the accurate predicting of celestial astronomy, it is important to recall that the cuneiform languages, dialects and scripts were used only by an elite. The scientific developments that form the focus of this study appear only in these scholarly languages. This is perhaps no more than a manifestation of the survivability of cuneiform, for it is conceivable that much theorising on methodology was conducted in Aramaic with only the results (the tables of observations or of calculations) being committed to clay precisely because clay lasted so well. However, as I discuss in Chapters 2 and 3, the cuneiform script itself contained much of the meaning of the celestial omens.[92] Given the dependence of the predictive techniques on such celestial divination, I am prepared to believe that the emergence of the PCP Paradigm was a largely cuneiform-only revolution perpetrated solely by these elite Scholars.

Very rarely some NA period texts are written in Sumerian (see for example Reiner, 1992 n124). More frequently, technical texts written in SB contain huge numbers of Sumerograms – signs which in Akkadian would take a syllabic reading or stand for an Akkadian word, standing instead for the Sumerian word, or part of a word. The variety of linguistic influences found in the texts from this period is important to keep in mind when attempting a translation. Undoubtedly the existence of many Sumerograms in the canonical series is related to the editing processes whereby they came together. They permit a greater condensation of the text – one Sumerogram sign can replace several in a syllabic rendition of the Akkadian equivalent. However, it also leads to greater ambiguity, as one sign can have a number of readings and a number of meanings.[93] This, combined with a development

[89] See Von Soden's articles (1966, 68 & 77), Tadmor (1982), and Greenfield (1982).

[90] Kinnier-Wilson (1972) 62f and Pl.20.

[91] See, for example, the discussion of cuneiform administrative documentation in the 7th century in Frame (1992) 12f.

[92] Nevertheless, EAE has been found translated into cursive scripts as a fragment from Ugarit reveals (App.1 §14), and see also Greenfield and Sokoloff (1989).

[93] Famously in 1972 p99f, Derrida describes how the word *pharmakon* used by Plato in *Phaedrus* has been translated as both "poison" and "cure" depending on the context – on what makes best apparent "sense" – a "sense" that emerges from the supposition that the signified (the supposed meaning) has priority over the sign (the letters that make up the word *pharmakon*). The translation that results from this supposition, Derrida argues, misses the full complexity of meaning implied by the sign. This observation applies readily to the translation of the cuneiform omina, and the supposition of the priority of the signified over the sign has led translators to ignore other nuances that the signs themselves may preserve. This is particularly important where the choice of apodosis may be based on graphic allusions to the protasis (on the very shape of the sign, e.g. see Livingstone, 1992) or on the sometimes many possible readings of a cuneiform sign (e.g. Bottéro, 1977). We shall return to this issue in Chapters 2 and 3. Derrida's work reminds us that the meaning of the texts discussed in this work are not absolute or ever-present,

in the dialects over the centuries, and indeed with the eventual demise of spoken Akkadian, probably led to many omen apodoses and protases becoming uninterpretable, misinterpreted, or reinterpreted. This is part of the richness of the omen series, its bilingual and multi-dialectical nature, and its character of repeated overlays and shifts of meaning, of misreadings, misunderstandings, and apparent rationalisations. It is for this reason that a repeated recourse to the signs themselves is required, as further and subtler corners of meaning are illuminated. The use of the celestial omen series during the 34 years which cover the Scholars' Letters and Reports allows uniquely for a study of the *then* meaning ascribed to some omens and names during a relatively short time-frame, and this study is presented in Chapters 2 and 3. The translated meanings will inevitably reflect this student's view of the intentions and motivations of the Scholars, but this will be explicitly stated.

---

but depend on the context in which the interpreter wishes to place them – that is, on the ideology of the writer.

CHAPTER 1

The astronomer-astrologers – the scholars

The aim of this chapter is to present the late Assyrian and Babylonian Scholars, many of whom were engaged in astrology-astronomy and to whom we must ascribe the developments in the techniques used to predict celestial phenomena. I will attempt to establish the extent to which family groups predominated amongst the Scholars, their locations, whether they were associated with temples, their relationship to the king, and their relationships with each other – whether competition and hierarchy existed between them.

"Scholars" was a term used by Oppenheim (1969) 97 to describe the authors of the Reports and Letters in an effort to move away from the more semantically loaded terms "magicians" and "astrologers" used formerly.[94] *Ummânu*, a term which appears in NA royal inscriptions and chronographic texts, is usually translated "Scholar".[95] The same term is used to describe some of the senders of the Reports and Letters and other experts associated with the royal courts. Parpola (SAAX xiv) defines Scholars at this time as practitioners of one or more of the five following disciplines:[96]

ṭupšarru "scribe/celestial diviner" (*ṭupšarru enūma Anu Ellil* e.g. 8499: r.5 = "celestial diviner" explicitly). Experts in interpreting celestial (and other) portents.
bārû "haruspex/extispicer/diviner". Experts in extispicy and lecanomancy.
āšipu "exorcist/healer-seer". Experts in magical manipulation of the supernatural.
asû "physicians". Experts in curing diseases by drugs and physical remedies.
kalû "lamentation chanters". Experts in soothing angered gods.

Other experts attested in the NA court include the augur (*dāgil iṣṣūrī*), the Egyptian scribe and magician (*harṭibi*),[97] the Aramaean scribe (*ṭupšarrū Arumu*),[98] "the wise man" (*hassu*),[99] the prophet and prophetess (*raggimu* and *raggintu*),[100] but their works are not relevant here. The presence of Egyptian and Aramaean scribes in the Assyrian capitals probably reflects the behaviour described in Daniel Ch.1: 4, where Nebuchadrezzar II has brought to Babylon the children of Jerusalem in whom:

94 E.g. Campbell-Thompson (1900).
95 E.g. in the king lists, Grayson (1980–3) §3.12, or in *Sargon's Eighth Campaign*. (Thureau-Dangin, 1973), l.428.
96 The briefest summary will given be here for convenience, as this subject has been treated at length by Parpola SAAX xiiif and in *LAS* II xivf. The term "Scholar" does not adequately describe the physical and mental perfection that the NA rituals indicate to be a prerequisite for a *bārû* (e.g. Lambert, 1967, 132 & Jeyes, 1991–2, 24–5), but these aspects are not relevant here and will be ignored.
97 SAA7 1 = ADD 851. See also Kinnier-Wilson NWL 75 for the same at Kalḫu in the early 8th C.
98 SAAXI 124: rev.ii4'.
99 SAAX xiv.
100 SAA9 xlv.

> "was no blemish, but well favoured, and skilful in all wisdom, and cunning in knowledge and understanding science, and such as had ability in them to stand in the king's palace, and whom they might teach the learning and the tongue of the Chaldaeans."

The NA kings similarly filled their courts with foreigners from whom and to whom knowledge could be imparted. They kept foreign princes as hostages in Nineveh (x112: r.3), in order to indoctrinate them with pro-Assyrian ideas.[101] This could well have formed one of the several functions of the Scholars.[102] The presence of (probably Anatolian) augurs, and Egyptian magicians at the Ninevite court indicates a royal interest in foreign technologies for dealing with the future and the supernatural. This undoubtedly led to some cross-fertilisation of ideas. Rather than a militarily-oriented cultural wasteland by comparison with Babylon, as it is sometimes characterised, the late Assyrian capital was more likely a hotbed of theological and scientific speculation and development. This has also been stressed by Oppenheim (1978) 650.

Defining a Scholar as the practitioner of one of the five disciplines differentiates them from simple scribes who served the secretarial needs of the court and the provincial governors. Parpola states in SAAX xiv that "not every scribe, diviner, exorcist, physician or chanter deserved the designation Scholar", and then points to x160, which lists 20 um.me.a.meš = *ummânus*. Parpola (loc. cit.) wishes to define an *ummânu* as someone proficient in more than one discipline, perhaps in all five. However, x160: r.1f makes it absolutely clear that those individuals who had mastered only one of the five disciplines still constituted an *ummânu*. Consequently, I equate *ummânu* and the "Scholar" as defined by Oppenheim, interpreting them as experts in one or more of the five disciplines described above while recognising that the other experts noted probably also deserved the designation *ummânu*.[103]

It is apparent that the *senior* Scholars[104] who wrote Letters and Reports to the Assyrian kings were familiar with more than one discipline, though perhaps not with them all, as Parpola suggests (1993b and SAAX xiv). Celestial diviners offered advice on rituals to avert portended evil (e.g. x010, 8022–3), normally the duty of the *āšipu*, and utilised omens drawn from *šumma ālu*, *šumma izbu*, and the hemerologies (see SAA8 xviii). Exorcists sent celestial omens and even Reports (e.g. 8160–3). *KAR* 44, the exorcist's manual found in Assur (see I.4), indicates that EAE formed part of the knowledge of these experts. The same was true of the *kalûs*. Urad-Ea sent Reports (8181–3), for example. The contents of the libraries

---

[101] See for example *ABL* 918 and Borger *Ash* 53: 15f noted in Parpola (1972) 34, n66, and Dietrich (1967–8) 245f. For the indoctrination of Arabian princesses in NA courts see Eph'al (1982) 126f.

[102] Cf. Parpola (1972) 33f and SAAXI 156.

[103] Lieberman (1990) 313 writes that *ummânu* is the word used by Babylonians to designate Scholars, but that in Assyria "the official bearing this title seems to have had a special status." However, Assyrian Balasî, for example, who never appears on a king-list or royal inscription calls himself *ummânu* in x039: r.8. Those *ummânus* who appear on the kinglists are simply *the* Scholars of the nation. They are no doubt given official status by the king (but not a unique title). They are perhaps the kings favourites, rather than the most senior Scholars, as is suggested by the Ninevite Scholars. Issar-šumu-ereš is less senior (at least in age) than his uncle Adad-šumu-uṣur, for example. Not all of *the* Scholars of the nation were celestial-diviner scribes, though all were literate, of course, if this is what is meant by *ṭupšarru* "scribe". It was possible to be *the ummânu* of a Mesopotamian king (i.e. on the king-list) without being chief-scribe (Nabû-zuqup-kēna was chief scribe at Kalḫu, but does not appear on the king-lists). The terms should not therefore be equated. Both in Assyria and Babylonia it was possible to be an *ummânu* of the king and neither appear on the king-list nor be a chief scribe.

[104] [lú]*Ummâni dannuti* (x294: 31).

and archives of the scribes, the exorcists, and the chanters in Assur fully confirm the interrelated interests of these three professions. It is further confirmed by the text x160 wherein Marduk-šapik-zeri informs the king on the skills of 20 other *ummânus*. It is not surprising that these three arts should overlap somewhat, for the exorcists sometimes enacted magical programmes designed to avert the evil portended (or implied through illness) by celestial and terrestrial phenomena, and the chanters performed before celestial bodies to the same end. They formed both sides of the same divinatory coin – the celestial diviners to warn of impending evil (e.g. signified by an eclipse), and the exorcists to avert it either before or after the phenomenon and the chanters to appease the supposedly angry gods.

The physicians and haruspices appear only to write about their own concerns, however. There is little evidence that the NA or NB *asû* or a *bārû* manifested an interest in celestial omens. This is perhaps explicable in the following way: a NA physician attempted to cure malaise through the use of herbs and remedies, where the exorcists tackled the supposed supernatural cause. This associated the latter, but not the former, with the appeasing and reading of that same supernatural cause. Also, there was a perceived difference between the evil portended by an extispicy and that by celestial and some other divination techniques. A mark in an entrail was often accounted for by the impurity of the diviner.[105] This cannot be done for an omen in the sky, say, which can be viewed by all. Many of the evil consequences of an ill-boding extispicy can be averted by repeating the operation. Furthermore, most of the texts composed by the NA *bārû* Scholars (SAA4) concerned queries to Šamaš on particular court and state issues. A response either way was not necessarily going to provoke an exorcistic ritual, or chanters' lament. That some crossover existed between celestial concerns and extispicy, however, is clear from the existence of prayers to the gods of the night for an extispicy to come out well, attested from the OB period on,[106] and one Kudurru who was apparently proficient in extispicy and had read EAE (x160: r31). Extispicies were also performed in order to confirm or decide between celestial omens as early as OB times.[107] However, this seemingly did not qualify haruspices to transmit celestial omens to the Ninevite kings. This itself may reflect an increasing specialisation of the Scholars during the period after c.750.

Interestingly, in the MAATs of the last few centuries BC from Babylonia the compilers signed themselves as scribes of EAE, *kalû*s or *āšipu*s. Haruspices and physicians are nowhere attested. The evidence concerning the intermingling of the first three professions seems clear, but I feel Parpola is incorrect in so closely associating the haruspex and the physician.

If some form of hierarchy existed between the five disciplines at the NA court in Nineveh it was not fundamental. SAA7 1 lists experts at Assurbanipal's court from about 650 BC and arranges the Scholars in the following order: scribes of EAE, exorcists, haruspices, physicians, and chanters, who are then followed by the foreign experts. The library records from the same period usually list EAE first (Parpola, 1983b 6, SAA7: 49–56). However, in the Ninevite *Catalogue of Texts and Authors* published by Lambert (1962), the exorcists' corpus (*āšipūtum*) is listed first (K2248: 1), followed by the chanters' lore, and then EAE.

105 E.g. the *ezib* formulations 2–4 (Starr SAA4 xxiii) "Disregard that an unclean person has performed extispicy in this place" etc. This difference between celestial and liver omens is one of the fundamental distinctions between provoked and unprovoked divination (Oppenheim, *AM*, 206f).

106 App.1 §11.

107 Starr SAA4 xxxii.

In the Letters jointly authored by those of different professions, (listed by Parpola *LAS* II 433 c,d,e,f,g,h,i,l,m,n) the order of the Scholars is apparently *not* determined by their professions.[108] The *ummânu*s listed against Assurnaṣirpal, Esarhaddon and Assurbanipal in the synchronistic king list[109] were all scribes – perhaps celestial diviner scribes. However, Nabû-šallim-šunu, *the ummânu* of Sargon II, seems to have been associated with the exorcists in Assur.[110] It is noteworthy, however, that in what remains of the Ninevite *Catalogue of Texts and Authors* not one of the historical Scholars listed was a *ṭupšarru*. Any notion that the profession of scribe (of EAE) was of higher standing than the others in the NA court must be considered unproven. It can be argued, however, that it was an Assyrian, perhaps even a Sargonid, innovation to raise the status of the scribe/celestial diviners to a level *equal* to that of the other four disciplines. I shall return to this point below. It is important to note in this context that the son of a NA and NB Scholar did not necessarily follow in his father's trade (see the following charts and *ALCA* II p47), which is again suggestive of the approximate equal worth of each discipline in the 7th century BC.

1.1 Familial and Ancestral Relationships

In Charts 1.1 and 1.2 I illustrate the family trees of as many of the Assyrian and Babylonian Scholars who composed astrological-astronomical texts during the period c. 750–612 BC as I could find. Chart 1.1 extends and improves upon some of the work on the Assyrian Scholars offered by Parpola in *LAS* II, but will undoubtedly be superseded by the results of the Prosopography of the Neo-Assyrian Empire project. I am unaware of any previous attempt to establish family or ancestral connections between Babylonian Scholars working between c.750 and 612 BC.[111]

A number of comments can be made immediately. Clearly, *families* of Scholars and scribes played a central rôle in the NA court. The Nabû-zuqup-kēna dynasty was particularly influential and over some 250 years produced at least three top Scholars (*the* senior *ummânu* – see n103) to three different NA kings. It is significant that despite his being located at Kalḫu, the children and grandchildren of Nabû-zuqup-kēna moved with the court to the new capital Nineveh. The Kiṣir-Aššur, Nabû-aḫu-iddina, and Bēl-kundi-alaia dynasties, in contrast, remained in Assur for many generations playing an important rôle in the Aššur temple until the very end of Assyrian hegemony over the Near East. It would appear that some scribal families were closely tied to temples, others were more intimately attached to the person of the king. Those closely linked to the king formed an "entourage" which I will discuss in more detail in Ch.1.3. The presence of the royal courts in cities that were less important religious centres, as say Kalḫu or Dūr-Šarkēn were by comparison with Assur, no

108 The author comes first, regardless of rank. In x001 the exorcist then follows, then a scribe, the chief chanter, and finally the chief scribe! In x232, the exorcist author is followed by the chief chanter and then the chief scribe. This contradicts Parpola's suggestion *LAS* II xvi and SAAX xxv that it can be argued that Nabû-naṣir out ranked Urad-Nanaya on the basis of the order of their names in *LAS* 222 = x297.

109 Grayson (1980–3) §3.12.

110 Also suggested by *CTN* 2 246 = ND 1120 which is a report concerning a ritual in Assur authored by Nabû-šallim-šunu.

111 The Scholars writing specifically on matters astrological-astronomical have been bolded, Babylonians are underlined. Most diacritics in the names have not been indicated, after the manner adopted in the SAA volumes. Arrows imply a father-son relationship, dotted arrows an insecure link, and dotted lines an association with an ancestor.

doubt encouraged the existence of these entourages. This late NA change in scribal practice, resulting in the concentration of Scholars around the king rather than in temples, played an important part in the developments then occurring in astronomy-astrology, I suggest. More on this later.

The Assyrian Urad-Ea family of Nineveh ascribed to itself a Babylonian ancestor. Similarly, Marduk-xxx, who wrote one of the tablets of EAE found in Kalḫu in Assyrian script,[112] belonged to a family of Babylonian Scholars with a long association with the Assyrian court. Apparently, Assyrian kings employed Babylonians long before they established direct control over their southern neighbour, and in several cases these families became assimilated in so far as they started to write in Assyrian script. One particular good example is afforded by the colophon of the MA tablet *KAV* 218 which contains a copy of *Astrolabe B* (see App.1 §16). The colophon states that the text is a copy of a Babylonian original made by a certain Marduk-balāṭsu-ēreš, son of Ninurta-uballiṭsu. As Horowitz (1998) 159 n17 argues, it seems most likely that these two were Babylonians, given the theophoric "Marduk". Apparently, Tiglath-Pileser I and/or his father employed Babylonian experts in astronomy-astrology to work in the capital's main temple as early as the 12th century BC. Even Nabû-zuqup-kena appended his name to one text written in Babylonian,[113] indicating that at the very least he could read Babylonian script and probably could write it. Perhaps he too had southern ancestry.

The rôle played by families in both the Assyrian and Babylonian scribal traditions had an impact on the development of astronomical-astrological wisdom from the OB period until c.612 BC, and in the case of the Babylonians thereafter as well. This impact will be discussed in Ch.5.1.2 in the light of Lloyd's 1996 study. Suffice it to note at the moment that there can be no doubt that being born into a good academic family assisted greatly a Scholars' chances of rapid advancement in both Assyria and Babylonia.[114] Being well-born did not, however, prevent a Scholar's rapid demise, as we see in x224 where Adad-šumu-uṣur petitioned Assurbanipal on behalf of his son Urad-Gula writing:

> "Nobody has reminded (the king) about Urad-Gula, the servant of the king, my lord. He is dying of a broken heart. He is shattered (from) falling out of the hands of the king, my lord."

Adad-šumu-uṣur petitioned again (x226), as did Urad-Gula himself (at length in x294), and all was well, for he was re-assimilated into the fold as Letter x227 shows.

---

112 *CTN* 4 pp5–6.

113 79–7–8,150, Hunger *Kolophone* No.20.

114 There are frequent comments about the Scholars' fathers performing the same functions for the kings' fathers. E.g. x221: 13, x182. Some administrative functions seem also to have been monopolised by particular families – Porter (1993) 36 n80.

Chart 1.1 Assyrian Scholars Writing on Astronomy-Astrology And Some Of Their Familial Relationships

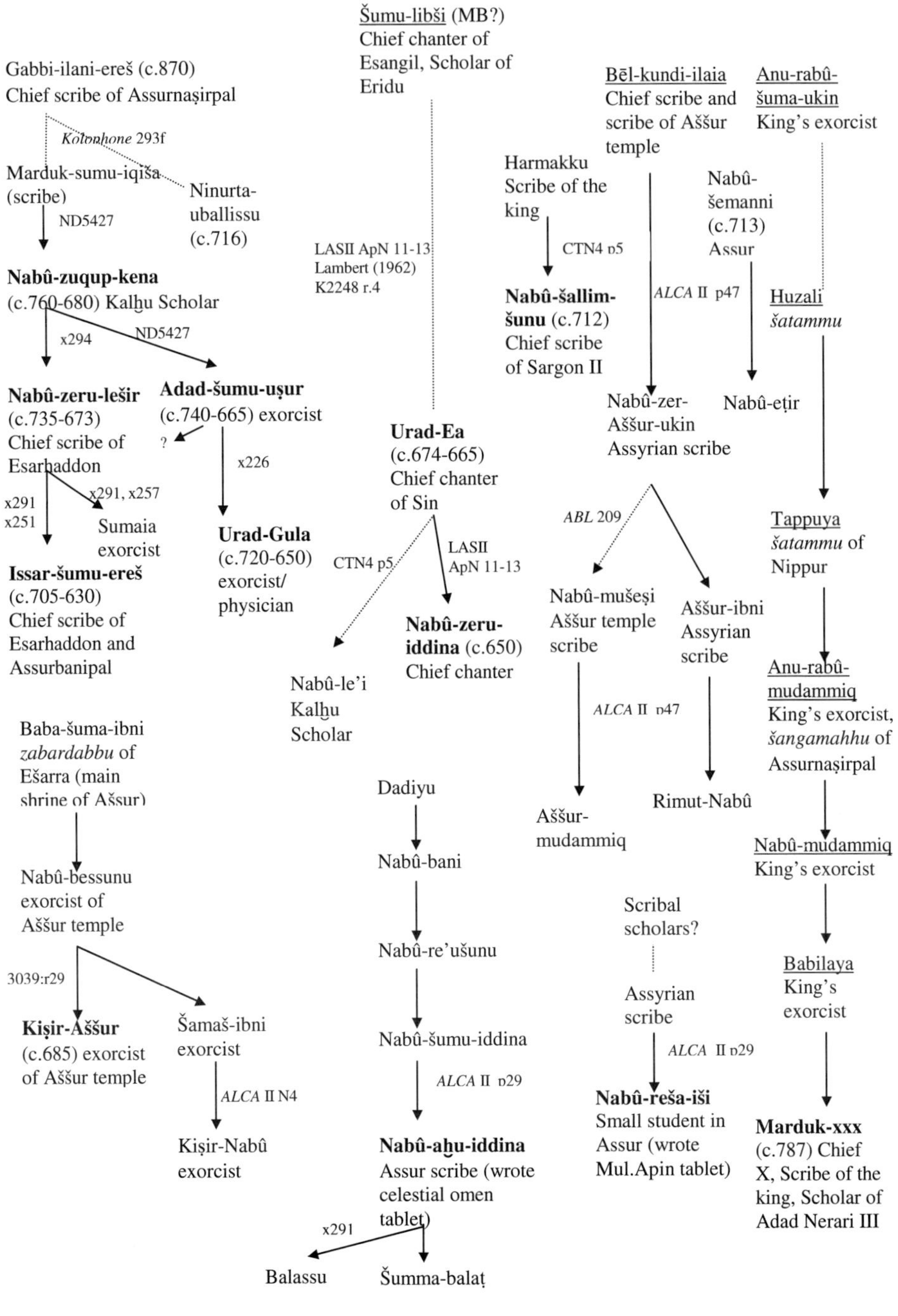
Šumu-libši (MB?)
Chief chanter of Esangil, Scholar of Eridu
Gabbi-ilani-ereš (c.870)
Chief scribe of Assurnaṣirpal
Kolophone 293f
Marduk-sumu-iqiša (scribe)
Ninurta-uballissu (c.716)
ND5427
Nabû-zuqup-kena
(c.760-680) Kalḫu Scholar
x294
ND5427
Nabû-zeru-lešir
(c.735-673) Chief scribe of Esarhaddon
Adad-šumu-uṣur
(c.740-665) exorcist
?
x226
x291, x257
x291
x251
Sumaia exorcist
Issar-šumu-ereš
(c.705-630) Chief scribe of Esarhaddon and Assurbanipal
Urad-Gula
(c.720-650) exorcist/ physician
LASII ApN 11-13
Lambert (1962)
K2248 r.4
Urad-Ea
(c.674-665) Chief chanter of Sin
CTN4 p5
LASII ApN 11-13
Nabû-zeru-iddina (c.650) Chief chanter
Nabû-le'i Kalḫu Scholar
Bēl-kundi-ilaia
Chief scribe and scribe of Aššur temple
Anu-rabû-šuma-ukin
King's exorcist
Harmakku
Scribe of the king
CTN4 p5
Nabû-šemanni (c.713) Assur
Nabû-šallim-šunu (c.712) Chief scribe of Sargon II
ALCA II p47
Huzali
šatammu
Nabû-zer-Aššur-ukin
Assyrian scribe
Nabû-eṭir
ABL 209
Tappuya
šatammu of Nippur
Nabû-mušeṣi
Aššur temple scribe
Aššur-ibni
Assyrian scribe
Anu-rabû-mudammiq
King's exorcist, šangamahhu of Assurnaṣirpal
ALCA II p47
Aššur-mudammiq
Rimut-Nabû
Nabû-mudammiq
King's exorcist
Baba-šuma-ibni
zabardabbu of Ešarra (main shrine of Aššur)
Nabû-bessunu
exorcist of Aššur temple
3039:r29
Kiṣir-Aššur
(c.685) exorcist of Aššur temple
Šamaš-ibni
exorcist
ALCA II N4
Kiṣir-Nabû
exorcist
Dadiyu
Nabû-bani
Nabû-re'ušunu
Nabû-šumu-iddina
ALCA II p29
Nabû-aḫu-iddina
Assur scribe (wrote celestial omen tablet)
x291
Balassu
Šumma-balaṭ
Scribal scholars?
Assyrian scribe
ALCA II p29
Nabû-reša-iši
Small student in Assur (wrote Mul.Apin tablet)
Babilaya
King's exorcist
Marduk-xxx
(c.787) Chief X, Scribe of the king, Scholar of Adad Nerari III

Chart 1.2 Babylonian Scholars Writing on Astronomy-Astrology And Some Of Their Familial Relationships

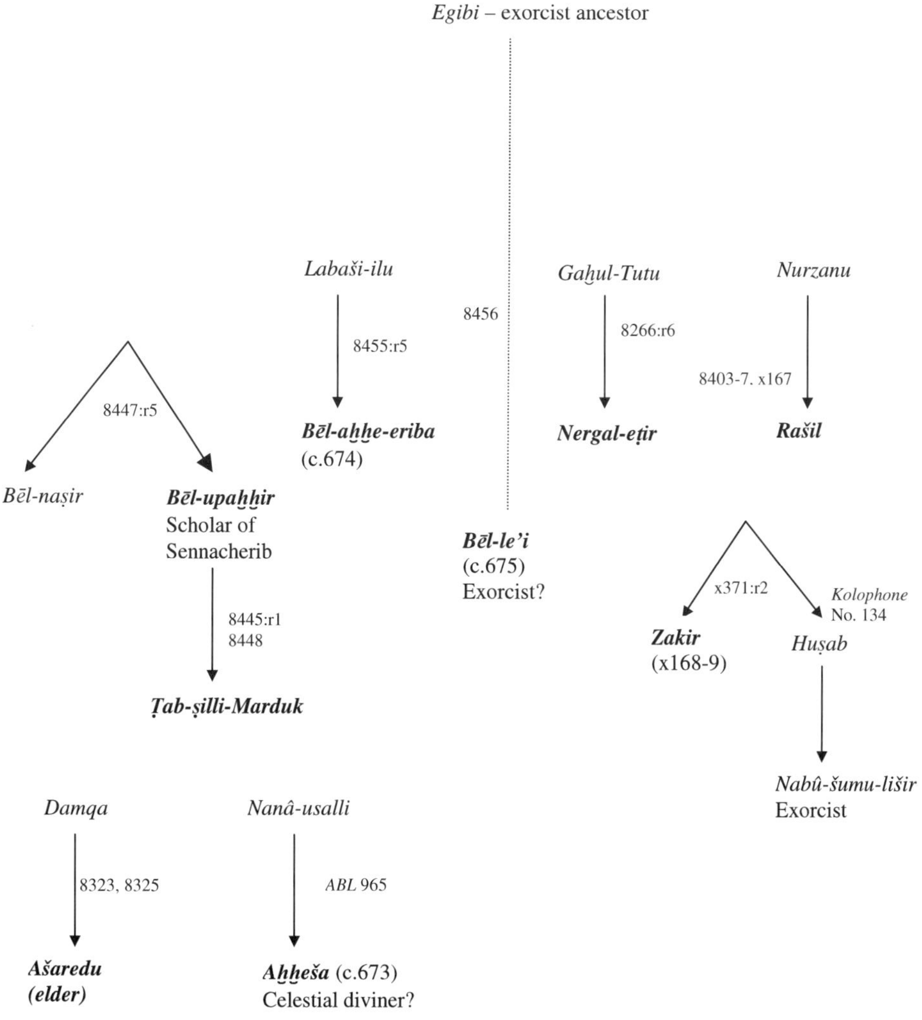

1.2 The Locations of the Neo-Assyrian and Neo-Babylonian Scholars

Assyrian Scholars who sent Reports and/or Letters to the kings in **Nineveh** were based in that city, **Kalḫu** (e.g. Babu-šumu-iddina in x134), **Assur** (e.g. Akkullanu in 8112: 1), **Arbail** (e.g. Issar-nadin-apli in x136–42), **Kilizi** (x143), probably in **Tarbiṣu** (x093: 8) and **Ekallate** (x294: r.16), and perhaps in **Kār-Mullissi** (8472: 8), **Kasappa** (x279: R.9), **Harran** (x013), **Dūr Šarkēn** and **Arrapha**. Babylonian Scholars sent Reports and Letters from the following locations:[115]

[115] This work improves on that undertaken by Oppenheim (1969) 101f.

Babylon

Ašaredu (younger) – suggested in x155.

Ašaredu (elder) – suggested in 8334, also the need to differentiate between elder and younger Ašaredus suggests that both worked at the same place.

Bēl-epuš – a *bārû* from Babylon in 8463: r.3.

Bēl-le'i – descendant of the ancestor from Babylon, Egibi.

Bēl-naṣir – Ṭāb-ṣilli-Marduk family. In 8463 he informs the king of a sick diviner in Babylon.

Bēl-upaḫḫir – being the Babylonian *ummânu* of Se., likely to have been from Babylon, and also in Bēl-naṣir family.

Damqa – in Ašaredu (elder) family.

Kudurru? – denounces? Bēl-naṣir in 8567.

Munnabitu – A witness against Zākir in 8309.

Nabû-šumu-līšir – in Zākir family.

Nabû-eṭir-napšati's father – meets people in Babylon (8517: r.9).

Šakin-šumi? – 8309: 2.

Ṭāb-ṣilli-Marduk/Ṭabiya – Ṭabiya is with Zākir in 8213: r.7. A Ṭab-ṣilli-Marduk is the son of Bēl-upaḫḫir and the nephew of Bēl-naṣir.

Zākir – informs the king about events in Babylon in x169.

Borsippa

Aplāia – calls himself Aplāia of Borsippa in 8356–8.

Bēl-aḫḫe-riba – said to be Borsippan by Bēl-ušezib in x118.

Labāši-ilu – the former's father.

Nabû-iqiša – calls himself Nabû-iqiša of Borsippa in 8288–99.

Šapiku – calls himself Šapiku of Borsippa in 8491.

Cutha

Nabû-iqbi – says he is of Cutha in 8416–7 etc. His father's house is in Cutha (x163).

Dilbat

Nabû-aḫḫe-iddin – says he is from Dilbat in 8481.

Nippur

None attested, but x114: 7 makes it clear an observation centre existed there.

Sippar

Urad-Gula, the Assyrian, was there for some time, since the king wrote to him in x295. This text was actually found in Sippar.

Ur

Šumāia – speaks to the king in Ur (8499: r.2). Is this the same Šumāia whom Kudurru sends to the king in x371?

Kudurru? – see above.

Uruk

Aḫḫeša – of Uruk in 8449–53.

It is immediately apparent that the Scholars (all male) identified themselves by city. These cities were presumably their birth places or where they learnt their trades. It would seem probable that each major city included at least one scribal school which was likely associated with one of its main temples. Many cities, both in the north and south of Mesopotamia, contributed Scholars to the entourage of the Assyrian king. Probably all major cities in Assyria and Babylonia produced them, though at any given time the Assyrian monarch might have favoured some cities over others. I suspect that not every minor city supported a scribal school. Some of the provincially located Scholars probably learned their trades elsewhere. I am thinking of the Scholar from Kilizi, for example. Some Assyrian Scholars from Nineveh were sent to Assyrian (e.g. Akkullanu) and Babylonian cities (e.g. Mar-Issar), where they kept an eye both on celestial and terrestrial happenings, and in particular on the proper functioning of the temples. No doubt the Scholars writing from the smaller towns were similarly seconded from the capital.

Many Babylonians worked in Nineveh, as others had done before them in Kalḫu, in Dūr Šarkēn, and in Assur. Some stayed on during Assurbanipal's reign, when Šamaš-šumu-ukīn was on the Babylonian throne, and Babylonian Scholars continued to send Reports to Assurbanipal from their native cities as Reports 8384, 8387, 8418, 8487, and Letter x172 demonstrate (see Table 1). They, or others, no doubt sent Reports and Letters to Šamaš-šumu-ukīn, too. One Babylonian Report, 8487, if dated correctly,[116] was sent to Assurbanipal during the time of the 652–648 BC revolt in Babylonia, though not necessarily from Babylonia.

It is difficult to assess if there were any perceived hierarchy between Babylonian and Assyrian Scholars, though it is interesting that Babylonian Bēl-ušezib recommended Adad-šumu-uṣur, son of the illustrious Nabû-zuqup-kēna in x110. Certainly, there is no evidence that the Assyrians looked down upon the Babylonian Scholars.[117] In general it appears as if relationships between Babylonian and Assyrian Scholars were good, as demonstrated by their joint authorship of some Letters (e.g. x176). This probably reflects the social background of the Babylonian Scholars. They formed part of the population of Babylonia that looked towards Assyria to hold off the advancing Chaldaeans.[118] This may have been due to the Chaldaeans having less use for their specialities, or simply a manifestation of *realpolitik* on the part of the Babylonian Scholars in the face of Assyrian might. Cooperation and mutual respect between Scholars in no way precluded fierce competition between them for royal favour, however.

From the Letters and Reports and other prosopographical material we know that many individual Scholars, although part of the royal entourage, were associated with temples either as temple personnel or as royal agents. In the latter category we find in particular Mar Issar, Esarhaddon's agent in Babylonia, who concerned himself with temples in Uruk, Der (x349), Borsippa (x353), Babylon (x354) and Akkad (x359), and Akkullanu who concerned himself with temples in Nineveh (x095), Kalḫu (x099) and particularly Assur (x096–8, x107). These two were clearly financed by the king and no doubt served to ensure that a certain degree of royal control over the temples was maintained. In the

116 De Meis & Hunger *ADABR* 82.

117 In x182 the Assyrian haruspex [Tabn]î compains that a foreign (*šaniti* = other – perhaps an over-translation by Parpola?) haruspex (– probably Aplāia or Naṣiru) has become the crown prince's favourite.

118 In x169 the Babylonian Zākir writes to the Assyrian king that Ubaru and the Babylonians (= the Chaldaeans) are making the land drift away from the Assyrian king. His loyalty does seem to be with Nineveh, rather than Ubaru. Bēl-ušezib assists Esarhaddon with the latter's military endeavours in Mannea (x111–112). In x118 he assures Esarhaddon of Bēl-aḫḫe-riba's loyalty to Assyria.

former category fall many of the Scholars based in Assur who were associated with the Aššur temple, those at Kalḫu connected to the Nabû temple, and *kalûs* such as Urad-Ea and Nabû-zeru-iddina who were connected to the Sîn temple in Nineveh. Some Babylonian Scholars were undoubtedly associated with temples, in particular the Marduk temple in Babylon where the "astronomical archive", containing material dating from the NA to the Christian period, was most likely situated.[119] Similarly, in Uruk the later MAATs and NMAATs were associated with the Reš temple sanctuary, though it is not certain that this temple played a significant rôle in the life of the Uruk-based Scholars in the 7th century BC.

The long-term connection of the Scholars who performed celestial divination with temples is undoubtedly important in the context of their discipline, whether this be through their education (see below) or their direct employment in these establishments. It helps explain the continuity of the tradition from OB times to the NA, and from NA times to the last centuries BC, despite the many changes of rule in the secular arena.[120] This continuity helps us to connect the Scholars' techniques of mathematical prediction of celestial phenomena with an older intellectual background that included both celestial divination and the wider concerns of the perceived creation and form of the universe. These issues are discussed in Ch.5.1. Such continuity is, to a limited extent, true for the entire scribal tradition. While writing for economic and administrative purposes would appear *a priori* to have remained useful at all times, it is by no means obvious that literary or divinatory texts would have continued to have been written, adapted and preserved *outside* of an institution in some way insulated from the changing fortunes of the land. Nevertheless, despite centuries of continuity, as a result of one particular circumstance forced upon the land by the late Assyrian kings, many senior NA and NB Scholars *were* employed directly by the king and not by a temple. Given that a development occurred in the art of astronomy-astrology practised by these same Scholars at this time, as we shall see, it would seem *prima facie* reasonable to suppose that the new, specifically royal nature of the employment of *many* of their number played some part in that change.

1.3 The Scholars' Relationship to the King

Assyria, throughout the 8th and 7th centuries was governed by absolute monarchs, whose power was checked only by religion, legal precedent, and by the mood of the noble classes that surrounded him.[121] The Assyrian king was the supreme human being in the state, nearer to the gods than anyone else. In ideology kingship descended from heaven, and the Assyrian king was the high priest (*šangû*) of the national god Aššur.[122] The royal inscriptions stress the intimate communication of rights and responsibilities from the gods to the person of the

---

119 Explicit evidence is hard to find, but Bēl-ušezib shows a keen interest in Marduk's temple in x109: 15' and the repeated sending by Babylonian Scholars of omens to the NA king with apodoses describing the rebuilding of the "temples of the land" suggests that they were not above promoting their own interests. See 8414: 5 *inter alia*.

120 Oppenheim (1969) 121f: "Babylonian interest in astral omens...continued after the fall of Assyria ...suggests that the entire practice...was basically a Babylonian institution."

121 Grayson (1991d) 196.

122 E.g. Sargon II in Luckenbill (1927) II §104, amongst many examples.

king. His position as the main object of divine interest is well known.[123] The society and to some extent the economy also revolved around him. He was during this period far and away the most significant power in the empire

His court has been compared to that which surrounded the Ottoman sultans,[124] with a harem, eunuch guards, the queen-mother's court, a "house of succession" (*bīt-redûti*) for the adult crown-prince, and an ethic of nepotism and patronage. The king's main concerns were military, but he played a very significant religious rôle in the community. The bureaucracy is described by Grayson (1991d) 199 as a pyramid, atop which sat the king, beneath whom came the major-domo, the field-marshal, and *the ummânu* – the most favoured Scholar of the nation. The only king of this period for whom there is any evidence that he was literate is Assurbanipal,[125] but all the NA kings had official inscriptions, records, and even works of "literature"[126] composed for them, collected together libraries, and to a greater or lesser extent were drawn to, or influenced by the intellectual achievements of the Scholars.

Brinkman (1991) 85, writes[127] that Tiglath-Pileser III had little time to spare for encouraging his subjects in cultural pursuits. Certainly, he and Sargon II campaigned ferociously, and more "intellectual" and "cultural" texts are attested from Esarhaddon and Assurbanipal's time than from their. However, much of this apparent transition in NA royal concerns may be due to the chances of discovery. In SAAX 109: r.1f Bēl-ušezib writes to Esarhaddon early in his reign, describing a time during Sennacherib's reign when the scribes and haruspices who ordinarily sent ill-boding omens to that king had arranged only to send him those whose prognoses were good. Clearly, a celestial divination industry was working powerfully under Sennacherib who also campaigned extensively, and as the presence in Kalḫu of early 8th century BC versions of EAE indicate, celestial divination was no new activity in Assyria at the time of Tiglath-Pileser III and his immediate successors. See also App.1 §§15–16 for MA texts concerned with celestial divination, though these were located in the Aššur temple and do not prove that the Assyrian kings then employed celestial diviners extensively, or at all.

The vagaries of recovery have also ensured that the texts known largely pertain to NA and NB palaces and temples. The few private archives found suggest, however, that the Scholarly pursuits in which the NA kings showed an interest were largely *not* followed in other less elevated social strata.[128] These arts were imported from Babylonia[129] in the cen-

123 The relationship of the king to the state religion, and of both to divination is a massive topic and will only be considered in passing here. I am of the opinion that divination does lie at the very heart of Mesopotamian religion, and that an understanding of the latter will be greatly enhanced by knowledge of the former. For a broad-based discussion of the issue see Saggs (1978) 137f. For a brief survey of the extent to which celestial bodies were seen by the Scholars as "representations" of the gods and their phenomena as "metaphors" for divine behaviour see Rochberg (1996).

124 Grayson (1991d) 198.

125 Assurbanipal boasts (Streck, 1916, 256: 18) "I study stone inscriptions from before the flood, which are difficult..." See also Grayson (1991c) 159.

126 E.g. SAA3 text 1: "Assurbanipal's hymn to Aššur", text 2: "The Nanaya hymn of Sargon II" etc.

127 Grayson (1991b) 103 suggests much the same thing.

128 Reiner (1991) 296 stresses the difficulty of separating the literature of the upper classes from that of the common people, the existence of which can only be inferred from scattered quotations. She also comments (ibid.): "The relation of literary texts to their cultural context (in Mesopotamia) is still little known."

129 Babylonian influences in religion and divination are enormous. Ellil appears in Assyrian religion in the OB period, followed by Marduk in about the 14th century BC. Nabû appears at the turn of the millennium. It is suggested (Livingstone, 1989, xxix *inter alios*) that by Sennacherib's time the cult of Marduk had become so popular in Assyria that (literary and other) measures were taken to reverse this situation. Similarly, it is generally assumed that

turies before the 8th century BC, and some Assyrians became expert in them as we have seen. However, these were very few in number and I suspect that the Babylonian cultural imports of celestial divination, chanting, extispicy and so forth were not used widely in the population as a whole. EAE-type celestial divination, in particular, was an art designed only for the king.[130] It was a study of celestial phenomena involving the decoding of divine messages to that king. It is hardly surprising that there is no evidence that it was used by ordinary people.[131]

As noted, those Babylonian Scholars expert in the arts of divining, exorcism and so forth belonged to an amalgam of the descendants of the Sumerians, Akkadians, Amorites and Kassites, who by c.750 BC shared a common culture broadly designated by the term *mārē āli* "citizens".[132] The more recently arrived tribesmen of the Aramaeans, Chaldaeans, and Arabs provided a major threat to Assyrian hegemony over Babylonia. Because of this and since "the venerable culture of Babylonia ...exerted a strong attraction for Assyria,"[133] the Assyrians increased efforts to establish solidarity between themselves and the Babylonian *mārē āli*, often giving them political and economic advantages.[134] The intentions of the Babylonian Scholars when working for the Assyrian kings should be understood largely in this light, I argue. It is more probable that they were easily persuaded rather than coerced into applying their skills to the supernatural protection of the king.[135] Indeed, Porter (1993) argues that Esarhaddon presented himself as the personification of the *Babylonian* concept of kingship through the adoption of traditional titles, and the performing of traditional functions and so forth. I suggest that Esarhaddon may have deliberately encouraged the legitimisation of this rôle that divination can bring, ensuring that many Babylonian Scholars worked for him. All this is significant in terms of the attempts to correlate scientific advancement and political transformation as Lloyd (1979), for example, has discussed in the light of the Greek experience. This question will be addressed further in Chapter 5.

The Scholars' status and power in the Ninevite court was probably significant, but was strictly limited by royal favour, which had to be curried at every opportunity.[136] The Scholars perhaps deceived their kings by manipulating what he was told.[137] That the kings were aware of this possibility is shown by x109: r.1 on Sennacherib's diviners, noted above. In theory the Scholars could influence state affairs by informing the king of auspicious and

all Assyrian astrological-astronomical texts, bar subtle variations, derive from Babylonian precursors. See App.1 §§ 16, 17, 21 & 30.

130 The intimate relationship between astrology and royalty is suggested in Barton (1994) 38 who argues that the discipline developed in Rome most powerfully after the demise of the Republic and the start of the Empire.

131 Even though its influence on 5th century BC and later personal astrology is clear. See Rochberg (1998) 11f.

132 Porter (1993) 36 n78.

133 Brinkman (1991) 16.

134 *Kiddinnūtu, zakûtu, andurāru*, and *šubarû* – forms of tax-exemption and protection were granted to the major Babylonian urban centres at different times. See Porter (1993) nn145–6.

135 It is noteworthy that the Assyrians' policy towards the Babylonian cities was often not successful. In 689 the Assyrians' frustrations at the fickleness of the *urban* Babylonians led to their destruction of Babylon and its main temple. There can be no doubt that this particular act of sacrilege alienated the urban Babylonians from the Assyrians, particularly those Scholars associated with the temple of Marduk.

136 The king's wrath may not have been quite that of Nebuchadrezzar in Daniel 2: 5: "if ye will not make known unto me the dream...ye shall be cut in pieces", but the position of servitude there implied for those Scholars probably compares well with that of the Assyrian king's *ummânus*.

137 Koch-Westenholz (1995) 56–73 provides a short description of the Scholars and their relationship to the monarch. She emphasises their aspect as trusted servants (p65 n1), their obligation to report (p66), and does not believe that the king was sceptical of divination (p67 *contra* Oppenheim, 1969, 120), only of diviners.

inauspicious times. In practice the number of different celestial diviners who watched the heavens and informed the king meant that these times were indeed regulated by the celestial phenomena, and not by the desire of particular Scholars. This may have been one of the reasons for employing experts based in many different cities, I suggest, and indeed a reason for the rise in the popularity of celestial divination itself, since it was apparently less easy than the other arts to manipulate. In principle, what was seen in the sky could only be interpreted largely through the extraction of the relevant omens from EAE, though as we shall see in Chapters 2 and 3, even this limitation could sometimes be overcome.[138] The Scholars' influence was, I suggest, mainly restricted to the extent and timing of apotropaic rituals, and to that which comes from being the king's informers on others' subversive acts. Just as in the Book of Daniel Ch.1: 18 where the chief eunuch brings Daniel and the other three children to an audience before the Babylonian king, so Sasî, the chief eunuch ([lú]gal-sag, *rab ša-rēši*) acts as an intermediary between the Scholars and the king (e.g. x113: r.11, x176: 8f, x270: 4, x377 & 8502: 5). Not even the most senior of the Scholars could expect to be received by the NA king without first writing or sending in a request via the chief eunuch. It is significant that even Issar-šumu-ereš, *ummânu* under both Esarhaddon and Assurbanipal, and mentioned in the Chronicles, wrote to both kings. His proximity to the kings in Nineveh, and his high status, did not mean that he could expect to deliver his reading of the heavens orally.

The clearest evidence for the absence of any real power in the hands of the Scholars in the royal "entourage" (see below) comes from their own comments on their remuneration. Theirs was, I noted, not a profession for which there was much call outside the court. Urad-Gula in x294: 16f, for example describes the good old days when his name was "mentioned among men of good fortune" and he repeatedly received a mule or an ox and annually a mina or two of silver from the king. These gifts he refers to as "leftovers".[139] Significantly, Urad-Gula, having fallen from royal favour, complains that he now receives and possesses nothing: "I cannot afford a pair of sandals or the wages of a tailor, I have not got a spare suit of clothes and I have incurred a debt of almost 6 minas, plus the interest", he writes in line rev.27f. In x163 and x164 Nabû-iqbi appeals to the king to help him in a legal case against the commandant of Cutha concerning property. It is apparent, then, that these Scholars were not, as is commonly argued,[140] powerful political agents in the NA court, but forced to depend heavily on royal favour for their livelihood. This aspect of their situation, I argue, led to extensive competition between them, and played a part in the development of techniques which predicted celestial phenomena and thereby gave one Scholar an advantage over another. Evidence for this can be found in the Letters and Reports and will be discussed in Chapter 5.2.

The Scholars were practitioners of what Parpola[141] has termed "Mesopotamian Wisdom", by which he means the magic, theology, divination, and scholarship that underpins the texts considered here, and to which predictive methods, hermeneutics and mathematics were harnessed. To what extent this is a wisdom that characterises Mesopotamia as a whole

[138] See also Koch-Westenholz's (1995) 140–51 case study.

[139] *rēhtu*, "what is left", though perhaps "profit share" gives a more accurate image than "leftovers".

[140] Lieberman (1990) 327 writes that "Assurbanipal collected his tablets in order to remove power from the hands of such consultants" (the Scholars). His article attempts to show how easy it could have been for the Scholars to manipulate the king. His view repeats that of von Soden (1954) 125 and Oppenheim (1969) 120 *inter alios*, and is often found in the secondary literature. Parpola *LAS* II xviii also provides a corrective to this view.

[141] Parpola SAAX, SAA9, 1993a, 1993b.

is hard to determine, but as a concept it suffices in the first instance to describe the materials written at all periods by the elite experts considered here. It will be the name I use to describe the knowledge used by the Scholars to protect their kings. I suspect that few if any Scholars were masters of all aspects of this wisdom, as the existence of specialisations immediately suggests.

The so-called *Catalogue of Texts and Authors*[142] leads from the legendary *apkallu* (sages) of legendary kings, to the historical *ummânus* of historical kings, indicating that the latter were perceived to have performed the same function as the former. The affiliation to ancestors that is so prominent in Babylonia, and attested in Assyria, indicates that in these Scholars' minds a connection to ancient Scholars and to sages formed part of their self-conception of their rôle and function – their self-validation. Just as the sages were thought to have behaved towards ancient kings, and indeed to have imparted all wisdom to the Mesopotamians, so the NA and NB Scholars behaved similarly and imparted this very same wisdom to Esarhaddon and the like. That "nothing is new under the Sun" was probably a familiar concept in Scholarly circles. See for example an author of antiquity quoting Berossus:[143] "(The sage) Oannes...taught them (the Mesopotamians) the knowledge of letters and science and crafts of all types...from that time...nothing further has been discovered."

This attitude has important ramifications for the reception of innovative methods and for conservative attitudes towards the established corpus of texts. New methods designed accurately to predict celestial phenomena were perhaps difficult to attribute to the corpus of standard texts (found in the NA libraries) because they had not been, or were not, attributed to ancient authors.[144] In Lambert (1962) K2248: 1–4 EAE is attributed to the god Ea. In Lambert (1967) K2486+: 18 *ṣâtu,* EAE and *arû* are attributed to the ancient king Enmeduranki.[145] For the Scholarly crafts of *ṣâtu* and *arû* see Ch.3.2.1.

Scholars engaged in sending omens to the king, performing rituals, chants, extispicies, applying medicines or indeed assisting him in the face of the supernatural with foreign technologies were said to be "standing before the king"[146] – they were in his "entourage" and no doubt handsomely rewarded for their efforts. If asked by the king to enter this "entourage" they were "summoned".[147] They protected him in a variety of ways including revealing intrigues to him (x199, x024), offering him wise advice (x111), and ensuring that he did not stray from the path laid out for him by the gods (x056). One aspect of this protection was known as "keeping the watch of the king".[148] In the case of celestial diviners this was done by observing and interpreting celestial signs, and by telling the king the manner in which he should respond to them. This might involve an apotropaic ritual (x206), or simply the

---

142 Lambert, 1962.

143 Burstein (1978).

144 The attribution of epics to "visionary experiences" was one way in which new texts could continue to be assigned divine origins, however. Cf. *Erra* V: 43f, and see Parpola SAAX n19.

145 *Šá* KI *ṣa-a-ti* ud an en.líl *u* a.rá (*arû*), which Lambert translates as " 'that with commentary', EAE and how to make mathematical calculations".

146 *ina pān šarri uzuzzu* (e.g. x226: r.9, x227: r.16). Someone belonging to the entourage of the king is a ^{lú}*ma-za-si pāni ša šarri*. In the NA text *ABL* 33: 10 we find "scribes, haruspecs, physicians, augurs, *manzāz ekalli* (and) citizens will take the oath" suggesting that the "entourage" may have been thought of as including many others employed by the king, besides the group of Scholars.

147 sag (*rēš*) *našû* = "to raise the head" = "to summon" (e.g. x160: 33 *rēša lišši* "let (the king) raise my head").

148 *maṣṣartu* (en.nun) *ša šarri naṣāru* (e.g. x118 r.8).

suspension of some policy or activity, however trivial (x192). It appears as if Scholars performed stints of "keeping the watch of the king" as Nabû-iqbi suggests in x163, perhaps on secondment from his temple, and the phrase should probably be equated with employment by the king within his "entourage". Some Scholars looked to the king's health (the *asû*), and checked for divine sanction of his decisions in the entrails of sheep (the *bārû*). The celestial diviners often sent good-boding omens, encouraging the king to act (x033), and at other times forbade him to act on unpropitious days (x038). This behaviour was much more then simply manipulating a gullible king, for as the Letters show the Scholars remained in constant fear of falling from royal favour (x160: 1f, x166, x173: 17f etc.). By looking after the king in this way the Scholars were in no small way playing a part in the state religion, for the king was thought to receive messages concerning all aspects of his behaviour directly from the gods. The Scholars' rôle was to decipher those messages.[149] As Balasî writes:

> "The god has (only) wanted to open the king's ears: He should pray to the god, perform the apotropaic ritual, and be on his guard" (x056: r.18f).

To summarise, the Scholars protected the NA kings in a manner believed, by them at least, to have been undertaken by their predecessors for at least a millennium before. Those "summoned" to the "entourage" of the king were specifically employed to this end. They included the top experts in the land, scribes who had learned their crafts in schools most, if not all of which were associated with temples. Protecting the king was not a solely secular activity, however, for his behaviour was thought to be of particular importance to the gods of the land. Using their temple-acquired skills the Scholars translated signs in the heavenly and terrestrial spheres into divine messages for their ward and responded to them accordingly. Performing the "watch of the king" and "standing before the king" were tantamount to being in direct royal employment, which was presumably more prestigious than temple employment, but did mean that the Scholars were entirely dependent on royal favour for their livelihood and advancement (x294: 13f). This established an atmosphere of dependency and resulted in intense competition and rivalry, as I indicate in Ch.5.2.

Consequently, I argue that the increase in the correspondence to Esarhaddon in his final years (noted in §I.5) was *not* because of that king's particular psychological bent, or superstitious nature, as is sometimes argued. Firstly, the NA kings were rather the passive recipients of Scholarly advice, the quantity of which reflected the Scholars' concern over their charge.[150] This concern might have grown due to the king's repeated ill-health, or due to unusual or excessive celestial or terrestrial phenomena. Secondly, under Esarhaddon, Assyria and Babylonia were at peace in part because he was presenting himself as the personification of a Babylonian monarch and therefore was of concern to Babylonian Scholars. Thirdly, his reign was accompanied by economic growth both North and South, and thus his ability to employ more Scholars and engage in more cultural pursuits was enhanced.

[149] As in the prayer to Sîn and Šamaš (PBS1/2 106: r.15f) quoted in Rochberg (1996) 476: "You stand by to let loose the omens of heaven and earth. I, your servant, who watches you, who looks upon your faces each day, who is attentive to your appearance...set down before me propitious and favourable omens".

[150] *Piqittu* "charge/ward" is used in the Letters only for the crown prince's baby, but it is this concept that I am asserting here governs the Scholars' attitude towards the king.

1.4 Scholarly Interrelationships

A Scholar's education was a protracted affair and a complex hierarchy of apprentices, students, experts and Scholars is apparent from the royal correspondence – a hierarchy summarised as "scribes great and small" in x171: r.1–2. In the case of Kiṣir-Aššur we have evidence in colophons of his path from apprentice to exorcist of the Aššur temple. This has been reconstructed by Pedersén *ALCA* II 45 in the following way, though I have translated *šamallû* = [lú]šamán.(mál).lá as "apprentice" rather than "student" following *CAD*:

Šamallû ṣeḫru – "young apprentice",
Šamallû – "apprentice",
Šamallû maš.maš *ṣeḫru* – "apprentice, young exorcist",
Maš.maš *ṣeḫru* – "young exorcist",
Maš.maš – "exorcist",
Maš.maš *bīt Aššur* – "exorcist of the Aššur temple".

"Apprentices", *šamallû,* were those studying to become specialised in one of the Scholarly arts. In x102: 6' Akkullanu remarks that Kiṣir-Aššur is copying (*šaṭāru*) the lexical series ur_5.ra. This was presumably the function of someone still lower in rank than a Scholar. Later, as exorcist of the Aššur temple, he composed the literary text on secret lore of the gods (SAA3 No.39) and was undoubtedly by then one of the top Scholars of the land.

It would appear from his colophons that when still a "young apprentice" Kiṣir-Aššur had already mastered cuneiform writing. This is also suggested by x143 where the scribes of Kilizi complain that they do not have time to keep the "watch of the king" (indicating that they were Scholars) or to teach the *didabû*s "pupils" the scribal craft (*ṭuppšarrūtu*). Presumably, while still learning cuneiform, future "apprentices" were known as "pupils".[151]

That an "apprentice" ([lú]*šamallû*) was taught more advanced subjects than a "pupil" ([lú]*didabû*) is also suggested by x171, where one Babylonian Scholar wishing to being "summoned" (n147) to the court with the other Scholars informs the king that he has taught his "apprentices" *Enūma Anu Ellil.*

Senior Scholars (*ummâni dannuti*) had "assistants" ([lú]*šaniu*) and in x294: 35 (if restored correctly) Urad-Gula states that one assistant exorcist had his own "apprentice" ([[lú]*š*]*amallû* ⌈*ša*⌉ [[lú]maš].maš 2-*i*, indicating that an "assistant" out-ranked an "apprentice" and was presumably soon to be an *ummânu*. Only once this final rung in the ladder had been reached,

151 *Didabû/didibû* are specifically NA terms, which I have translated as "pupil" in order to differentiate them from "apprentices". This is not done in the glossary to SAAX. In x097 it is made clear that a [lú]*didibû* when raised in status is "shaved" (*gallubu/gullubu*), though in this case it may apply only to those pupils becoming priests and not Scholars. See also x096 and compare the phrase "standing with hair" (x096: r.25, x097: 10'), which means serving in the entourage of the king only as a pupil, with "standing with the king" performed by the Scholars (n146). Perhaps those graduating from scribal pupilage were also shaved, becoming young apprentices and beginning on the road to becoming an *ummânu*. I note on this basis that both scribes represented in the famous relief from the central palace of Tiglath-Pileser III in Kalḫu are without beards but with hair on their heads (Moortgat, 1969, Plate 272). These may indicate "apprentices" rather than eunuchs, as has previously been thought, provided the status-raising shaving of "pupils" applied only to their beards. From the evidence thus far apparent this seems possible (see *CAD* G p130). That some Scholars were not eunuchs is clear from Ch.1.1. The winged sages (if this is what they are) shown anointing Aššurnaṣirpal II in the NW Palace relief from Kalḫu (Moortgat, 1969, Plate 257) are bearded, however, which does perhaps suggest that the Scholars who played their part in later times were also bearded.

could the Scholar be called into the royal "entourage", I presume.

In x160 Marduk-šapik-zeri attempts to convince the king of his standing, arguing that he has completed his "apprenticeship" (*šamallūtu* – l.31) and has "mastered" (*gummuru*) his father's profession, the art of lamentation, that he has "studied/inspected" (*murruru*) and "chanted" (*zamāru*) the (lamentation) series (of tablets), is "competent" (*le'û*) in the profession of his father and has also "read" (*šasû*) EAE, made astronomical observations (mul.meš an-*e ṣubbû*), read the anomaly series and various other works. He goes on to comment on his fellows, describing them as useful to the king since they "master", are "competent" in, or have "read" a discipline or series. He refers to all twenty as "Scholars", yet I suggest that distinctions can be drawn between "mastering", "competence" and "reading". In general a Scholar was one considered to have "mastered" a discipline, which may have involved in some cases the ability to chant it. In addition to "mastering" one discipline, some Scholars were competent in or had read the works relating to other disiciplines, though at a lower level of expertise. No doubt "mastering" a discipline went well beyond simple familiarity, and as we shall see in Chapter 2, some Scholars felt able to adapt and comment on extracts from the tablet series themselves. Scholars this confident presumably saw themselves as being at or near the same level of expertise as the supposed composers of the series, the mythical sages.

Some, but by no means all top Scholars were chief chanters, or exorcists of this or that temple or even chief scribes. Balasî, for example, was the Scholar appointed to teach the crown prince (x039), but apparently he held no temple post, and never appeared on a king list. As noted above, it was, I suggest, the king's preferred Scholar who became *the ummânu* of the king and so appeared on king lists. Holding one particular temple post, such as chief exorcist of the Aššur temple, say, was not the precondition of this position.

In this vein Parpola has argued[152] for the presence of an "inner circle" of the highest ranking Scholars based in Nineveh, and an "outer circle" made up of (still important) Scholars generally not resident in Nineveh. Correspondence from the outer circle, he argues, was more sporadic and "owed its existence to the need to calibrate inconclusive lunar (and solar) observations made at the capital".[153] There are some problems with Parpola's argument,[154]

152 *LAS* II xivf and SAAX xxvf, now followed by Koch-Westenholz (1995) 68f, who also argues that only in Assyria did a formalised network of observers exist, and that the relationship of the Babylonian Scholars (in Babylonia) to the king was "opportunistic" (p71). Considering that more Babylonian than Assyrian Reports were found in Nineveh, this is highly unlikely, though we accept that the Babylonian Scholars (in Babylonia) *may* have existed independently of the Assyrian king, perhaps funded by their temples.

153 Note that this is a move away from Oppenheim's (1969, 122) suggestion (after Neugebauer) that the network of observation posts was due to the need to regulate the calendar. It was clearly also to check on the observation of ominous phenomena. I agree.

154 Parpola's choice of "inner-circle" Scholars is based on their residence in Nineveh, high ranking titles, and the recovery of a large number of their Letters, but not their Reports. I suggest that using familial ties (Charts 1.1 and 1.2) and co-authorship (*LAS* II App. M) as criteria would alter the picture somewhat, swelling the ranks of the inner circle with Sumāia (Issar-šumu-ereš's (half) brother), Nabû-mušeṣi (x205), Naṣiru (x176), Aqarāia (x176), and Tabnî (x177). The haruspices attested in SAA4 should be added, since they were clearly resident in Nineveh, particularly Nīnuāiu, the chief haruspex (4326: r3). If Babylonians (Bēl-ušezib) can be members of Parpola's inner circle, as can Assyrians posted in Babylonia (Mār-Issār), then surely senior Babylonians resident in Babylonia can be too. Nergal-eṭir, Aplaya, Ašaredu the younger, Bēl-naṣir, Munnabitu, Nabû-iqbi, Rašil, and Zakir sent both Letters and Reports to the Assyrian kings, and even to other members of the royal family (x154). Marduk-šapik-zeri, as x160 shows, was significant enough to inform Assurbanipal of, and evaluate, other Scholars in the empire. Similarly, the location of some of the Assyrian Scholars who sent Reports cannot (as yet) be ascertained (e.g. Nabû-mušeṣi). They may well have been resident in Nineveh. Finally, the list of experts at Assurbanipal's court (SAA7

though it is natural to assume that those Scholars based closest to the king would have had most influence over him. As we have seen, however, access to the monarch was extremely limited. I would prefer to do away with the concept of inner and outer circles altogether, and replace it with the notion of a larger circle, the "entourage" of those "performing the watch of the king", which included at one time or other all the Scholars. Their entry into this circle depended on the extent of their education and their chances were obviously enhanced by being members of certain families. Their survival in the circle was dependent on the whim of the king. In my opinion no hierarchy existed between the five disciplines, nor between Babylonians and Assyrians, but a hierarchical continuum of age (and family) dominated, tempered by the obvious difficulties of communication experienced by those Scholars who lived far from the capital. It must not be forgotten that Esarhaddon was king of both Assyria and Babylonia, and was no doubt obliged to treat his Babylonian Scholars equally.

Parpola's view of a distinction between inner circle and outer circle Scholars was, no doubt, influenced by Oppenheim's (1969) article. Oppenheim argued (p113–5) that those Scholars who authored the largest number of Letters authored the smallest number of Reports, that the similarity of the nature of the Reports indicated that they demonstrated the workings of a well-established empire-wide institution, and that the Scholars who sent Reports from Babylon included in their missives various requests for and complaints about pay, *because* Reports were their only means of communication.[155] In other words Oppenheim has argued that in general only Reports were sent by Babylonians in Babylonia, while Parpola has argued that those Scholars formed part of the outer circle. The supposition of this combined position is that the Reports were *less important* pieces of communication than the Letters.

In fact since Oppenheim's publication many more Letters from Babylonian Scholars have emerged. Also, the high-ranking, Ninevite Scholar, Issar-šumu-ereš, sent as many Reports as he did Letters, even at the height of his powers in 669 BC (8004 and 8005). Reports must *now* be understood only to be a form of communication designed for celestial omens, and that even those without interpretation[156] sent by Nabû'a from Assur, say, were not in any way inferior in import to the Letters sent by Akkullanu, say, from the same city. Reports were not solely sent by scribes of EAE, as mentioned, but it is clear that there would be little call in Nineveh for the correspondence of physicians, haruspices (and perhaps also for chanters and exorcists) based in cities distant from the capital. The use of Reports as opposed to Letters reveals *nothing* about the relative standing of the authors, only that they were sending interpretations drawn from a study of the heavens.

As Ašaredu says in 8338: 7f:

---

1) offers perhaps the best guide to the inner circle at this time, but it includes many names not on Parpola's list.

155 "It seems that the scholars stationed there (Babylonia) did not communicate in any other way with the king, while a complaint of the Assyrian Balasî ...is contained in a letter." (op.cit. 115). Also, Hunger agrees in SAA8 xvii: "As noted by Oppenheim, almost all complaints (in Reports) come from Babylonian Scholars who probably had no other way of approaching the king."

156 The distinction between Reports with and without omens is made by Chadwick (1992) 13–15, who asserts that the latter were made to assist in regulating the luni-solar year. He consequently designates them "calendar Reports". I note, however, that the interpretations of the observations recorded in the calendar Reports were *obvious* (e.g. compare 8136 and 8266 etc.), and it cannot be excluded that they were *not* distinct in purpose from those Reports containing omens.

"The scribal art is not heard of in the market place. Let the lord of kings summon me on a day which is convenient to him and I will investigate and speak to the king my lord."

On the contrary, *ṭupšarrūtu* took years of training, involving long, subordinate association on the part of the pupil, then apprentice, then assistant, and finally the Scholar with a temple institution of learning and with other experts – Scholars of the *previous* generation, many of whom would have been blood relatives. The products of just such an academic environment might be expected to adhere to texts written centuries earlier, and it is all the more remarkable that so much development in the ability to predict celestial phenomena can be discerned in this period when the authority of one's elders still mattered so much. More on this in Ch.5.1.3. At the same time, regardless perhaps of the participants' extended educations, the new, large, royal "entourages" of the NA kings encouraged fierce competition for royal favour between members of the *same* generation, and the extent to which the invention of new predictive methods can be accounted for on this basis is discussed in Ch.5.2.

To conclude this chapter, I make the following observations:

Although celestial divination was used in Assyria before the mid-8th century BC, I argue that an increased interest in that and other Scholarly disciplines took place in royal circles thereafter. It has been suggested that under Sargon II the goddess Ištar (NA Issār) "protrectress of kings, reappears in Mesopotamian ideology".[157] This may have been prompted by Assyrian military expansion, for Issar was the goddess of war, or by the legitimating ambitions of the royal usurper Sargon.[158] It is the Ištar section of EAE that deals with the planets and stars, and Ištar herself is "associated" with the planet Venus. I noted in I.3 that the planets appear for the first time in the royal inscriptions of Sargon II and particularly in those of his grandson Esarhaddon. I suggest that their presence in these royal inscriptions attest to an increased interest in what the planets were thought to portend for these kings, which was enhanced by an increased concern with Issar, and in the case of Sargon II with an intention to copy such motifs in works describing the exploits of his illustrious predecessor Sargon of Agade (see n43, above). Since the Scholarly discipline – "wisdom" – of which celestial divination was but a part was a southern Mesopotamian invention, numbers of Babylonian Scholars had long since been brought to the Assyrian capitals. With increasing Assyrian success more resources were devoted to cultural matters, particular to those matters which reflected on the glory of the Assyrian king and told of his destiny in the stars. At the same time, despite the Babylonian origins of this wisdom, the Assyrians themselves became "competent" in it, indeed they soon "mastered" the series and large numbers of Assyrian Scholars swelled the ranks of the "entourages" of Esarhaddon and Assurbanipal.

It is argued that under Sennacherib a re-assertion of Assyrian religion took place[159] and increased Assyrian independence in the associated arts of divination may well have gone hand in hand with this. By Esarhaddon's time the Assyrians felt themselves to be so able to ape Babylonian religious behaviour that they considered this to be a means by which to control the south other than by military force (Porter, 1993). By Assurbanipal's

[157] Reiner (1985a) 22..

[158] It was under Sargon II that a NA revival of texts concerning the third millennium BC Sargon of Agade took place. See n43, above. Ištar was the city goddess of Agade.

[159] E.g. Livingstone (1997) 167f with references to the "Marduk Ordeal" text SAA3 34/35.

time, the Ninevite libraries had undoubtedly become the main cultural repositories in all Mesopotamia, and Assyria, I suggest, had taken a lead in intellectual matters, in particular celestial divination. Assyrian nationalism had encouraged Assyrian religious fervour, militarism an interest in Issar, absolute monarchy the interest of an empire's worth of the finest Akkadian minds who for the first time had a single object of concern – the king. Collectively, they protected their charge, and took great interest in what the gods, via the heavens, had to say about him. Methods to assist in this developed rapidly and circulated amongst the Scholars, obliterating any clear distinctions between Babylonian and Assyrian schools. To some extent this new knowledge was profane, perhaps esoteric, and only rarely found its way into the institutional libraries. Nevertheless, enough clues remain in the correspondence of the Scholars, in the copies of astronomical observations from Babylon, and in the occasional library text from Nineveh to provide us with the evidence to reconstruct at least some of it.

CHAPTER 2

The Planets and Their Ominous Phenomena c. 750–612 BC – Names and Terms

The aim of this chapter is to study in detail the names given in Mesopotamia during the period c.750–612 BC to the seven planets visible to the naked eye, and to determine the phenomena they manifest which were at that time considered ominous. I have tried to be as comprehensive in my coverage as possible in order to produce an analysis of the state *of celestial divination in the late NA period. This is the period when, it is suggested here, a rapid development in the discipline took place resulting in the use of and interest in methodologies that can predict some planetary phenomena to an accuracy useful for divination. Efforts will be directed towards elucidating this suggestion.*

No study of this form has been attempted before, to my knowledge. Other analyses treat "Mesopotamian" celestial divination as a unified and broadly unchanging body of knowledge, and consider it legitimate to make comparisons between texts separated by more than a thousand years without first establishing, as far as it is possible, what the names and terms meant at any given period. I consider this to be a methodologically unsound approach, as I noted in §I.2, for continuity in meaning cannot be assumed, but must be demonstrated. Also, the non-mathematical astronomical-astrological texts (NMAATs) dating to this period have only very occasionally been compared with the contemporary Reports, Letters and other related texts found in the Assyrian capitals. This has perhaps been because the NMAATs have mostly been thought to have been the result of ambitions quite distinct from celestial prognostication, and have been treated by an almost entirely different group of modern students. I argue in Chapter 4 that the NMAATs formed part of the predictive (PCP-) Paradigm, while here in §2.2 I reveal that their connection to the divinatory material is extremely close.

Much has been done in earlier studies in gathering together planetary names, in working out which planets were assigned to which gods, and so forth. Of greatest significance is the 1950 work of Gössmann. Also of immense help were the computer generated Indices in the SAA series, and the works of Reiner & Pingree *BPO*2, *BPO*3, Hunger & Pingree *Mul.Apin*, and (the partly outdated) Bezold (1916). Early contributors include Jensen (1890), Jastrow (1898) & (1912), Virolleaud *ACh.*, and Jeremias (1929). Parpola's *LAS* II is also full of important insights relevant to this study. Similarly, the elucidation of the terms used to describe the heliacal risings, the stations, and so forth of the planets has been the work of many students from the time of Sayce (1874) onwards. Of most importance are perhaps the several studies of Kugler, Schaumberger, and Weidner to be listed where relevant. The SAA glossaries were once again very useful, as were *LAS* II and the two dictionaries, *CAD* and *AHw*.

Many names are shared between planets, and between planets and constellations or stars in this period. Why is this, and under what circumstances does it take place? Some names are used for the planets under certain circumstances. What are these circumstances? Some names are both unique to the planets, and can be used for them under any circumstances. Does their usage imply something different from that of the others? These are some of the questions that this study will attempt to answer in part. Only then can issues such as the possibility that the meanings of names were lost over time, or that texts had became "corrupted" by the NA period, and so forth be approached.

Much is said about planets "representing" or "standing for" gods or constellations. A new terminology will be developed in an effort to distinguish between different types of "association", and clear up some of the complications found in the secondary literature.[160] This study is a potential minefield, with as many "rationalisations" of the associations between planets and constellations as there are Assyriologists. Every effort has been made here not to become bogged down in attempting to "explain" certain names, as if the ancient astronomer-astrologers and modern students really do share the same mind-set.[161] Nevertheless, the several methods by which planets and stars were associated will be looked at in order to see the extent to which they were typical of celestial divination more generally and whether or not they were still being used during the late NA period. It will be shown that these means of associating names form part of the normal practice of the EAE Paradigm outlined in Chapter 3.

2.1 The Planet-names in Cuneiform, c.750–612 BC

I have found that all the names attested for the seven planets in the period c. 750–612 BC can be placed into five categories. For example, the names Sagmegar, Delebat, Ṣalbatānu, Šiḫṭu, Kaiamānu, Šamšu and Sîn are unique to Jupiter, Venus, Mars, Mercury, Saturn, the Sun and the Moon respectively. They are never used for any other celestial bodies. They are what I am terming the "A-names" for these planets. The Marduk planet however, is a B-name for Mercury, which means that this name is shared only by other planets. In this case it is also used of Jupiter. Nēbiru is a C-name for Jupiter, Venus and Mercury, for it can be used only for these planets and only when one of them is located near the horizon. D-names, such as Nīru, are used for constellations or particular stars as well as for more than one planet. Finally some names refer only to *one* planet and to a constellation or individual star. These are the E-names, for example Šēlebu, used for the fox constellation and Mars,

[160] E.g. Hunger writes in SAA8 xvif "Saturn is considered *equivalent* with the Sun...scholars can *replace* one by another, *interchanged*...any planet can be *intended*...by constellations. Boll found the explanation for these *substitutions*...a planet could *take the place* (of a star)."

[161] As early as 1961 (English 1967) Foucault argued that the methods by which the "rational populace" asserted control over the insane were dependent on the society in question, that they did not manifest cultural universals, nor could they be understood in terms of an *a priori* human essence. He argued in 1969 (English 1972) against the tendency of historians to analyse the past in terms of categories like "the general will of the people", and more significantly for the topics herein covered, he criticised the modern tendency to classify into categories of "rational" and "non-rational". There is, indeed, a strong temptation to "rationalise" the omen corpus, the assumption being that the compilers of the omen series ought to have formed part of the same "culture of the sane" as do we – that their mentalities are "commensurable" – see Rochberg (1992) 549. When an omen seems inexplicable to us, some scholars (see below) have resorted to the notion that the omen must once have been "rational", and that it has subsequently become "corrupted". Much that is merely speculative can lead from this. E.g. completing the quote from above, Hunger SAA8 xvi writes: "There are many omens which speak of movements of fixed stars relative to each other... It is unclear what these protases may have originally meant... The scholars considered the names of constellations in such cases to be substitute names for the planets (on the basis that) if a planet had the same color as a fixed star, it could take the place of the other in the interpretation of the omen." This last "rationalisation" was the work of Bezold (1916, after Boll), but is both untestable scientifically and cannot be corroborated on the basis of the Scholars' own comments. (See also the remarks by Koch-Westenholz, 1995 131–2.) The effort here will be to discuss the "rationalisations" actually attested in the 7th and 8th centuries BC, and not those which seem familiar to us and which we (perhaps understandably) feel to be universal.

but for no other planets.

Many of the names used for heavenly bodies are those also applied to gods and their attributes. This is not the place to discuss the extent to which the gods and the planets were equated, and yet it is quite clear that the relationship was often close.[162] For my purposes it is sufficient that the phenomena manifested by the stars, constellations and planets were understood by the astrologer-astronomers to be messages, binding or otherwise, from the gods, and that they and the king whom they guarded against supernatural misfortune acted on these signs accordingly. The celestial bodies were "bearers of signs to the inhabited world" – see §2.1.1 below.

The following is a comprehensive list of all the names known to have been *used* in the period studied. The following section, wherein many of the names are analysed, refers back to this list. The results are displayed in Chart 2.1.

A *Names unique to the planet (amongst celestial bodies) and which can be used under any circumstances*[163]

JUPITER	$^{d/mul}$**sag.me.gar** = Sagmegar (reading uncertain, meaning unknown). Used in all text-groups. *Planetarium* No.334. SAG.ME.GAR is written in the -567 Diary and sàg.me.gar in the -651 and the -418 Diaries. It is used less frequently in the late periods, where it is replaced by the A-name $^{mul/múl}$babbar (which is quicker to write). ^{mul}**en.gišgal.an.na** = Engišgalanna (reading uncertain). In 8254 it is equated with Sagmegar, and in Assurbanipal's acrostic hymn SAA3 2: 43 it is said to be *mamlu šūpû* "noble, illustrious, the lord [who...] the (celestial) positions (*manzāzu*) of the Anunnakkī, [who...] lustration rites [...] rituals, and offerings [...]"
VENUS	$^{d/mul}$***dele-bat*** = *Delebat* (meaning unknown). Used in all text groups in all periods. It appears as ***dele-bat*** in the -651, and all subsequent Diaries. *Planetarium* No.109. ^{d}**u.dar/**^{d}**iš.tar** = Ištār (NA Issār, the goddess of war and love). E.g. 8051: 4/8461: 3. The deity can also be written ^{d}15 and $^{d(+)}$Innin, but I have been unable to find either spelling used to refer to the planet in the texts herein considered. This is presumably only by chance.[164]
MARS	$^{d/mul}$***ṣal-bat-a(an)-nu*** = *Ṣalbatānu* (meaning unsure, though Lambert (1996) has suggested seeing *ṣalbatānu* as a variant of *ṣarbatānu*, a rare adjectival form derived from *ṣarbû* "pertaining to the poplar", an epithet of Nergal). It is used in all text groups. *Planetarium* No.360. *Ṣalbatānu* does not appear in the Diaries where Mars is always referred to by the single sign

[162] Rochberg (1996).

[163] References, when *not* given, are very frequent. The planets, for which the names are attested in the texts of interest here, are italicised *and* bolded.

[164] For references to the use of 30 and 15 for Sîn and Ištar from the OB period on, see Lieberman (1987) n202. See also Parpola (1993a) nn87–9.

an (-651 Diary: 10). *Planetarium* No.21. It is found in the SB *ACh.* 2Supp.80: r.9. An is not usually written as the "star An" (though in the Hellenistic period new year ritual, *RAcc.* p138 l.308, ^{múl}an is attested), nor as the "god Anu". Simply the cuneiform sign AN is written.

^{d}u.gur = Nergal (lord of the Underworld, linked to Erra, the war and plague god, sometimes called Meslamta-ea and identified with Lugal-irra). In 8114: 8, 8284: 2, 8502: 11, 8541: 12 Nergal is used as a name for Mars. The few references in the SB texts listed in *Planetarium* No.302 do not change Nergal's status as an A-name for Mars in the period of concern here. See von Weiher (1971) 76f.

^{mul}***sa-ar-ri*** = *Sarru* (false planet). In 8288: 3 Mars is referred to in a protasis "If Jupiter and the false planet meet." Also written ^{mul}lul.la in the SB texts, there is no textual evidence that *sarru* is the name of any heavenly body except Mars, as *Planetarium* Nos. 249 & 342 also show.

mul *šá* ^{kur}**su.bir$_4$.ki** = Planet of Subartu (a region at this time often synonymous with Assyria[165]). In 8491: r.7 Mars is said to be the Planet of Subartu.

MERCURY

$^{d/mul}$**udu.idim.gu$_4$.ud** and

$^{d/mul}$**gu$_4$.ud** (x051: s1) = *Šiḫṭu* (jumping planet). Used in all text groups, including the -651 Diary, and all subsequent Diaries where the form **gu$_4$.ud** is used without determinative.

^{mul}***Na-bu-ú***/$^{d(+)}$**ag**/^{d}**pa** = Nabû (Biblical Nebo, god of wisdom/scribes, son of Marduk and god of Borsippa). Attested (probably) as a name for Mercury in *Sargon's 8th Campaign* l.317 (§I.3 n41), and perhaps as a planet in x064: 5. It is found as a name for Mercury in the SB text *ACh.*1Supp.8: 7 (*Planetarium* No.290).

dumu-lugal = *Mār šarri* (Crown prince). In x052: r.9, x073: r.7–8 & x074: r.6 ^{mul}udu.idim.gu$_4$.ud is equated with the crown prince.

SATURN

$^{d/mul}$**udu.idim.sag.uš** and

^{d}**sag.uš** = *Kaiamānu* = (steady/normal/constant planet). (*Planetarium* No. 333.) Used in all texts including the -651 Diary: 8 where sag.uš is written, and in the -567 Diary: 2 where ^{d}sag.uš is used. In this and in all subsequent Diaries the name genna is used for Saturn, with the -567 Diary using a divine determinative.

SUN

^{d}**utu**/^{d}**20/20/*šamšu*/*šá-maš*** = Sun god. The distinction between *šamšu* "the Sun" and the vocative Šamaš is made in the inscriptions,[166] but the need repeatedly to mention the Sun, made the use of the signs ^{d}utu, ^{d}20, and 20 more common in the majority of texts under consideration. Since utu and babbar share a sign, the use in the Diaries of ^{múl}babbar and ^{d}utu (with their respective determinatives) is understandable.

---

165 Subartu was the land of the Subareans, nomads based somewhere north of Sumer and Akkad in the late 3M and early 2M BC and traditionally part of Sargon of Agade's empire. By the NA period Assyria itself was sometimes referred to as Subartu, though in *The Sargon Geography* the region appears to have been considered to belong to the empire of another king. For details see Horowitz (1998) 79.

166 E.g. see the glossary at the end of Borger (1956).

MOON **$^{(d)}$30/nanna/zu.en/en.zu**[167] = Sîn = (Moon god). Its characteristic number 30 is, no doubt, derived from the length of an "ideal" (see Ch.3.1.2) month in days. $^{(d)}$30 is used most frequently in the texts herein considered.

B Names shared only with other planets and usable under any circumstances

JUP/MERC **$^{mul\text{-}d}$amar.utu** = Marduk (God of Babylon) planet. = *Mercury* in (8093: r3, 8454: 3, 8486: 6, 8503: 1), *Jupiter* in (8147, 8244, 8326). *Planetarium* No.260 offers no evidence that it was ever a name for any other planet. In the mystical work SAA3039: r.5 it states that the inside of ^{d}utu is ^{d}amar.utu.

ME/MA/SA/J/V **$^{d/mul}$udu.IDIM/*bi-ib-bu*** = *Bibbu* (planet – the Sumerian and Akkadian names have usually both been interpreted as meaning "wild sheep" thereby invoking the errant nature of these bodies' movements against the background stars, also implicit in the name "planet". However, doubt has been cast on the meaning of the Sumerian word udu.idim by Reiner, 1995, n22 though see also Horowitz, 1998, 153 n5). Bibbu is only attested as a name for Saturn, Mercury and Mars in the texts under consideration here. *Me* (8051: r3, 8113: r.1, 8157: 2/4, 8158: r.6 etc.), *Sa* (8039: 6, 8082: 4 etc.), *Ma* (8102: r.6, 8288: 8, 8311: 3, 8341: 3 etc.) ^{d}*Bi-ib-bu* in 3032: r.19 "The Underworld Vision of an Assyrian Prince" is the hangman of the underworld.

MERC/SAT **mul-gi$_6$** = Black planet. Probably Mercury in 8180: 4, but as *Planetarium* No.86 indicates, the black planet could mean Saturn. See also *LAS* II p343.

SAT/MERC **^{d}nin.urta/^{d}maš** = Ninurta (war and farmer god whose local form prior to the OB period in Girsu was Niñgirsu. It is rendered Inurta in the NA texts). In 8154: 8 the protasis: "If the Moon is surrounded by a halo and Ninurta stands in it" describes Saturn. *Planetarium* Nos. 323 and 316 suggests that in the SB texts both Niñgirsu and Ninurta were A-names for Saturn. However, in Mul.Apin Iii22 Niñgirsu appears as a star, and in Iii16, IIi38 & Iii54 ^{d}maš and in IIi5 & IIi66 ^{d}nin.urta are said to be names of Mercury. Ninurta was probably a B-name of Saturn and Mercury at this time.

SUN/SATURN **^{d}utu/20** = Šamaš – see above. These signs (and no others) still mostly designate the Sun's A-name, but occasionally Saturn is meant. This is the case in 8082: 5, 8095: 1, 8110: r.1f, 8166: 1, 8168: 6, 8297: r.1, 8301: 3, 8317: 3, 8350: 3, 8383: 3, 8416: 5 and x113: 3f etc. where Saturn is being described, but the omen protases use the B-name ^{d}UTU or 20. In 8095: r.3f Saturn is said to be the star of ^{d}utu. Often the C-name, aš.me, is used for Saturn.

---

167 Free organisation of writing: Beaulieu (1995) 3–4. Also found in ^{d}BIL.GI for Gibil. Rarely attested before the 1st Mill.

SUN/SATURN **mul-*šar-ri/ša* lugal** = Planet of the king. In 8095: r.7 the Sun is said to be the planet of the king. In x051: r.8–9 it is Saturn. Note: mul *ša* lugal is not the same as ^{mul}lugal.

MOON/JUP ^{d}**30** = Sîn – see above. In x043: r5 Sagmegar is said to be the Moon. This association is unique in the texts under consideration, and is explained by the context of the Letter, but is not unprecedented in the SB material (see Parpola *LAS* II p59).

MARS/MERC ^{mul}**mín-*ma*** = *Šanumma* (strange planet). A name for Mars in 8101: 7, 8288: r.5, 8341: 5, 8452: 6 and probably in 8125 & 8064: 7. However, in 8503: 6 the strange planet is a name for Mercury.

C Names shared with other planets that can only be used under certain circumstances

JUP/ME/VE? $^{\text{d/mul}}$***né-bi-ru*** = *Nēbiru* (ferry/ford). In the texts under consideration this name is only applied to Jupiter. In 8147: r.1 Nēbiru is said to be a name of the Marduk planet (J/Me) when the latter stands in the middle of the sky (*ina* murub$_4$ an-*e* gub-*ma*). In 8254: r2 and 8323: 7, Nēbiru is said to kur-*ḫa-ma* or sar-*ma* = *ippuḫamma* (8069: 1g), to have "lit up", which suggests its rising and proximity to the horizon. In x362: r3 a Nēbiru omen is used by Mar-Issar for Jupiter five days after its heliacal rising. The evidence from texts not dating to the period 747–612 confuses rather than clarifies the issue concerning Nēbiru. It is discussed by Koch (1991).[168] Most importantly the older evidence shows that Mercury (and perhaps even Venus) could be called Nēbiru.

JUP/MERC $^{\text{d/mul}}$**šul.pa.è** = Šulpae (brilliant youth – a Sumerian god attested from the Early Dynastic period on). The secondary literature assumes Šulpae to be a name for Jupiter alone,[169] but this is incorrect. It is a name given to the Marduk planet (J+Me) when it is rising heliacally – stated explicitly in 8147: 7. *J* (8147, 8212, 8214, 8288, 8398, 8438), *Me* (8093, 8114 by calculation). In 8438: 4, and perhaps in 8398: 3, it appears as if Šulpae is being used when Jupiter is near the Western horizon, but this may be a misinterpretation. In the vast majority of attested cases the name is used only when the Marduk planet is near the eastern horizon.

SAT/SUN **aš.me** = *Šamšatu* (Sun disk). It is used for Saturn when it is near the Moon (8095: 3, 8297: 3) and for the Sun, if the Sun and Moon are seen together on the 14th (8426: r6, 8501: 4, 8521: 4).

[168] As with so many of these attempts to rationalise the apparently contradictory evidence concerning the naming of the planets, very little interest is shown in the different eras from which the documentation emerges, and the differing purposes to which they were put. Koch's aim is to provide a solution to a problem that he himself (Schott (1936), Schaumberger, *SSB* ErgIII 313f, and Hunger & Pingree *Mul.Apin* 126, likewise) has generated by bringing together every reference to Nēbiru he can find; material which spans hundreds of years and many different text types. His basic assumptions are that the Nēbiru omens can be derived from original observations, and that the tradition concerning Nēbiru neither changed over time, nor over text group. Both of these assumptions are unfounded. The former will be discussed in Ch.3, the latter here in §2.1.2.

[169] E.g. Black & Green (1992) 173, Gössmann *Planetarium* 383, and the SAA8 index.

D Names shared with planets and constellations or stars

MA/SA/ME/J? ^{mul}**apin** = *Epinnu* (Plough constellation = Triangulum Boreale). In all the references in the texts herein considered Epinnu is a name for Mars (8049: r.4, 8219: 1, 8452: 1 & 8502: r.1). *Planetarium* No.39 indicates that it is a D-name for Saturn and Mercury and perhaps Jupiter, as well as for the constellation. Triangulum Boreale is non-ecliptic and is not close to the constellations Scorpius or Cancer. However, omens in the texts cited concern its approach (te = *ṭehû*) to those constellations and its reaching (kur = *kašādu*) the ecliptic. Consequently, I argue that ^{mul}apin was likely already a name used for the constellation *and* for planets by the time these omens were compiled, and thus long before c.750 BC.

MA/VE/ME/J ^{mul}**aš.gan** = *Ikû* (Field constellation = Pegasus+Andromedae). The constellation is meant in 8357: 6 and 8537: 3, but Mars is meant in 8082: r.6, and 8311: 7–9, and perhaps in 8072: 7. Ikû is also equated with ^{mul}ab.sín in 8082: r.8. In *Planetarium* No.193 IIIA3 the suggestion is made that Ikû is a Mars name when the latter is sitting in Virgo. It is equally possible that Ikû is a Mars name when the planet is near the Moon. I suggest the name is generally applicable to Mars. *Planetarium* No.193 shows that in the SB texts Ikû is also a D-name for Venus, Mercury, and Jupiter.

ME/MA/MO? ^{mul}**bir** = *Kalītu* (Kidney constellation = Puppis). A name for Mercury in 8325: 5, in the SB texts it is a D-name for Mars and the constellation. "Kidney" is the name given to the Moon on the 7th day in l.284 of the *GSL.*[170]

ME/V/SA?/J? ^{mul}**en.te.ne.bar.guz** = *Habaṣīrānu* < *ḫumṣiru* (the Sumerian reads "the shaggy winter constellation", and the Akkadian "mouse-like", and according to *CAD* is Centaurus, which is non-ecliptic). In 8158: 7f it is made clear that the protasis, "if Centaurus *flickers* (*mulluḫ*) when it comes out", applies to Mercury's appearance. *Planetarium* No.123 indicates that Centaurus is in the SB texts a D-name for Venus, and possibly also for Saturn and Jupiter.

***J/METEOR*/MO/SI/SA?** **mul-gal** = *Kakkabu rabû* (Big star). This name is attested in 8288: r.1, where Jupiter is meant, and in 8334: 2, where a meteor[171] is meant. Although not found in the texts herein considered, mul-gal can also act as a D-name for the Moon, Sirius, and possibly for Saturn, as *Planetarium* No.62 shows. The Big star appears in ll.9 & 12 of *The 12 Names of the Marduk Planet* (K5990, Weidner *Hdb.* p24). In 8339: 1.

^{mul}***ṣallummû*** (unknown meaning), the usual word for a comet, is perhaps referring to Jupiter in 8339 and 8456.

---

[170] *GSL* = The "Great Star List", K250+ (App.1 §29), now in Koch-Westenholz (1995) App.B, which provides some of the code of the EAE Paradigm (Ch.3.2.2). The text may have been first composed during the period of interest, but this cannot be established as yet.

[171] See most recently Chadwick (1993) who shows that mul-gal is one of five names that identify meteors, where ^{mul}*ṣallummû* (8456: 1) identifies a comet.

SAT/VE/MA? ^{mul}**ka.muš.ì.kú.e** = *Pašittu* (Deleter star = β-Andromedae). It is a D-name for Saturn in 8491: r.3, and *Planetarium* No.215 shows that in the SB texts it is also a D-name for Venus and possibly for Mars.

V/ME/MA/J/SA ^{mul}**ku**$_6$**(-an-*e*)** = *Nūn* (*šamê*) (fish (of the sky) constellation = Piscis austrinus, which is not ecliptic, but is close to Capricornus). It is probably a name for the constellation in 8073: r.1. It is attested seemingly as a D-name of Venus in 8055: r.10. In 8325: 4 the Fish constellation is attested as a name for Mercury.[172] This is fully confirmed by *Planetarium* No.218 §III where the Fish constellation is also apparently attested as a D-name for Jupiter, Saturn, and Mars in the SB texts.

MARS/SAT **mul** ^{kur}**mar.(tu).ki** Star of Amurru (*Māt Amurrî*) is a D-name for Mars as stated in 8412: r.2. Ṣalbatānu is listed under the stars of Amurru in *GSL* 1.219. However, in 8491: r.9, Saturn is said to be mul *ša* ^{kur}mar.tu.

VENUS/MARS **mul** ^{kur}**nim.ma.ki** Star of Elam (*Elamtu*). Venus is a star of Elam in 8302: r.2 and is also found in the list of Elam stars (*GSL* 1.201). *Planetarium* No.318 points to several references which indicate that Mars was also referred to by the title "star of Elam" (e.g. *GSL* 1.94). Perhaps, it was *the* star of Elam. None of the references can be shown to have been composed or used in the period of interest, however.

SATURN/JUP **mul** ^{kur}**uri.ki** Star of Akkad (*Māt Akkadî*). Saturn is equated with this name in 8383: r.7, but no A-name for Saturn appears in the *GSL* list of Akkad stars. In that list, the D-name of Saturn, Zibānîtum appears, as do the names Nēbiru, and UD.AL.TAR.

ME/MA/JUP ^{mul}**šudun/šu-du-un**[173] = *Nīru* (Yoke constellation = Bootes). It is a Mercury name in 8073: 4, and a Mars name in 8383 r.4 The omens used are different. In 8546 Jupiter is probably meant, as *Planetarium* No.379 B suggests. The Yoke is a recognised constellation, which also takes the name ^{mul}šu.pa (Reiner & Pingree *BPO*2 p15). Under this name it appears in the list of Ellil stars in Mul.Apin. Yoke constellation omens are attested in Mul.Apin IIB7-iv8 and IIiii43, where it could well be acting as a D-name for the planets.

MARS/ME/J? ^{mul}**udu.idim sa**$_5$ = *Pelû/Sâmu* (red/red-brown planet cf. 8252). Mars is called the red planet in omen protases from texts 8274: r.4–6, 8288: r.2, 8419: 4. In 8281: 5 Mercury is meant. Jupiter is described as sa$_5$ in 8004: 12, 8115: 3, 8170: 5, 8211: 3 & 8326: 5, but is never explicitly called the "red planet". ^{mul}sa$_5$ "red star" and ^{mul}*makrû* (*Planetarium* No.255) "fiery red star" are Mars names in *GSL* 1.85–6. *Planetarium* No.114 shows that ^{mul}sa$_5$ can describe Mars and Jupiter, but also constellations and stars in the SB material. *CAD makrû* and *BPO*2 take $^{\text{mul/d}}$*makrû* to be a name for Mars alone. This is incorrect. Although, the "Red planet (*bibbu)*" and not the "Red star (*kakkabu*)" is attested in the texts discussed here, it has been decided to categorise the former as

[172] In 8325 Mercury in Virgo is paralleled by the fish star approaching the Bow star. The Bow star is a name for Virgo in *ACh.*1Supp.50: 14.

[173] This is the syllabic writing of the Sumerogram found as a gloss in 8073. The syllabic writing of *nīru* is found in Mul.Apin IIiii43, but is not attested in the texts considered here.

a D-name for Mars, Mercury and perhaps Jupiter, because of its clear connection to the SB "Red star".

ME/MARS/SA ^{mul}**uga(mušen)/ú-ga** = *āribu* (Raven constellation = Corvus). It is used as a D-name for Mercury in 8073: r.1 where the protasis: "if the Fish constellation stands close to the Raven constellation" describes the situation of Mercury's proximity to Capricorn. The Fish constellation is located next to Capricorn, and its name is sometimes used to designate that ecliptic constellation (*Planetarium* No.218 I). Mercury is perhaps also meant in 8414 where the Raven constellation appears in broken context. In 8082: 5 ^{mul}uga.mušen is a D-name for Mars for "if the Raven constellation reaches the path of the Sun", describes the near conjunction of Mars and Saturn (designated by its B-name). *Planetarium* No.132 indicates that the name ^{mul}uga is also used in the SB material to indicate the constellation Corvus and the planet Saturn.

SAT/MA ^{mul}**zi.ba.an.na/*zi-ba-ni-tum*** = *Zibānītum* (Scales constellation = precursor to Libra). In 8547: 5', probably in 8039: 3, and perhaps in 8544, it is a name for Saturn. All other attestations in the texts under consideration describe the constellation. Note the Sumerian spelling in x172: 6'. In *Planetarium* No.176 it is shown that in the SB texts the Scales constellation is also a D-name for Mars.

Generic **mul** = *Kakkabu* (Heavenly body = star, constellation, planet, meteor, comet). In the lists above and below I have used the appropriate name depending on the categorisation. The sign mul derived from a pictogram of three stars. In later times the signs TE, GÁN and ÁB were used. They are read by modern students as múl, mul_4 and mul_x.[174]

dingir = *Ilu* (god, though in this context it is equivalent to mul). In the same text one heavenly body might be designated by dingir and mul (e.g. 8005), or indeed without any determinative. The sign AN for dingir evolved from the pictogram representing one star. The divine nature of all the heavenly bodies is apparent, not least from the sign for mul, though I will not be considering this issue further. See the second paragraph to §2.1.

E Names shared only with constellations or stars

JUPITER ^{d}**gàm** (read zubi, by Parpola SAAX) = *Gamlu* (Crook constellation = Auriga). An Ellil star in Mul.Apin and apparently ecliptic according to Mul.Apin I iv 34 . In 8115 and 8170 Jupiter is described as rising heliacally in month III, in the path of the Anu stars. In 8115: r.4 and in 8170: r.1 Auriga is said to "carry radiance" še.er.zi íl. This protasis could be accounted for by saying that Jupiter is rising near to Auriga in both instances. Alternatively, Auriga is acting as an E-name for Jupiter. This is suggested by the text *The 12 Names of the Marduk Planet* where we find (1.10) ^{mul}gàm listed. Also, in x160: 13 the same *Gamlu* omen is re-

174 Reiner (1995) 5.

peated where Jupiter is not near to Auriga (see also App. 2 below, year 660).

^{mul}lugal = *Šarru* (King star = Regulus). Šarru designates the star in 8040, 8041 etc., but in 8170: r.3 the omen "If Regulus carries radiance" is used when Jupiter is meant (i.e. when the text was written no planet was located near to Regulus). As with Auriga above, an omen concerned with radiance is linked with Jupiter's brightness. In 8489 Jupiter's awesome radiance or *melammu* is explained in r.4–5 as meaning "Regulus stands either to the right or to the left of Jupiter". Jupiter is perhaps meant in 8205 and in x160: 23. That ^{mul}lugal is an E-name for Jupiter is confirmed by the text *The 12 Names of the Marduk Planet* l.11.

VENUS

^{mul}a.edin = *Eru'a* (Frond star = γ-Comæ-Berenices?) "If the frond star reaches the Pleiades; explanation, Venus stands in the Pleiades" – 8536: 3f. In 8055: r9 the same protasis is quoted, and yet Venus is at that time located in Scorpius. The Pleiades in r.9 is the E-name of Mars, as r.10 shows. 8055: r9 indicates that this Assyrian Scholar was using ^{mul}a.edin as an E-name for Venus, not only when Venus stood in the Pleiades, but also when it stood near Mars. That is, the Scholar appears to have extracted from the omen: "If the Frond star approaches the Pleiades: Adad will devastate" an association of Venus to the Frond star which applies *only* in the vicinity of that which can be called by the name "Pleiades".

^{mul}*tul-tum* = *Tūltu* (Worm star/constellation?). It is connected in some way to Anunītu, an eastern part of Pisces as *GSL* l.165 suggests. It is attested in this context in 8357: 3 and 8538: 3. Perhaps, the Worm star is not so much an E-name for Venus, as the name Anunītu takes when Venus stands inside it.

^{mul}ùz = *Enzu* (Goat constellation = Lyra). The constellation is probably meant in 8074, but in 8175: r.1, x088: r.5, and in 8247: 8 we find the omen: "If the Goat star comes close to Cancer". Constellations cannot approach each other, so either the Goat star or Cancer refers to a planet. In 8175: r.7 ^{mul}ùz and Venus are explicitly equated. When Venus approaches Cancer, a Goat star omen (whose apodosis is positive) is used by the Scholar in question by associating Venus with that particular star. That this association is also found in EAE is clear from *ACh.* Išt. 8: 3.

MARS

$^{\text{mul/d}}$im.dugud.mušen/*an-zu* = Anzû (a lion-headed bird, thief of the tablet of destinies) star. Mars is meant in 8064: r.2f and 8114: r.1. *Planetarium* No.196 points to VR 46 20ab where the Anzû star is equated with the horse star, which appears in Mul.Apin Ii30.

^{mul}ka$_5$.a = *Šēlebu* (Fox constellation = Ursa Major?) In 8049: r.9 the Fox constellation probably denotes Mars, also suggested by the evidence cited in *Planetarium* No.205 in which it is attested only as a constellation and as a name for that planet.

mul.mul = "The Stars" = *Zappu* "bristle"/ **^{d}7.BI** = *Sebetti* (e.g. 8507: r.1) "The Seven". All names designate the Pleiades. It appears as if The Seven (generally beneficent) gods, represented by the seven dots, became identified with the constellation at least by the NA period.[175] Mostly, the references to The Stars in the texts herein considered are to the Pleiades, but in x063: r.8 the following line is given: udu.idim mul.mul ^{d}*ṣal-bat-a-nu* "(as) a planet, the stars (=) Mars", which demonstrates explicit awareness of the possibility of constellation names acting as planet names. No other planets use The Stars as a name. Mul.mul is an E-name for Mars in 8050: 7, 8055: r.9, 8072: r.2, 8376: 6f & 8491: r.3.

^{mul}**ud.ka.du$_8$.a** = "Demon with the gaping mouth constellation" = *Nimru* "Panther constellation" = Cygnus. In 8284: 2 the Panther is apparently equated with Nergal – an A-name for Mars. In 8415: r.1 it may be acting as a name either for Mars or for the constellation, as in 8507: 5f. *Planetarium* No.144 tentatively suggests that the Panther can act as a name for Jupiter (*ACh*.2Supp.64II: 13 is fragmentary). In the absence of any clear examples, I have listed the Panther as an E-name of Mars.

^{mul}**ur.bar.ra** = *Barbaru* (Wolf star = α-trianguli). Mars is meant in 8048: 5. *Planetarium* 161 shows that, thus far, it is only attested as an E-name for Mars.

2.1.1 Discussion of the Associations between the Planets and their Names

This section discusses some of the above listed names used for the planets. Those that are not discussed are those for which no "rational" connection between the use of the name and the planet and its properties has been discerned (here, or by others, to my knowledge). There may have been none, but further research may well cast light on the processes by which certain of the names became associated with the planets.

By "associated name" is meant no more than that the planet as celestial object is sometimes referred to by it. Some names are *strongly* associated to the planets. For example Delebat is associated with the planet Venus throughout most of the Akkadian cuneiform period. It is, in effect, *the* name given by the Mesopotamians to that particular heavenly body. The same ought perhaps to be said for Sagmegar, except that in the late period *the* name for the planet we know as Jupiter was ^{mul}babbar. Other names can refer to more than one heavenly body, they are perhaps associated with more than one planet, star, or constellation. These names are more *weakly* associated with the respective planets. When any of the names are used in the texts herein considered, they only ever refer to one celestial body (by which I may mean constellation) at a time. When ^{mul}lugal is used, for example, either Regulus or Jupiter is meant. No single use of ^{mul}lugal refers to both simultaneously. It is consequently a name associated with *both* heavenly bodies.

Five main kinds of association between the planets and their many names have been discerned. A ***basic*** association is one between the planet and a deity. For example, the association between Marduk and the planet Jupiter is basic, or traditional/fundamental – inexact terms which serve to cloak the impossibility of finding the origin of the association. A ***theo-***

[175] Black & Green (1992) p162.

logical association is one based on a knowledge of the Mesopotamian gods, a small part of which we have gained. For example, Nēbiru is a C-name of Jupiter and Mercury. The basis of the association of the name to the planets is via the Marduk planet, for Nēbiru is a name given to Marduk in the religious text *Enūma Eliš* VII 124. A ***learned*** association between a planet and a name is one that could only have been drawn by those expert in the field of celestial divination – the Scholars. Two main types of learned association have been discovered, one based on the many possible readings of the words themselves, and another based on the collection of omens and associated works. Learned association will be discussed in §2.1.2. An ***observational*** association is simply one drawn from the phenomena manifested by the heavenly body. For example, Saturn is associated with the name "the steady, constant planet" because it moves more slowly than any other planet against the background stars. A ***symbolic*** association is one made between the generally recognised representation of a deity (the symbol) and the planet. For example, the C-name for Saturn, aš.me, is the name given to a symbol of the Sun god, who is also associated with Saturn.

In assigning an association to one of these five categories I have tried not to be overly speculative. In particular, I have considered the evidence to be found in the Letters and Reports to be of greater value than that found in the omen series and associated texts (the so-called Standard Babylonian material) or "literary" texts, even when the latter are known to have been composed in the period of concern here. This is because the Letters and Reports were written by and for known individuals whose intentions are broadly understood. The same cannot be said for those who wrote some of the literary texts or for those who put together the divination series. I am wary of the positivist assumption that the SB material or the NA literary texts *necessarily* tell us about the attitudes of the NA Scholars to the heavenly bodies and their phenomena, without first assessing what beliefs can be gleaned from texts that they *themselves* wrote. Literature, after all, is 'literary' and its intentions may not always be to display society's views as to religion, science or whatever.[176] Equally, the intentions of the 2nd millennium BC authors and compilers of EAE may not have been the same as those who used the series in the 7th century BC. Nevertheless, both types of material will be considered in passing in the following, which discusses, in turn, the associations behind the names known to have been applied to each of the seven planets in the period c.750–612 BC.

JUPITER

Sag.me.gar's attested epithet is found in the SB text VR46: r3 and elsewhere. It reads:

> *na-áš ṣa-ad-du a-na da-da-mu*[. . .]
> "the bearer of signs to the inhabited world. . . "[177]

In Craig *ABRT* 1 30: 42 the following is found:

> "(O Marduk) your name when you are visible (as the planet Jupiter) is ^{d}sag.me.gar, the foremost god (*ilû rēštû*), the leader of [. . .] (*ašarē*[*d*..]) who when he shines forth, shows a sign (*ṣaddu*). . . "

176 "Near Eastern scholarship – the positivist approach . . . .treats 'literary texts' exactly as any other form of 'historical text', discarding . . . any attempt to account for precisely those distinctive qualities that make literature 'literary'. . . The result has been that literary works have been trawled for evidence of social conditions or historical facts, as sources for the history of thought or religion. . . " Black (1998) 6–7.

177 For variants see *CAD* Ṣ p56.

^{D}sag in Akkadian is *ilû rēštû*. Sag also takes the Akkadian *ašarēdu*. Me can be read *têrtu* = "oracle/omen/instruction" which is semantically linked to *ṣaddu*. The epithet has been derived from the signs used to make up the name SAG.ME.GAR.[178] Perhaps, the epithet points to why sag.me.gar and Marduk were associated, for Marduk was the foremost god of Babylonia from the Kassite period on. It is perhaps a ***theological*** association.

^{Mul}en.gišgal.an.na means "star (mul) of the lord (en) of the *manzāzu*s (gišgal) of the Anuna gods/sky (AN)", just as in the broken epithet from 3002: 43 listed above. The Anuna are the 600 gods of the sky and/or underworld over whom Marduk had power.[179] This A-name for Jupiter is presumably based, therefore, on ***theological*** associations.

As noted, the association between Jupiter and Marduk must be understood to be ***basic***, and it is tempting to speculate that with the assignation of gods to the Moon, Sun and Venus taking place at least by the third millennium BC, the next brightest body in the sky would likely have been linked with the new top god of southern Mesopotamia.

The single association in the texts herein considered of the Moon and Jupiter in x043: r5 is paralleled in two other cases cited in Parpola *LAS* II p59 nr.5. Parpola argues that the association of Jupiter and the Moon is a "logical" corollary of the association of Saturn and the Sun. It is consequently a ***learned*** association. Incidentally, Sîn shares the same epithet as Sag.me.gar, "the bearer of signs".

As discussed above, the name Nēbiru is found in the ***theology*** of Marduk. Similarly, the god Šulpae's association is probably ***theological***. In the MA *Astrolabe B*:B: III: 23 Šulpae is described as the *sukkallu* "herald, or similar" of Marduk.[180] Perhaps this was in some way visualised[181] as describing Jupiter near the eastern horizon. Some form of syncretism between Marduk and the "brilliant youth" perhaps took place, since the possibility cannot be excluded that in a different (earlier) tradition the association between Šulpae and the planet was ***basic***.

The reason for the association between mul-gal "Big star" and Jupiter is not clear to me. Certainly, that Jupiter is the largest planet was not known by the Mesopotamians. In Mul.Apin Ii37 it is written that "mul-gal, although its light is dim, divides the sky and stands there: Marduk, Nēbiru, Sag.me.gar." Was Marduk's star large, because Marduk was the top god to the Babylonians? Did the name only later come to be used to describe meteors and the like?

^{Mul}ku$_{6}$ an-*e* "Fish of the sky constellation" is also known by the name ^{mul}ku$_{6}$-^{d}Ea in the text *The 12 Names of the Marduk Planet*. Ea was thought, in some circles at least, to be Marduk's father.[182] The association of Marduk and the Fish constellation is thus likely ***theological***.

---

178 This is typical of Mesopotamian literature, and will be discussed in Ch.3.2.2. Livingstone *MMEW*, on VR46: r.42: ^{mul}*ṣal-bat-a-nu* | *muš-ta-bar-ru-ú mu-ta-nu* notes that ZAL = *muštabarrû* and *batānu* = *mutānu* since *bat* = UG$_{5}$ = *mut*.

179 *Enūma eliš* VI: 40–44. In some later texts, e.g. *KAR* 307: 37, they are only located in the lower earth. See Horowitz (1998) *sub* Anunnaki in the index. *Manzāzu* in this context is probably referring to the "stations" which mark the start of the year. In *Enūma eliš* V: 6 and 127, Marduk "set fast the station (*manzāzu*) of Nēbiru...who holds the turning point" and in the astrolabes Nēbiru is often positioned in month XII or month I. En.gišgal.an.na is a name describing the rôle marked out for Nēbiru by Marduk in *Enūma eliš*. For further details see Horowitz (1998) 116.

180 For other instances of gods acting as a *sukkallu* of other gods see *CAD* S pp358–9.

181 As a "metaphor" – see Rochberg 1996.

182 Marduk was syncretised with Asarluhi who was regarded as the son of Ea; Black & Green (1992) 128.

Mul ^{kur}uri.ki "Star of Akkad" was probably associated with Jupiter through Marduk's status in Babylonia, sometimes designated by the name Akkad.

^{Mul}gàm "Crook star" in VR 64: 3 is said to be "the weapon of the hand of Marduk", so the association with Jupiter was again probably ***theological***. The evidence gathered in *Planetarium* No.64 shows that the Crook star and Jupiter were linked long before the period of interest here, though the use of the ^{mul}gàm omen with omens describing Jupiter's brightness was probably determined in part by its particularly positive apodosis: "the foundation of the king's throne will be everlasting".[183]

^{Mul}lugal "King star" omens also concern Jupiter's brightness.[184] Both ^{mul}lugal and ^{mul}gàm are ecliptic.[185] They lie close to the path of the planets, and when it is said they "carry radiance" (e.g. 8029: 1) this probably meant, in the period of interest here, that a planet was positioned nearby. The similarity of the conditions under which they both act as E-names for Jupiter suggests that ^{mul}lugal's association with the planet may have been based solely on that protasis key-word "radiance" še.er.zi, *šarūru*. This would make the association ***learned***.

In those Astronomical Diaries dating from -463, -453, and -440, Jupiter is given the name ^{múl}babbar = *peṣû* "White planet" for the first time. This name is a late A-name for Jupiter, and is not attested in the texts of interest here. Nevertheless, whiteness is associated with Jupiter in EAE, its commentaries, and in other explanatory works. In *ACh.*1Supp.36: 9 (EAE 61) it is written that if Venus is wearing a white crown, then Jupiter is standing before her. See also *ACh.*1Supp.4: 11–12. In K2346+:r.26' (*BPO*3 Group F pp248–9) it states explicitly "the white star is Jupiter, the red star is Mars, the green star is Venus, the black star is Saturn – variant Mercury". I suggest that it is merely chance that Jupiter's whiteness is not attested in texts known to have been used in the period c.750–612 BC. This is because I argue that there existed a strong connection between the astronomical-astrological material written in Assyria before 612 BC and that written in Babylonia thereafter. In *GSL* I.168 the white star is Mars (Ṣalbatānu). Perhaps, ^{mul}babbar was a B-name for Jupiter and Mars prior to 612 BC and only later became an A-name for Jupiter. The association between the white planet and Jupiter is perhaps ***observational***, though all the planets can appear white to the naked eye.

Another well known name for Jupiter attested only in SB texts is ^{mul}ud.al.tar = *dāpinu* "Heroic/martial planet". Again, I presume it to be only chance that this name is not attested in texts known to have been composed during the period of interest, since the later A-name for Jupiter, ^{múl}babbar, probably derived from the name ^{mul}ud.al.tar. That is, ^{mul}babbar derives from ^{mul}ud.(al.tar),[186] for babbar is a reading of the sign UD. In looking for a succinct way of designating Jupiter, the later Scholars chose a shortened form of an attested name. The association of ^{mul}babbar to Jupiter is thus also ***learned***.

---

183 Note Koch-Westenholz (1995) 144 "*– A certain tendency to see things from the bright side*".

184 The protasis with which the Crook star omen is found: "If Jupiter becomes steady in the morning" is explained in 8184: 7 and *ACh.* Išt. 4: 34 as " "morning" means "to become bright"- it carries radiance."

185 They are both "gods who stand in the path of the Sun" in Mul.Apin I iv 33–9, even though what we know as Auriga was quite some way from the ecliptic in the period of interest. Identifications made by Reiner & Pingree *BPO*2 10f and 2.1.2.4.1, Parpola *LAS* II App.B, Hunger & Pingree *Mul.Apin* p137f and Koch (1989) are still somewhat in dispute, it must be noted.

186 MUL.AL.TAR is also attested in *ACh.*Išt 36: 14, where it is identified with Šulpae, a Jupiter C-name.

VENUS

The association of Inana/Ištar and Venus was made very early. It has to be considered ***basic*** to all further associations. Inana, the name of the Sumerian deity, probably derives from Nin-ana "lady of heaven", and one may surmise that a certain logic underpinned the assignment of the next brightest celestial body to the Moon to the deity thought (in one tradition) to have been the daughter of the Moon god. In a unique seal[187] from the former Erlenmeyer collection dating to the early third millennium BC a series of five Sumerian signs refer to the "festival (ezen) Inana, evening/morning star (celestial body)" which suggests very strongly that at this very early time Inana was identified with a heavenly body that was known to appear *both* in the morning and the evening. In other words the morning and evening "stars" were not thought to be different. This single heavenly body was, of course, Venus.

It is not clear from any Sumerian evidence whether the evening and morning manifestations of Venus were thought of as male and female, but gender in Sumerian *is* often very hard to determine. In Akkadian, however, Venus was sometimes described as being male in the evening and female in the morning, and sometimes vice versa. Indeed, Ištar was sometimes represented bearded. The name, Eštar/Ištar is related to the *male* South Arabian god Athtar, and this may have played a part in the later male-female aspect of the goddess, which was then reconciled by the Scholars with the evening and morning manifestations of Venus.[188]

The omen "if ^{mul}a.edin (Frond star) reaches the Pleiades; Adad will devastate" is used in 8055: r.9 to describe the close approach of Venus and Mars, and in 8536: 3 to describe Venus's proximity to the Pleiades themselves. The fact that the names were associated with planets makes this omen applicable to observable celestial phenomena. In 8055 Nabû-aḫḫe-eriba is also warning the king about eclipses (l.5) and Venus's approaching Scorpius, and is sending him ill-boding apodoses as a consequence. One in particular – "Adad will give his rains to the Gutian lands" – closely mirrors the apodosis above. Perhaps Venus and ^{mul}a.edin were associated by this Scholar because of the particular warning he wished to impart. If so, this association was ***learned***.

One omen is attested for ^{mul}*tūltu* in the texts herein considered: "If the Worm star is very massive: there will be mercy and peace in the land" (another is attested in *BPO*2 XV: 24). Both Reports (8357 & 8538) in which it is used include other favourable prognostications. Since this omen is attested in EAE 51 (*BPO*2 XV: 25), which does *not* include the other Venus omens found in the two Reports, it seems likely that the association between ^{mul}*tūltu* and the planet was made by the authors of the Reports, rather than by the earlier compilers of EAE. Again, the association seems to have been driven by a desire to match apodoses (compare "the harvest of the land will prosper" and "abandoned pastures will be resettled" in 8357: 2 & r.1f). This ***learned*** association between Venus and the worm star may thus date only from the late NA period.

MERCURY

Gu_4.ud, *Šiḫṭu* "Jumping" no doubt reflects the varying length bewteen Mercury's appearances and non-appearances.[189] Its origin is ***observational***. Mercury's association with Nabû is ***basic***. Nabû is Marduk's son and I am tempted to speculate that much as Venus was as-

[187] Nissen ed. (1993) 17.

[188] For some references to the male and female manifestations of Venus in Akkadian context see Reiner (1995) 6.

[189] Mercury has a relatively large ellipse anomaly and inclination to the ecliptic.

signed to the Moon god's offspring, so the fainter Mercury was linked to Marduk's son, once Nabû had become established in that rôle in the second millennium BC. This happened perhaps only soon after Mercury was identified as a planet (which perhaps occurred only in the late OB period, see App.1 §13 and below), and presumably took place at the same time as Marduk acquired dominion over Babylonia and Jupiter/Šulpae. Similarly, the association of Mercury and the crown prince mirrors the relationship between Marduk and Nabû. It is both ***learned*** and ***theological***.[190] The B-name for Mercury, the Marduk-planet, can be accounted for on the basis of this same ***theological*** association, as can the names Nēbiru and Šulpae.

Udu.idim, *bibbu* "wild sheep" = planet is used in the late NA period for Mercury, Mars and Saturn, though in the SB material it was applied to all seven wandering heavenly bodies.[191] As noted above, it seems reasonable to posit an ***observational*** source for the word's association with the planets, particular for Mercury whose behaviour is the least regular, the most "wild". Perhaps this is one reason why its name gu_4.ud is often accompanied by udu.idim, unlike the names for Jupiter, Venus and Mars. Saturn's name is also often prefaced by udu.idim, again perhaps because it was discovered later than the other planets. Indeed, it's A-name contrasts so markedly with the Jumping planet, it suggests that both were discovered to be planets at around the same time and differentiated by their behaviour. I further suggest that because *bibbu* also meant "plague" and "Hangman of the Underworld"[192] it was only (or mostly) applied to the often ill-boding Mercury, Mars and Saturn, making them ***learned*** associations.

The black planet, and the property of turning heavenly bodies black by their proximity, applies both to Mercury and to Saturn. Mars can also turn other heavenly bodies "dark" in the late NA period – see n228. The simplest explanation for Mercury and Saturn's association with blackness is that they are the dimmest of the planets – an ***observational*** association, though a ***learned*** connection between black and ill-boding is not impossible, but cannot be assumed. Their association with the black feathered ^{mul}uga.mušen "the raven (star)" is perhaps related. Also, Gössmann in *Planetarium* p48 suggests that ^{mul}uga.mušen is a D-name for Mercury because the constellation Corvus is located south of ^{mul}ab.sín, "the furrow" – precursor to Virgo, and the so-called *ašar niṣirti*[193] of Mercury. This would make the association ***learned*** and ***observational***. It is also noteworthy that the father-son relationship of Jupiter and Mercury might underlie their white-planet black-planet opposition.

SATURN

Kaiamānu, the adverb which shares the same spelling as the A-name for Saturn, derives

[190] It is interesting that Marduk-Jupiter is not the planet of the king, though the mystical text SAA3039: r.5, where it states that the inside of the Moon is Nabû (Mercury=crown prince) and the inside of the Sun (=king) is Marduk (Jupiter), ought perhaps to be understood in this light. Saturn and the Sun are the planets of the king. In K148+:16 (*ACh.* 1Supp.36, referred to in Hunger & Pingree *Mul.Apin* p147, now *BPO*3 56f), a commentary to EAE 61, the equations between Mercury and the Moon, and Saturn and the Sun are made. (The context is slightly fragmentary, but the restoration seems to be sound). The latter equation is explained below. The former equation derives perhaps from the Nabû-Moon association in the mystical text, or perhaps from the association of Jupiter and the Moon via the common association that the two planets have with Marduk.

[191] Also, in the LB Anu temple ritual from Uruk (*RAcc.*79: 33) and *GSL* 241–3 there are said to be seven planets. The planets are distinguished from the stars in *ACh.* Išt. 25: 46 and op.cit. 1.38 "three or four" are said to rise.

[192] *CAD* bibbu 3 and SAA3032: r.19.

[193] "Secret place", also *bīt niṣirti* "secret house" – a location in the heavens wherein the planet in question bodes well, precursor to the Greek hypsoma or "exaltation", attested in EAE and used in the late NA period. See Weidner (1919), Hunger & Pingree *Mul.Apin* 146, Rochberg (1998) 46f, *BPO*3 14 (for Venus) and Koch (1999).

from *kânu* "to be firmly in place/to be stationary (said of planets)" (*CAD* K 159) and *kittu* "legal truth/justice" (*CAD* K 468 & SAA3 xxiii). This etymology is suggested further in 8547 where a Babylonian Scholar writes 5'- r.1:

> [^{mul}udu.idi]m.sag.uš [^{mul}*zi-ba-ni*]-*tum* "Saturn (=) [the Scal]es"
> obliterated line
> [1 ^{mul}*zi-ba*]-*ni-tum* ki.gub-*su* gi.na "[if the Sc]ales' 'station' is stable:"

No constellation can become stationary, so what does gi.na = *kânu* imply here? It could imply that Saturn, called by its D-name, is stationary, or that Saturn is lying within the Scales, which would explain the ki.gub.[194] Regardless of explanation, the etymological link between *kaiamānu* and *kânu* is being alluded to here. This word play is part of the normal practice of the diviners, part of the "normal science" of the EAE Paradigm, I argue later. *Kaiamānu* was no doubt a name given to Saturn to reflect the planet's slow, indeed saturnine (though not necessarily gloomy), movement – an ***observational*** association.

Saturn's B-name, Šamaš "Sun god" (and by extension its C-name Aš.me "Sun disk"), could also have arisen from the following associations. Hunger & Pingree *Mul.Apin* p147 suggest that the association between Saturn and the Sun is ***observational*** – that is, Saturn's "secret house" (the Scales – n193 above) rises as the Sun's sets. Parpola (*LAS* II pp342–3) suggests that the association is mostly ***learned***, in that the etymological connection between *kaiamānu* and *kittu* links Saturn with Šamaš, as stated. (Parpola's suggestion that the Sun and Saturn were associated because of the latter's steady motion can be ignored.) He also suggests (op.cit.) a connection via the word *ṣalmu* which means both "black" and "image". One of Saturn's C-names is "the black planet " and the Sumerian word an.dùl which means "protection", a quality of the Sun god, can also be read *ṣalmu* "image" in Akkadian (Ṣalmu and Šamaš were also connected theologically, see below). In the late NA version of the lexical series Ur$_5$-gud, we find the line ^{mul}gi$_6$ = an$^{ṣa\text{-}lam}$dùl = ^{d}sag.uš ^{d}utu (Civil & Reiner *MSL* XI p40 lines 39–41) which confirms that this ***learned*** etymological link between the black planet, ^{d}sag.uš and ^{d}utu was *at that time* thought to justify the association between Saturn and the Sun. This meant that omens pertaining to either body could be used to interpret particular celestial configurations.

Parpola argues, without evidence, that Saturn is the black star through the planet's ***basic*** association with Ninurta. But war = evil = black are white eurocentric equations.[195] Instead the association between Saturn and "black" may be ***observational*** for it is the dimmest planet, and ***symbolic*** as the following suggests:

Ṣalmu is used as the name of a Sun deity in the MA/MB period (*CAD* ṣalmu (a)d') and is identified as the father of the minor god Bunene, worshipped at Assur in the period of interest. Šamaš was similarly identified as the father of Bunene,[196] and a form of syncretism must have taken place between Ṣalmu and Šamaš. The association between ^{d}utu and Ṣalmu

---

194 In *BPO*2 17, Reiner & Pingree propose that ki.gub = *manzāzu* referred to a planet or star's location when first seen, and in *BPO*3 they propose that it refers to the place on the horizon above which a heavenly body rises or sets. See also n179 above.

195 Goody (1977) 65f reminds us that even the geo-political North associates "black" with some good things, e.g. soil etc. The Akkadians referred to themselves as the "black-headed people" *ṣalmāt qaqqadi*. Also, a war-god is not, to my mind, obviously associated with the colour black. Ištar is a war-goddess, for example, and Ninurta is said to "illuminate" the apsû in *AKA* 257: 18.

196 Black & Green (1992) 159 and 184.

is thus as much ***theological*** as learned, and their learned association via an.dùl (above) no doubt confirmed this. It has also been asserted that *ṣalmu* may be a specific name for the well-known symbol, the winged disk,[197] This winged disk may possibly have symbolised Ninurta in the glyptic of the 9th century BC[198] and only later came to symbolise Šamaš and Aššur. It is thus perhaps a ***symbolic*** association that lies behind Ninurta's link with *ṣalmu* rather than war god = black.

The association of mul *ša* lugal and Zibānītum "Scales" with Saturn is via the planet's association with the Sun. For the latter's association with the star of the King, see below. The Scales were no doubt ***symbolically*** associated with administering justice and therefore with Šamaš, god of justice.[199] Their association with Saturn is thus ***learned***.

The late name Genna takes the meaning *šerru* "child' (*Planetarium* No.69), but it is made up of the signs TUR + DIŠ. TUR takes the Akkadian *ṣeḫru* "small" or *īṣu* "little". DIŠ can take the Akkadian adverb *ginâ* "constantly'. Clearly then, the description offered by the signs that make up Genna is appropriate to Saturn – small and constant. The association between Sag.uš and Genna is thus graphically ***learned***.

MARS

In §2.1 I noted that Lambert (1996) offered a ***learned*** derivation of Mars's early A-name, ṢAL-batānu, from an epithet of Nergal whose association with this planet must be considered ***basic***. The later A-name for Mars, An, is perhaps derived from the end of the name, for "ānu" in the late period was homophonous with the Akkadian name for the sky god Anu, written in Sumerian as An. A variant rendition of the name in 8102; dṢAL-*bat-a-ni* indicates a genetival ending "of Anu" which suggests that ṢAL-BAT-ānu may have been understood retrospectively to allude to *ṣarbat-Anim* "the poplar of Anu" with perhaps the word *ṣarpu* "red" (see below) resonating as well (reading *ṣar*$_x$, *pá* and *pát* etc. where necessary). If so, the association of *ṣalbatānu* and AN is, in part, ***learned***.

The False planet, Strange planet, as with the Anzû star, the Fox and Panther constellations, Wolf star, and the stars pertaining to Babylonia's enemy countries (Elam, Amurru, and Subartu) were probably associated with Mars because of their common ill-boding names. To this extent they are associated with the ***theology*** of Nergal, whose intimate connection with death, plague, fire, and war is well attested. I am assuming, without justification, that animals such as the fox, wolf and panther were thought of as evil in some way. It is apparent from the celestial omens and elsewhere that Mars almost always bodes ill, and it would be surprising if some of the planet's names were of creatures real or supernatural who were not feared in some way. Equally, Mars's association with the black feathered mul uga.mušen "Raven" is perhaps because of that animal's often ill-boding nature. Alternatively, it may be because Mars, like Saturn and Mercury, can turn celestial bodies dark (n228), even though the body itself appears relatively bright in the night sky. This "darkening" is perhaps metaphorical, though see n195 above.

I assume that the grounds for associating Mars with the name red star were ***observational***. Mars does indeed often appear red to the naked eye. However, the very fact that other stars and planets are described as red (see above, *BPO*2: 2.2.6 and *BPO*3 19) shows that the colours produced by horizon effects were recognised as well as those colours produced

197 Black & Green loc. cit. 159.
198 Black & Green loc. cit. 143, 185, Fig.155.
199 Reiner (1995) 4.

by qualities intrinsic to the heavenly bodies, as is the case with Mars. The name ^{mul}*makrû* "Fiery red star" is unattested in the material under consideration, but this is undoubtedly purely by chance. It is noteworthy that *makrû* sounds like $^{(mul)}$*Nakru* "Enemy star" another well known SB name for Mars, suggesting further ***learned*** resonances between "red" and "ill-portending".

The collection of ^{mul}šudun "Yoke" omens in Mul.Apin and in *ACh.*Išt. 21 offers an interesting example of the data base of omens from which the Scholars could select omens that best suited their needs. Within the parameters established by the Yoke acting as a D-name for Mars, Mercury, and Jupiter, and the phenomena observed, a certain amount of flexibility was still possible. The apodoses of the collection of omens are both good and ill-boding. For example, in 8383 Mars is located in a lunar halo. The omen: "If the Moon is surrounded by a halo and Ṣalbatānu stands in it," is found in 8082: 3, for example. The apodosis of that omen concerns the destruction of cattle and crops, and the diminution of the Westland. In 8383 Rašil decided not to use that omen, but the ^{mul}šudun omen: "If the Moon is surrounded by a halo, and the Yoke stands in it," the apodosis of which concerns the death of the king of Elam and thus bodes well!

It is also worth noting that several of the attested ^{mul}šudun omens describe phenomena in their protases which cannot apply to constellations. E.g. "If the Yoke is low and dark when it comes out;" (8073: 1) "If the Yoke keeps flaring up like fire when it comes out;" (Mul.Apin IIiv1). Unlike the technical terms such as "gaining radiance" (the meaning of which could once have meant something other than the proximity of Venus or Jupiter), these ^{mul}šudun protases cannot possibly be understood to be using terms describing constellation phenomena. The protases can only be describing the behaviour of *planets*. Consequently, some of the ^{mul}šudun omens were *always* intended to describe planetary phenomena. They were not describing constellation phenomena and only later were applied to planets. This suggests that ^{mul}šudun was the D-name (or perhaps even the A-name) of a planet(s) *before* these particular omens were constructed.

In contrast the protases: "If the Yoke is turned towards Sunrise when it comes out," and "If the Yoke is turned towards Sunset," (8546) describe phenomena which can *only* apply to the constellation - a group of stars whose arrangement can swivel relative to the horizons. It appears then that the Yoke was considered to be the name of both a planet and a constellation *before* the creation of the omens discussed. Two different traditions were subsequently gathered together, with the result that even those omens that described the behaviour of constellations could by NA times be applied to planetary happenings. To some extent, then, the association of Mars and ^{mul}šudun is ***learned***, whilst the earliest assignation of the name "Yoke" to a planet was ***basic***. Several of the apodoses relating to Martian phenomena concern the destruction of cattle herds.[200] Reiner (1995) 7 notes that the omen "Mars will rise and destroy the herd," is the only non-eclipse celestial omen to appear in collections of omens whose main concerns are not the phenomena of the sky. Mars was seemingly closely associated with the destiny of cattle, hence perhaps the names "Yoke", Epinnu "Plough" and even Ikû "Field".

In all the omens where ^{mul}min-*ma* is a B-name for Mars, the apodoses are ill- boding. In the one example where Mercury is being described, the omen apodosis is good-boding This suggests that the choice of planet for whom Šanumma "Strange planet" could act as a

200 8049: r.4, 8081: 3, 8288: r.1.

B-name was apodosis-dependent - that is the choice was made *after* the omens were formed. The association would consequently be termed ***learned***. This is fully confirmed by the SB material recorded in *Planetarium* No.374 II. The associations long predated the NA and NB Scholars, however.

The learned commentaries which derive the epithet *muštabarrû mūtānu* "he who keeps plague constant" from the syllabic components of ZAL-BAD-*anu* (n178) were alluding to Mars's association with Nergal.

SUN

Šamaš/Utu's association with the physical body the Sun is ***basic***. The association with aš.me is ***symbolic***, as explained. The equation of the Sun and the star of the king is on one level perhaps ***theological*** - the concept of the roi-soleil etc., and at another level ***learned*** in that the logogram 20 is both man = king and utu = the Sun god. See Parpola *LAS* II p130 note to r4f.

MOON

The heavenly body's association with Nanna/Suen/Sîn is ***basic***, its association with Jupiter ***learned*** as noted above. The use of the logogram 30 derives both from ***observation*** (about half of all months have thirty days) and from the concept of the "ideal" month, for which see Ch.3.1.2.

Chart 2.1 Planet-Name Associations In The Period c.750–612 BC[201]

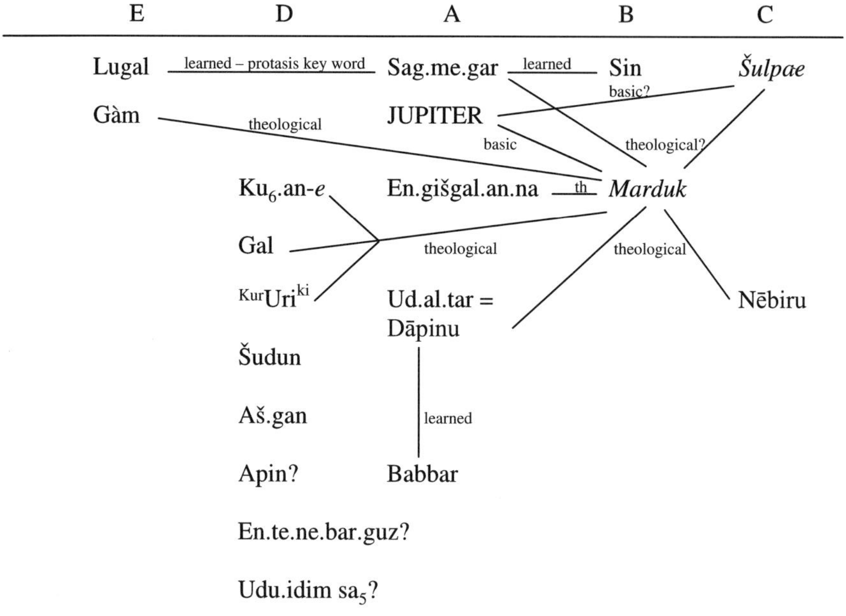

[201] I have not included **udu.idim** in Chart 2.1, since it is associated with all 7 planets through ***observation***. Its particular association with Mercury, Saturn and Mars is, perhaps, ***learned***.

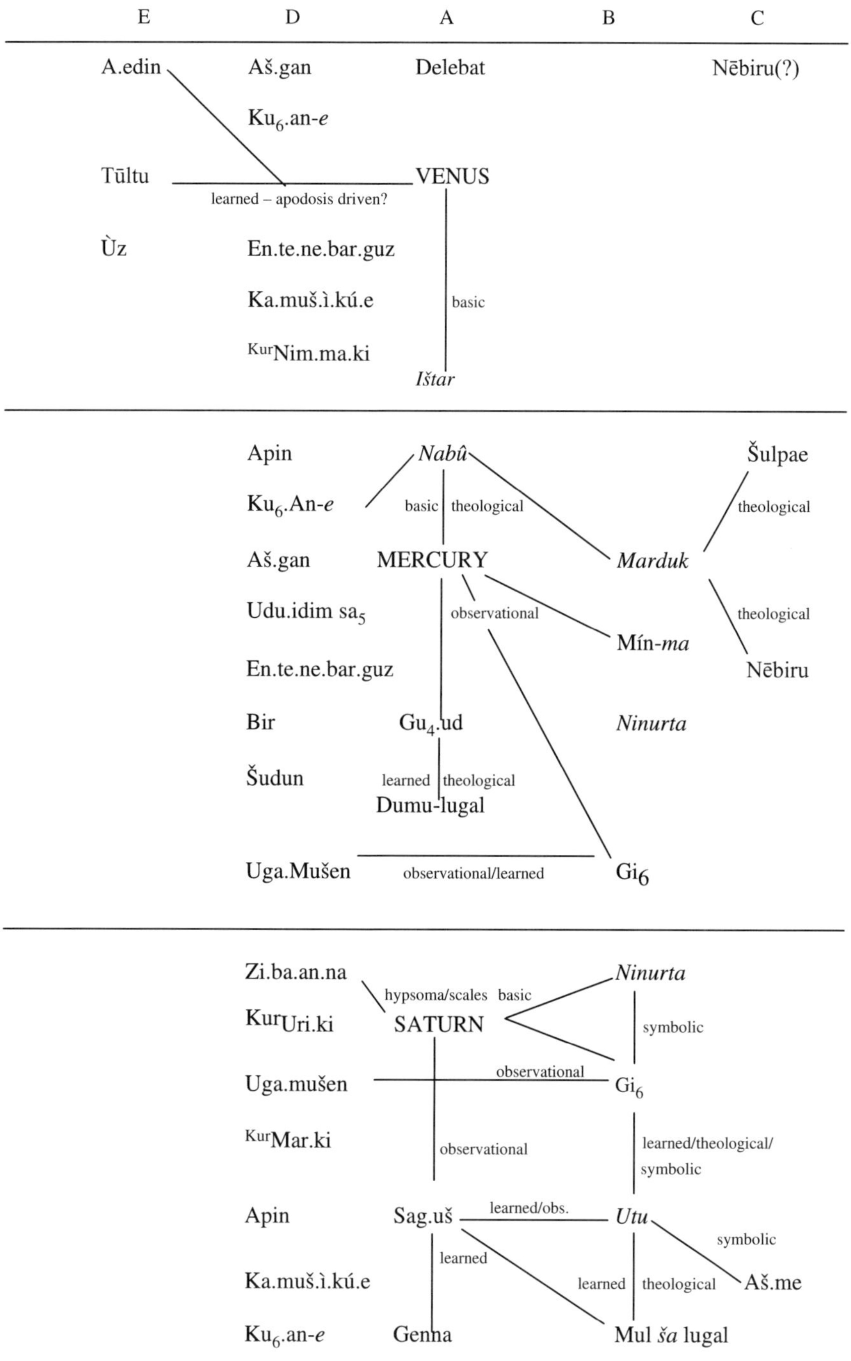
E
D
A
B
C
A.edin
Aš.gan
Delebat
Nēbiru(?)
Ku6.an-e
Tūltu
VENUS
learned – apodosis driven?
Ùz
En.te.ne.bar.guz
Ka.muš.ì.kú.e
basic
KurNim.ma.ki
Ištar
Apin
Nabû
Šulpae
Ku6.An-e
basic
theological
theological
Aš.gan
MERCURY
Marduk
Udu.idim sa5
observational
theological
Mín-ma
En.te.ne.bar.guz
Nēbiru
Bir
Gu4.ud
Ninurta
Šudun
learned
theological
Dumu-lugal
Uga.Mušen
observational/learned
Gi6
Zi.ba.an.na
Ninurta
hypsoma/scales
basic
KurUri.ki
SATURN
symbolic
Uga.mušen
observational
Gi6
KurMar.ki
observational
learned/theological/
symbolic
Apin
Sag.uš
learned/obs.
Utu
symbolic
learned
Ka.muš.ì.kú.e
learned
theological
Aš.me
Ku6.an-e
Genna
Mul ša lugal

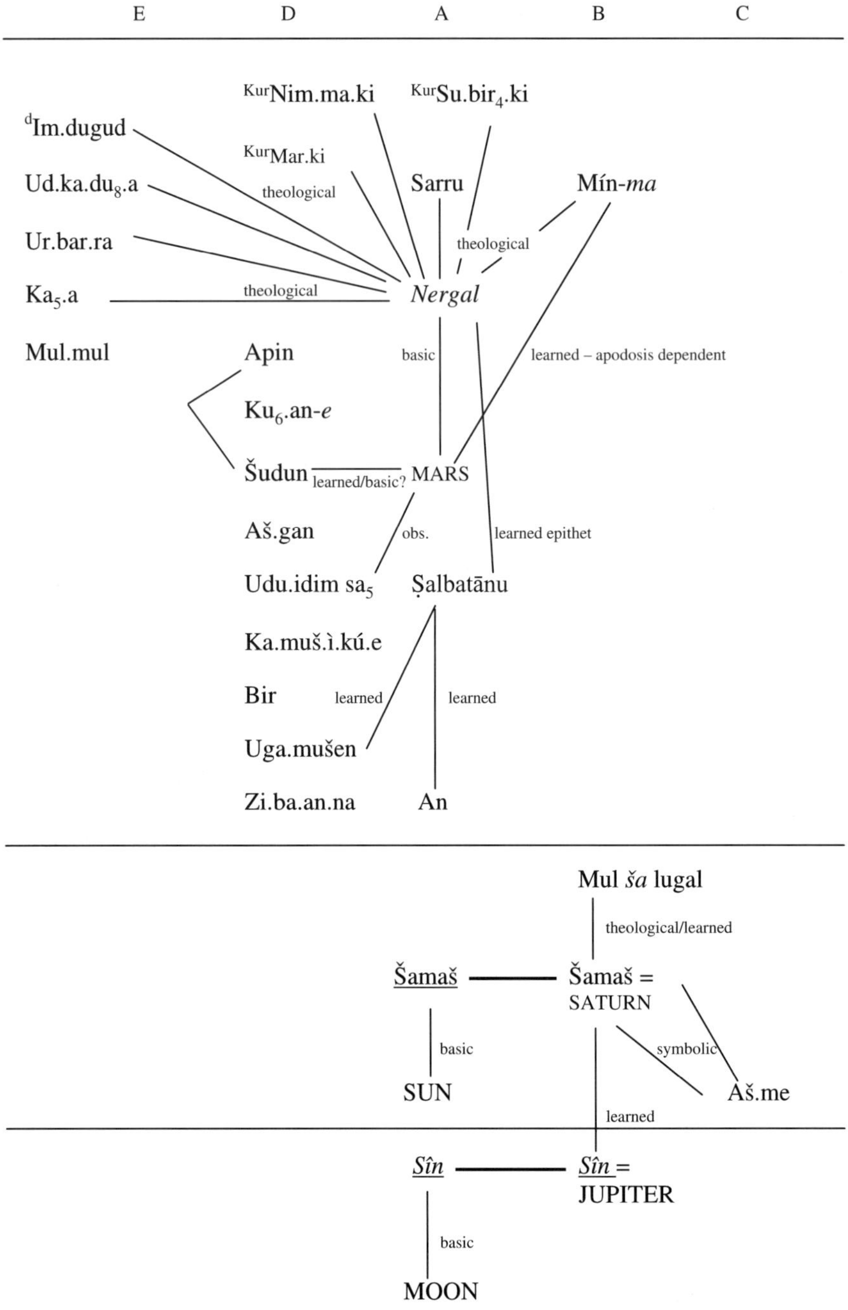
E
D
A
B
C
KurNim.ma.ki
KurSu.bir4.ki
dIm.dugud
KurMar.ki
Ud.ka.du8.a
theological
Sarru
Mín-ma
Ur.bar.ra
theological
Ka5.a
theological
Nergal
Mul.mul
Apin
basic
learned – apodosis dependent
Ku6.an-e
Šudun
learned/basic?
MARS
Aš.gan
obs.
learned epithet
Udu.idim sa5
Ṣalbatānu
Ka.muš.ì.kú.e
Bir
learned
learned
Uga.mušen
Zi.ba.an.na
An
Mul ša lugal
theological/learned
Šamaš
Šamaš =
SATURN
basic
symbolic
SUN
Aš.me
learned
Sîn
Sîn =
JUPITER
basic
MOON

2.1.2 "Learned" Associations and Interpreting Chart 2.1

Whilst I accept that the commentaries which accompany the divinatory series are not the same as literary texts (n176, above), in that they purport *to illuminate a recognised genre, and are not simply flights of fantasy, nevertheless it becomes apparent from a close reading of the material composed in the late NA period that the Scholars interpreted the omen series in order to elicit prognoses which favoured their personal agendas. It will be shown here and in Ch.3 that although these means of interpretation are commonly found in the Letters, Reports, and commentaries, they were not unique to the Scholars who authored them, but also lie at the heart of the "official" iškaru series.*

Chart 2.1 offers a view of the names and associations that were in use by a known group of scribes during a relatively short period of time. Undoubtedly, many more names and associations were recognised by the Scholars. Many others are attested in the versions of EAE and its commentaries found in the archives of Nineveh and elsewhere. Many of these have been listed in Gössmann's *Planetarium* and elsewhere. However, it is dangerous to assert that the names' associations, which might be found in a canonical text, would inevitably continue to be used in the period under discussion. Instead the aim here has been to locate all the names and associations in texts that were known to have been used between c. 750 and 612 BC (and most between 680 and 648 BC), and then to see if they are attested in any of the canonical material (n30, above). The canonical or reference material serves to confirm suspicions, and also to alert the translator to other possible nuances or associations that might otherwise have been missed. It would, however, be unsatisfactory to list every attested association known from Mesopotamia, because associations recognised in one generation might well have been dropped in the next, and so forth. The consistency of "the stream of tradition" needs to be justified, not assumed.[202] The picture (Chart 2.1) provided by the textual material considered here is incomplete, but consistent and comprehensive within the confines of the texts considered, and the discoveries which derive from it relate directly to a known group of individuals. Almost all the associations derive from the class of documents described as "correspondence" in I.3, for which both the temporal and spatial locations are known, and whose authorship is attested.

It is apparent from Chart 2.1 that no E-names are attested for Mercury and Saturn. This means that there are no names which these two planets share with constellations, and yet do not also share with another planet. This implies that (being the dimmest of the planets) they were not discovered until after all the constellation names had already been associated with the three brighter planets (not including the Sun and Moon, which are special cases). If this

202 The "stream of tradition" is Oppenheim's *AM* p13 term (see also n30, above) for that which unites material composed in the OB period with that "preserved" in the later centuries. Is there a continuum in the tradition of these names? To use material from texts separated by hundreds of years, and by genres as diverse as omen texts and literature in order to construct a definition of Nēbiru, say, and then to use this definition in order to see a continuous tradition transcending the texts, is a circular argument, yet this is exactly what is done by many contemporary students working in the field (n168). Data from other eras and other text types must be used cautiously, and only to suggest other possible interpretations. To reconstruct some of the ideology concerning a planet during the period of the NA Scholars, say, provides useful information, which can then be tested against older or later texts, and against texts of different genres in order to assess, for example, the continuity and spread of the EAE Paradigm. To do otherwise is merely speculative. Elman (1975) 23 similarly warns that the "apparent homogeneity (of "Mesopotamian" culture) stems from our lack of data."

is the case then it suggests immediately the antiquity of many of the associations. I discussed in the previous section how the known omens pertaining to ^{mul}šudun, the Yoke constellation, suggest that it was associated with planets *before* those omens were created. The reverse is the case for other associations. Goody (1977) 102 comments that the technique of listing or mapping categorises a literate creation well:

> "The implication of so-called 'primitive classifications' are, in part at least, the simplicities produced by the reduction of speech to lists and tables, devices that typically belong to early literacy rather than yet earlier orality."

The associations of many of the B, C, D & E names to the planets were perhaps made only after they had been written down. They may have been made and recorded by the original compilers of EAE or by its later users – the Scholars.

In general terms Chart 2.1 indicates that during the period of interest each planet was most closely associated with one god. Many of the planets' associated names derived from the theology and symbolism surrounding the respective gods. Other associations appear to have been derived from the observation of phenomena characteristic of the planets. Many of these associations were *not* the result of the activities of a literate few, but may have been more widely appreciated. Those associations that I have termed "learned", however, were undoubtedly the result of "book work". By this I mean that connections were drawn (A) from the words themselves (whether this be etymologically, ideographically, semantically, or graphically) and (B) connections were drawn from what I term "Listenwissenschaft"[203] or "the technology of listing", noted above. The meaning of this terminology will become apparent in due course. Some examples were noted in §2.1.1:

[203] This term was coined by von Soden (1936, now 1965) 29f, and is now familiar in descriptions of the Mesopotamian literate bequest to the world. Essentially "Listenwissenschaft" denotes the techniques of ordering and association used in the extensive lexical materials found in Mesopotamia. These list at great length objects, concepts, and linguistic particles in one or more languages. For details see Civil (1976) and idem (1987). The Mesopotamian mind which created these lists was motivated by an "ordnungswille", so von Soden op.cit. argued. Such ideas are still current – for example Steinkeller (1995/6) 212 writes: "Rather than being a mere superstructure of writing "lexical" (logical) speculation appears to have contributed significantly to its very invention: by mapping out semantic fields and setting out their boundaries it very likely helped to determine the content and scope of the sign repertoire itself". The influence of the lexical material on the omen series is significant. It is discussed in Leichty (1993). Here "Listenwissenschaft" is meant in the sense of a more general literate phenomenon. It is also discussed in this sense by Goody (1977) 81f. Goody stresses the impositions created by listing, from (loc. cit. 57) "the tendency to seek fixed, linear sets of associations seems likewise to be connected with the written mode of communication", to a discussion of how tables and lists require the compiler to fill in the gaps. It leads to the pressure to be *complete* and to notions of *equality*, *analogy* and *polarity*, concepts which are often not clear when describing objects. E.g. what is the opposite of the Sun? The Moon? Goody asks (loc. cit. 105) "the question, is a tomato a fruit or a vegetable...is the kind of question generated by written lists," and (loc. cit. 106) "when graphic representations are drawn, it often leads to binary divisions". Goody goes further, noting (loc. cit. 61) that a list or table is more likely to be "a specialist's elaboration, rather than a fundamental cultural code." As stated before (I.2), I am wary of assigning concepts such as "ordnungswille" to a trans-generational Mesopotamian mind. The use of the literate-listing techniques by the compilers of EAE and the NA and NB Scholars in finding parallels, opposites, and analogies, fulfilled their particular needs. It was a "new technology of the intellect" (Goody, loc. cit. 81) which was probably only available to the literate, but it should not be argued, then, that the literate *had* to think that way. The fact that the Scholars *did* use the technology of listing both with regard to names and also to the omens (see Ch.3.2.1) is a manifestation of their interest in the products of that technology. These products were the *wider applicability* of the names, protases, and apodoses, and a greater flexibility in EAE.

A.1 The association of Saturn and the Sun is perhaps based on, or justified by, the common *etymology* of *kittu* and *kaiamānu*.

A.2 Mulbabbar was, I argue, associated with Jupiter through the first sign of the name UD.al.tar – ud and babbar are two different Sumerian readings of the same cuneiform sign, each with a distinct meaning,[204] each a distinct *ideogram*. Similarly, utu and lugal are associated via the shared sign 20=MAN.[205]

A.3 Mulgenna is made up of the signs TUR and DIŠ, which together occupy the same *semantic* arena as the name *kaiamānu*. The connection is made *graphically*, which means that only the signs themselves could convey the whole meaning of the name genna (see n93, above).

B.1 Jupiter and the Moon are associated through a *parallelism* with the Saturn-Sun link. The crown-prince is associated by *analogy* with Nabû because Nabû is Marduk's son.

B.2 Mercury is perhaps associated with muluga because of the constellation's proximity to the hypsoma of the planet. This is an association that could probably only have been drawn by an initiate familiar with a broad range of cuneiform astronomical-astrological writings – a Scholar, perhaps.

"Learned" associations are those which the experts, the Scholars, were able to draw on the basis of their familiarity with the written corpus of texts concerned with the sky. They are distinct from traditional or common associations, and those based on a generally accepted theology. It would seem plausible to hypothesise that learned associations are more flexible and more rapidly changing than those that are basic or theological in origin. Is there any evidence for this hypothesis?

Within the correspondence of the Scholars are found many omens with more than one apodosis. The subject index to SAA8 under "variant" reveals how frequently alternative apodoses were submitted. Mostly, the variants are separated by ki.min, sometimes by *šanīš* "alternatively", sometimes by *ša pî ṭuppi šanî* "according to an alternative tablet". Usually they are not mutually contradictory, though frequently they seem to be unrelated.[206]

The presence of these variants is important to recognise when attempting to understand the scope and purpose of Mesopotamian celestial divination. They demonstrate that even at this late stage in the development of the discipline they had not been excluded by any

---

204 This is not dissimilar from the scholarly technique, attested in the exegesis of Hebrew scripture and elsewhere, known as *notarikon* – the use of words as abbreviations of phrases, and letters or syllables as abbreviations for words. MulUD (=BABBAR) is an abbreviation for mulUD.al.tar. See Lieberman (1987) 179f and Tigay (1983) 176f for further Mesopotamian precursors to this technique of exegesis. Lieberman (op.cit.) 201–3 argues for *notarikon*'s ancient OB roots, indeed for its intimate connection to the earliest stages of writing systems.

205 The technique whereby the numerical values of the letters of a word are used is known as *gematria*. (Hebrew *gēmaṭriah*, Greek γεωμετρία > geometry) Words might be interchanged if the sums of the numbers corresponding to their letters were equal. With *notarikon*, *gematria* forms part of Onomancy (ref. in OED) or "divination by names", also part of *Aggadic* hermeneutics which in addition includes paronomasia, allegory, and the substitution of letters, all of which can be found in cuneiform writing (see Lieberman, 1987, 161f). A pleasing example of *gematria* is described by Tolstoy in *War and Peace* Book 9 Ch.XIX, where the sum of the letters assigned values as in the Hebrew of "L'Empereur Napoléon" came to 666! The use of "20" for the Moon, "30" for the Sun, "15" for Ištar in the texts herein considered is perhaps a precursor to this technique. See Röllig (1957–71) and above n164. Parpola (1993a) n88 argues that the technique of writing the gods with numbers was an Assyrian innovation whose origin can be traced to the 13th century BC, though Lieberman (1987) nn200–202 argues that the numbers were associated with the gods in the OB period, and that cuneiform *gematria* dates from this period.

206 8147: r.2 "If the Moon is surrounded by a halo, and Scorpius stands in it; *entu*-priestesses will be made pregnant; men, ki.min: lions will rage and block the traffic of the land."

"canonising" process, for many variants appear in the "official series" itself. They perhaps derived from different cities' editions of EAE and other celestial divinatory texts that were brought together by its original redactors.[207] Equally, variant apodoses may have been generated according to what I am here terming the "normal, puzzle-solving practice of the EAE Paradigm" (see Ch.3.2.1) using "learned" techniques such as those just outlined. Whatever the means, one celestial event was clearly not considered by the compilers of EAE to spawn one single interpretation and certainly not by the late NA Scholars. Variant apodoses were not eliminated in their correspondence, they were perfectly acceptable to them and by extension to their kings. On the one hand this makes good sense, for if for any one given celestial configuration two or more prognostications were deemed to be valid, this increased the chances of the prediction "coming true". Indeed, the often general nature of the prognostications assisted in this. On the other hand, if the Scholar wished to impart one particular theme to his king, variants culled from the official series might dilute his message.

"Learned" elaboration of names provided for both these scenarios. If for some particular celestial phenomenon more than one apodosis might be sent to the king, it is entirely reasonable that a *further* prognostication might also be sent which derived from another complete omen which used alternative names for the celestial bodies in question. Provided that the *rules* governing the association of names were not broken, different omens culled from the series which *could* describe the same celestial phenomenon were acceptable material to send to the king. Within the parameters set by the celestial phenomenon itself and by the list of possible name associations, the omens sent were limited only by the ingenuity of the Scholar. The original editors of EAE provided for some variation in the interpretation of any given celestial event, but the possibilities for further exegesis on the signs in the sky were still great for those who had truly mastered the multi-lingual writing systems and who were familiar with the full range of relevant divinatory material. The omens sent were those whose apodoses matched the message intended by the Scholar, I suggest.

In 8320: r.1 the following omen is sent by a Babylonian Scholar:

> "If on the 16th day the Moon and the Sun are seen together; one king will send messages of hostility to another; the king will be shut up in his palace for the length [of a month]; the step of the enemy will be set towards his land... variant; the king of Subartu will become strong and have no rival."

From the Babylonian point of view these variant apodoses are not contradictory, for Subartu (n165, above) is identified in EAE as an enemy nation to Akkad=Babylonia. However, this omen was sent to the king of Assyria, which was identified with Subartu.[208] We have no evidence as to how these mutually contradictory apodoses were understood by him. To us it reveals the Babylonian origin of EAE, and the limitations of its applicability to foreign royalty. To the Assyrian king and his guardian Scholars, though, this omen must have presented a confusing message from the gods. Significantly, the same contradictory apodoses are sent by the top Assyrian Scholar Issar-šumu-ereš in 8025. Not even the Assyrians edited out this inconsistency. EAE must have been consulted without the expectation that only one prognostication would emerge. This reflects both on the "origin" of the omens, and on the

[207] This important field is discussed a little by Weidner (1941/4a) 175f and by Rochberg-Halton *ABCD* Ch.2. Much further work remains to be done.
[208] As is clear from x048, amongst many examples.

concept of causality considered to have linked protasis and apodosis, issues discussed in Ch.3.1.1.

Mars was the star of Subartu (see A-names, above). To the Babylonians, Mars almost invariably portended misfortune (see also *Planetarium* No.360 III). No doubt this was why it was associated with the enemy country Subartu. For the Assyrians, however, the situation was reversed. It is for this reason, I argue, that the Babylonian Šapiku writes in 8491: r.7 that when Mars is bright, and carries radiance, this is good for Subartu. It contradicts the statement made in 8114: r.3 by the Assyrian Bulluṭu that "if Mars becomes faint, it is good; if it becomes bright, misfortune." The latter is an omen and applies to Akkad, the former is Šapiku's re-interpretation. He has attempted to reconcile the contradiction by noting, on the basis of his familiarity with the series, that a bright Mars portended well for Subartu, and thus Assyria, when he must have known it portended ill for Akkad=Babylonia. His Report can also convincingly be shown to date from the end of Esarhaddon's reign. Esarhaddon was king of both Assyria and Babylonia. Presumably, Šapiku felt that an omen portending good for Assyria was best complemented by the one portending ill for Babylonia being left unsaid.

Similarly, it is interesting that Akkullanu, writing in 657 BC to Assurbanipal, when the latter was king only of Assyria, states in x100: 33:

> "Assyria is the land of Akkad of the king, my lord."

This is said in the context of a series of omens giving the prognostications for the situation when Mars was faint and small in appearance. The omens portend victory and prosperity for Akkad (these are among the only good-boding Mars omens). Akkullanu has asserted that they apply also to Assyria. Again, this Scholar's desire to send good-boding omens has lead him to read the texts in a particular way. Clearly, if the interpretations of Akkullanu and Šapiku were taken together Mars's appearance whether dim or bright would portend well for Assyria.

Thus, in order to reconcile contradictions in the predictions of EAE (in this case brought about by it being applied to Assyria and not to Babylonia) some Scholars sometimes left out apodoses that they thought were unsuitable, or made connections (Akkad-Assyria, Mars-Assyria) in order to widen the arena of applicability of the omens. To this extent differences in the usage of EAE by individual Scholars can be discerned. These differences form part of the on-going "learned" development of EAE. They show that although EAE was religiously consulted, and omens extracted and sent to the king, incremental changes, brought about by the individual intentions of the Scholars, were taking place. This is the "normal, puzzle-solving science" of the EAE Paradigm.

Further to this, I noted in §2.1.1 that some names appear to have been associated *because* of similarities in the apodoses or protases of the omens that describe phenomena connected with them. I suggested that Venus may have been associated with the Worm star (^{mul}*tūltu*) and with the Frond star (^{mul}a.edin) because of similarities in their attested omen apodoses. If this were the case, it implies that the associations were made after the omens were formed, perhaps during the compiling of the omens into large series, but perhaps later still by those NA experts. It was also suggested above that Jupiter and ^{mul}lugal, the King star, and ^{mul}gàm, the Crook star, were associated through the key phrase "še.er.zi" in protases attested for them all. In both cases the associations could only have been drawn by scribes familiar with the omen corpus. Again, I noted how the association of Mars with ^{mul}šudun permitted Rašil in 8383 to use a '^{mul}šudun in a lunar halo' omen whose prognosis differed from that of the omen 'Ṣalbatānu in a lunar halo'. Where the latter omen predicted loss of cattle in all lands, the former predicted the death of the enemy king of Elam. Lastly, in 8245 Nergal-eṭir interprets Mercury's proximity to the star Regulus, which lies inside Leo, using the omen: "If a planet (a Mercury B-name) comes close to Regulus:" the prognosis of which includes the line "he will restore the temples and establish sacrifices of the gods; he will provide jointly for (all) the temples". He chose not to use the equally applicable omen: "If Leo is black; the land will become unhappy", attested in 8337: 3.[209] In both cases the agenda of the Scholar in question was playing a part in determining which omens were sent and which names were being associated – the former in order to send good-boding messages, the latter in order to plead the cause of his alma mater. An *active* and not merely *passive* use of EAE was being made by these experts, and very good at it they clearly were. It is from the ranks of these learned types that the developments in the prediction of celestial phenomena emerged, I believe.

A glance at the indices to SAA8 and SAAX reveal only a few names of celestial bodies whose association with the planets are not known to us. This is because, in almost every case, an A-name of the planet is also provided by the Scholar in his Letters or Reports. The correspondence was not meant to be obscure. It was, after all, written for a non-expert – the king. The correspondence did not need to be re-interpreted, which suggests that it was read directly to (or even by) the king.

The use of the non-ambiguous A-names is interesting. In the majority of cases only the seven following were used: Sagmegar, Delebat, Ṣalbatānu, Šiḫṭu, Kaiamānu, Šamaš, and Sîn. It hints on the one hand at the Scholars' need to demonstrate to the king that they were sending in omens that pertained to *actual* celestial happenings, and on the other it suggests a certain degree of uniformity of approach on the part of the Scholars. There is no evidence, for example, that one school of celestial diviners regularly referred to Mars, say, by another of its A-names. This perhaps indicates the extent of the unifying influence the late Assyrian kings had on the tradition through his direct employment of the Scholars. This is in contrast to the variant editions of EAE attested from various locations in the empire.[210] The importance of this for the development of predictive techniques is unquantifiable, but was probably significant.

The use of unambiguous names also suggests that the Scholars knew precisely which planets they were observing. There is no evidence of Scholars hiding their ignorance of which of two planets they were observing by using names shared by both. One might expect that the consistent use in the royal correspondence of these seven A-names reflects their use in the raw observational data – the data for which omens from EAE were subsequently selected. More on this below.

---

209 Compare 8175 where Šumaya interprets Venus's approach to the constellation Cancer, ^{mul}al.lul using the omen "if the Goat constellation (^{mul}ùz = a Venus E-name) approaches Cancer", the prognosis of which includes the line "the temples of the land will be restored". Cf. 8247 & 8414.

210 Weidner (1941/4a) 181. The theory of different editions of EAE based on different schools has been criticised by Koch-Westenholz (1995) 80f. It remains unclear to what extent the variant editions use different names, in any case.

Throughout many hundreds of years prior to the 7th century BC many scholarly name associations must have been accreted to the main body of EAE in the gradual manner described. Whilst learned techniques typify the compiling of EAE, these same techniques were flourishing at precisely the same time as methods for calculating the locations and times of celestial events were being developed. The first enabled the Scholars to extend the arena of applicability of the EAE Paradigm and to invest their missives with their own agendas, no doubt designed to promote their personal causes in the royal courts. The second, I suggest, was developed for much the same purpose – see Ch.5.2.

2.2 The Ominous Phenomena and Configuration

My intention in this section is to list those celestial happenings that were deemed to have been significant in the texts herein considered, and then to compare that list with the list of celestial events recorded in the earliest Astronomical Diaries, Eclipse and Planetary Records, and consequently with those events predicted in the later NMAATs and the MAATs.[211]

I have explicitly avoided using the remains of EAE as a source for what was considered ominous in the sky, relying only on material known to have been used between c. 750 and 612 BC. For example, a constellation "gaining radiance" in an EAE protasis may once have described the luminosity of the stars. In the period of interest it most often meant the proximity of a planet. It is the latter that can usefully be compared with the observations recorded in the NMAATs. Most of the earliest records of celestial observations and some of the correspondence to the Ninevite kings derive from the city of Babylon. It is consequently possible to test whether or not the observations recorded in the oldest NMAATs match those considered ominous by Scholars writing in the same place and at the same time.[212] If the answer is in the positive it becomes highly likely that the Scholars authored, or arranged to have authored, these NMAATs as well. The following questions then emerge. If at a later date there is good evidence that these observational-record NMAATs provided some, at least, of the raw material for theories designed to predict celestial phenomena, was this their intention in the late NA period also? Is there any evidence that the Scholars had succeeded in part in this activity by this time? If so, as I argue and attempt to demonstrate in Ch.4.2, then if the original intention of the accurate prediction of celestial phenomena was in order to facilitate the work of celestial diviners, should not the late MAATs also be considered (at least in part) in this light. By the time *they* appeared, celestial divination had undoubtedly moved on, but the premises which underpinned them were established between c. 750 and 612 BC when celestial divination was still dominated by the EAE Paradigm.

2.2.1 A Description of the Celestial Phenomena Afforded by the Planets

The following outlines which planetary celestial phenomena occur in order that we might interpret the phenomena described in omen protases and isolate those omens which describe

211 For details on these texts groups see App.1.

212 The relationship of the Diaries to EAE is considered cursorily by Rochberg-Halton (1991a) 330–1, who points to a selection of phenomena and terminologies found in both.

planetary phenomena that can never *occur (see Ch.3.2.1). I also describe the periods that exist between phenomena of the same type, the knowledge of which enabled the Scholars to predict them (see Ch.4.1.2).*

The superior planets (those further from the Sun than the Earth) are Saturn, Jupiter, and Mars, in the order of the length of their periods. From the point of view of a naked-eye terrestrial observer, on one particular day a superior planet first appears over the eastern horizon in the morning just before the Sun rises. This is its *heliacal rising* or *first appearance*. It is located somewhere within a narrow band in the sky that stretches 360° around the earth. Within that band move all the planets. Its central line is known as the *ecliptic*, and it passes through and near a set of constellations known as the zodiacal constellations. At least by the 5th century BC twelve constellation names were chosen to represent 30° each of the ecliptic.[213] Each 30° arc is located near to the constellation from which it derived its name, but in order to distinguish the arc from the star group the former is known as the *zodiacal sign*, the latter as the constellation. In the texts from the main period of interest herein discussed, the constellation is always meant (so far as we know), but in the later Diaries, for example, a planet can be said to be in the constellation Aries, and/or in the sign Aries.

The following days the planet rises with its stretch of the ecliptic earlier and earlier than the Sun. At the same time the superior planet moves slightly faster than the background stars which mark out the ecliptic. It is said to be moving forward in the ecliptic. In the case of Jupiter, about 130 days after its heliacal rising the planet no longer moves forward in the ecliptic, but becomes *stationary* with respect to the background stars. It then starts to move in the opposite direction (*retrograding*) along the ecliptic, which it does so for another 120 days or so, at which point Jupiter stops once again (*second station*), and then starts to move forward once more. Exactly between the two stations the superior planet is in *opposition*, which means that it lies 180° away from the Sun. Superior planets are at their brightest in opposition. Near opposition there is a morning upon which the superior planet sets in the west just as the Sun is about to rise in the east, and an evening when the Sun sets in the west just as the superior planet rises in the east. They are known as *morning setting* and *acronychal (evening) rising* respectively. The time between and the order in which opposition, acronychal rising and morning setting occur depends on the season and what is known as the *latitude* of the planet. This is its angular distance from the ecliptic. It is also possible for bad weather or horizon effects to hide the superior planet from view at the point at which acronychal rising or morning setting might be expected. This applies, of course, to any planet's rising or setting.

After its second station a superior planet continues to move forward in the ecliptic, moving closer and closer to the points in the sky where the Sun sets, until, in the case of Jupiter, some 370 days after its heliacal rising the planet is no longer visible in the evening after Sunset. Its *heliacal setting* or *last appearance* has occurred for it has moved out of sight behind the Sun. It remains behind the Sun for about a month (the middle of which is known as *conjunction*) until heliacal rising takes places once again. In one cycle, known as a *synodic interval*, since the interval is measured between configurations relative to the Sun, Jupiter moves forward in the ecliptic on average by some $30^1/_3$° , during which time it also retrograded about 10° . The synodic interval is approximately 398 days. After an average of

213 App.1 §42.

11.86 years Jupiter travels once around the ecliptic belt. This is known as its *sidereal interval*. Since they are not a whole number of years, Jupiter will not be in the same place in the ecliptic on the same date (according to a well-regulated calendar) after one sidereal interval, however. Also, since 11.86 years are not a whole number multiple of 398 days, Jupiter will not be in the same configuration with the Sun after one sidereal interval – it will not be demonstrating the same heliacal phenomenon. Longer periods are needed to reconcile these three things. In 427 years 391 heliacal phenomena of the same type take place for Jupiter, and 36 sidereal rotations, which means that opposition, say, will occur once again in the same place in the sky *and* at the same "time of year", or on the same date provided a well-regulated calendar is being used. It is worth noting here that the Mesopotamian calendar was lunar and needed regulating against either the solar (seasonal) or sidereal (stellar) year.

Equivalent long periods are required for Mars (sid. interval = 687d) and Saturn (sid. interval 29.46y) to reproduce their heliacal phenomena on the same date and in the same place in the sky. Values for just these periods are found in both the NMAATs and the MAATs.

The superior planets move slowly along the entire ecliptic, passing into and out of constellations and by individual stars, repeatedly stopping and retrograding, and disappearing for a time behind the Sun. All the planets move along the same path in the sky, so occasionally they will meet. If the latitude of the planets is similar at that time they will approach very closely. This is also called *conjunction*, though when it is with the Moon, and the latitudes are such that the Moon blocks out the planet, it is known as *occultation*.

The inferior planets, Venus and Mercury, similarly move through the entire ecliptic, only more rapidly than the earth. Also, being closer to the Sun than the earth, from a terrestrial point of view they are never located far from the Sun. This means they appear only in the evening, shortly after Sunset, and in the morning, shortly before Sunrise. They never appear in opposition to the Sun, of course.

An inferior planet rises heliacally in the east in the morning (*morning first* = mf), after what is called inferior conjunction. It moves backwards along the ecliptic to those stars that rose heliacally some time earlier, until it reaches its stationary point 2 weeks later, in the case of Venus. Its *elongation* (distance along the arc of the ecliptic) from the Sun at this morning station is about 28° . It moves forward in the zodiac and fifty days later Venus reaches its greatest westerly elongation. This is just over 46° . 180 days later and Venus has moved closer and closer to the morning rays of the Sun, until it finally disappears (*morning last* = ml). Superior conjunction with the Sun lasts about 80 days (it depends on the latitude of the planet, the seasonal angle of the ecliptic, and weather conditions) and then Venus rises (*evening first* = ef) in the west in the evening, right after Sunset. 180 days later it has moved rapidly through the zodiac and is at its furthest elongation east (46°) and moving at about 1° per day. It slows until its *evening station* 50 days later, and then retrogrades into inferior conjunction a few weeks later. This is its *evening last* = el. Inferior conjunction lasts but a few days.

The synodic interval of Venus is 584 days, during which time it has travelled right around the ecliptic 2.6 times. It takes only 225 days to travel around the Sun. The earth, in the meantime has also travelled 1.6 times around the ecliptic. In 8 years Venus travels almost exactly 13 times around the Sun, which is 5 times more than the earth. This means that the same heliacal phenomenon will take place once again on virtually the same (solar) date, and in the same place in the sky. In those 8 years 5 synodic intervals of Venus occur. One *characteristic* period for Venus lasts 8 years.

Mercury travels even more rapidly. It takes only 88 days to travel around the Sun. From the point of view of the earth this means that its synodic interval is on average 116 days.

The Moon appears on the first of the month in the west just as the Sun is about to set. It sets shortly afterwards. As it waxes so it appears, once daylight has gone, higher in the sky each night (for the first week, the second week it reaches its greatest height after Sunset) and sets progressively later after Sunset. Around mid-month it appears near the eastern horizon as the Sun sets and disappears in the west as the Sun rises in the morning. It may be visible all night, if the nights are not too long. The Moon wanes in the last two weeks appearing later and later into the night until it only appears just before dawn. The following morning it does not appear at all, nor does it appear in the evening. The following evening, or possibly the one after, it appears once again at Sunset, and a new month begins.

The interval from new Moon to new Moon lasts 29 or 30 days,[214] though due to bad weather other lengths were occasionally recorded . This again is its *synodic interval*, and the average is about 29.53 days. During this month the Sun has itself moved by about 30°, so the new Moon is no longer positioned against the same star. The *sidereal interval* is consequently only 27.32 days. In fact the appearance of the new Moon provides only a poor way of establishing the synodic interval between similar luni-solar configurations, for its appearance depends very much on the latitude of the Moon and night length, as well as being dependent on the seasonal inclination of the ecliptic, weather conditions and the velocities at which both bodies are moving.[215] Much better is an eclipse, lunar or solar, for at that point the three planets (earth, Moon, Sun) are in a line (a *syzygy*) whose orientation with a distant star, and the time of which, can be established with high accuracy. When an eclipse occurs, not only must the Moon be at the right elongation along the ecliptic, but it also must have a small latitude relative to the ecliptic. In fact the Moon undulates in and out of the plane (or line as viewed from earth) of the ecliptic. The points at which it crosses the plane are called the nodes (ascending and descending). The interval between two nodes of the same type is the nodal or *draconitic month*. This is approximately 27.21 days. For an eclipse to occur, something near to a whole or half number of synodic and draconitic months should have passed since the previous eclipse. For an eclipse to happen at the same point in the sky a whole number of sidereal months need to have passed.

Six synodic months are c.177.2 days. This is close to 6 1/2 draconitic months.
Five synodic months are c.147.7 days. This is close to 5 1/2 draconitic months.

This indicates that eclipses can be separated by 5 or 6 months.[216]

---

214 Stephenson & Baolin (1991).

215 That is, the period between new Moons is partially dependent on the solar and lunar anomalies (the effect of the ellipses in which the earth travels around the Sun, and the Moon around the earth, which results in the Sun and Moon both appearing to a terrestrial observer to be travelling at slightly different velocities through different parts of the ecliptic). Interestingly, the *interval between* any lunar heliacal phenomena of the same type (first visibility, opposition, conjunction) is barely dependent on the lunar anomaly, but mostly on the changing solar velocity, and therefore on the location of the phenomenon in the zodiac. This important simplifying result was discovered by Brack-Bernsen (1969) and bears on the development of cuneiform lunar MAATs.

216 For a more complex derivation based on similar principles see Aaboe (1972). See also Beaulieu & Britton

223 synodic months are about 6585.3 days. This is very close to 242 draconitic months, and also to 239 *anomalistic months*. These last about 27.55 days and are the intervals between lunar returns to the same velocity. 223 synodic months is also about 18 years, and thus restores both lunar and solar anomalies, meaning that eclipses separated by this period are "of equal circumstance".[217] Thus, eclipses will often be separated by 6585.3 days, or 18 years made up of 12 lunations each and 7 additional intercalary months, since 12*18 + 7 = 223. This period was familiar to the authors of the late MAATs and NMAATs and is known to modern students as the "Saros" period. See App.1 §39. In Ch.4.2.2 and 4.2.4.3 I discuss the evidence for a NA knowledge of it. The .3 day in the length of 223 months means that sometimes an eclipse will not be *seen* to occur at one given location on earth a Saros period after another had been observed, because it has become impossible to see either the Sun or the Moon due to the 0.3 of a day of terrestrial rotation. Better is the triple Saros or "Exeligmos" period, which eliminates this problem.

235 synodic months are 254 sidereal months, so after this period the Moon will appear at the same longitude (position on the ecliptic) on the same date of the month (within a day), since this is determined by its configuration with the Sun. It thus implies that the Sun, too, is at the same longitude it was 235 months earlier and that a whole number of years (reckoned by the stars) have passed – in this case 19. Were an eclipse to take place at each end of this interval (which equals one Saros + 12 months) its length and significance could be established directly from observation.[218] If in 235 months the intercalations are arranged so that the *same* month appears at the beginning and at the end, then the sidereal and *lunar year* (simply 12 months) would be reconciled. This occurs, for example, in the Metonic cycle.[219] 19 sidereal years are accurately equivalent to 235 months, and in the Metonic cycle they are spaced in such a way that 12 years each have 12 months each, and 7 years have 13 months each. The cycle thereby links the months (and therefore any dates) to the Sun (and therefore to heliacal phenomena) if one ignores the small difference between the solar year and the sidereal. The Metonic Calendar regulated the luni-solar year and permitted intervals between phenomena to be recorded in *years* and not merely in months.

The Sun appears directly in the east at the equinoxes, to the north in summer and to the south in winter. It is mainly recorded at its rising and setting, when its colours were considered to be important.

2.2.2 From c. 750 to 612 BC the following Planetary Phenomena were Ominous:

It has proved useful to assign the phenomena to 18 categories in order to facilitate comparison with the Diaries and related texts. Nothing is intended by the order in which the categories have been listed. The translations are mostly those provided by Hunger and Parpola in SAA8 and SAAX respectively.

(1994) 78f and *HAMA* 497f. See also Hartner (1969) Table 1.

217 That is, with similar entrance angles, magnitudes and lasting similar amounts of time. See Neugebauer, *HAMA* 502.

218 Moesgaard (1980)'s argument.

219 See Sachs (1952b), Bowen & Goldstein (1988).

1 *Heliacal rising*

JU "If Jupiter becomes visible (igi = *innammar*) in *aiāru* (II)" (8254: r.1)
"If Jupiter becomes steady in the morning" (8184: 5)
"If Jupiter in *simānu* (III) approaches and stands where the Sun shines forth" (8170: 3f)
"[Last year] it (Jupiter) became visible on the 22nd of *aiāru* (II) in mul[ ]šu.gi]. It disappeared in *nisannu* (I) of the [present] year, on the 29th. It appeared on the 6th of *simānu* (III)" (x362: 3f)
"If Jupiter becomes visible in the path of the Anu stars" (8115: r.6)

MA "If Mars lights up (kur-*ḫa-ma*) faintly (x100: 18)
"If Mars stands in the east" (8114: 4)
"If at his appearance (igi.du$_8$.a-*šu*) Nergal is small" (x100: 20)
"If Mars becomes visible in VI" (8491: 1)
"Mars has appeared in the path of the Ellil stars, at the feet of [mul]šu.gi." (x100: 5)

SA "If a planet (Saturn) lights up in V (8324: r.1)

VE "If Venus becomes visible in XI" (8357: 1)
"Venus became visible in the west in the path of the Ellil stars" (8175: 1) (ef)

ME "Mercury became visible on the 16th" (8050: r.3)
"If a planet (Mercury) becomes visible at the start of a month" (8157: 2)
"If the Marduk planet (Mercury) becomes visible at the start of the year" (8486: 6)
"Mercury became visible in the west" (8486: 1) (ef)
"Mercury is going beyond its (normal) position and ascends" (8093: r.3)

SUN "If the Sun rises in the path of the Anu stars" (x079: r.12)

MOON 28th "If the Moon at its appearance becomes visible on the 28th as if on the 1st" (8014: 1)
"If the Moon becomes visible on the 28th" (8014: 3)
29th "[If the Moon] becomes visible [on the 29]th" (8457: 1)
"If the day is short compared to its normal length" (8457: 4)
30th "If the Moon becomes visible on the 30th" (8011: 1)
"If the Moon becomes visible in month (1–12) on the 30th" (8304: 1, 8191, 8192…)
1st "If the Moon becomes visible on the 1st" (8290: 4)
"If the day reaches its normal length (*ana minātišu erik*) – the 30th completes the measure (*mināt*) of a month" (8007: 3, 8290: 1)
"If the Moon's position at its appearance is stable/true (gi.na[220]) – it becomes visible on the 1st" (8506: 3)

2 *Heliacal Setting*

"If Jupiter passes (*itiq*) to the west" (8456: 6)
"If the Sun rises in a *nīdu*[221]" (8456: 3, 8339: r.3)
"If Mercury disappears in the west" (8274: r.1)

220 Gi.na = *kânu* "to be firm/true". The implications of this phrase and those in the previous line to the notion of the "ideal" month are discussed in Ch.3.2.3

221 *Nīdu* is a cloud formation (*CAD nīdu* B), however there is more than reasonable evidence in the texts herein considered that it also refers in some way to a planetary absence. Parpola *LAS* II p310.

"If Venus disappears in month I from the 1st to the 30th" (8403: 7)
"If Venus disappears in the east" (8056: 1) (ml)
"If the day of the disappearance of the Moon is at an inappropriate time...- the Moon disappears on the 27th" (8346: 1)

3 *Visibility period*

"Jupiter [becomes visible in the east and stands] in the sky all year" (8167: r.1)
"Jupiter remained steady in the sky for a month of days" (8339: r.5)
"Jupiter stands in the sky for excessive days" (8329: r.5)
"Jupiter stood there one month over its period (*adannišu*)" (8456: 5/8)
"Jupiter retained its position; it was present for 15 more days. That is propitious" (x100: 30)
"Venus...became visible quickly... If Venus stays in her position for long... If the rising of [Venus] is seen early" (8027: 6f)
"If Venus gets a *flare* (*ṣirḫu*[222]) – she does not complete her days (of visibility), but sets" (8145: 2)

4 *Invisibility period*

"Jupiter [*may* remain invisible] from 20 to 30 days; now it has kept itself back from the sky for 35 days." (x362: 5')
"If Nēbiru drags (*išdudma*)" (x362: r.3)
"If the rising of Venus is seen early" (8247: 6)
"Venus is [not] yet visible" (x072: 14 see *LAS* II pp72–3)
"If a planet (Mercury) becomes visible within a month" (8281: 3)
"3 days (the Moon) [stayed] inside the sky...the Moon disappeared on the 27th; the 28th and the 29th it stayed inside the sky, and was seen on the 30th, when else should it have been seen? It should stay inside the sky less than 4 days; it never stayed 4 days" (8346: 6f)

5 *Retrogradation and Stationary points*[223]

"If it (Jupiter) turns back out of the breast of Leo, this is ominous. It is written in the series as follows: If Jupiter reaches and passes Regulus, and gets ahead of it, afterwards Regulus, which Jupiter had passed and got ahead of, reaches and passes Jupiter, moves to/stays with it (the stationary point) in its setting. This aforesaid is the only area which is taken as bad if Jupiter retrogrades (*isaḫur*) there, where ever else it might turn, it may freely do so" (x008: r.16, 8502: r.4f)
"When the planet Mars comes out from Scorpius, turns and re-enters Scorpius, its interpretation is: If Mars, retrograding (*itūra*), enters Scorpius... This omen is not from the series

---

222 See *BPO* III 14 §7.2

223 The phrases "is stable in the morning" *ina šēreti ikūn* (x072: 18) and "keeps a stable position" ki.gub-*sa raksat* (8357: r.3) are said of Venus. Jupiter's "becoming steady in the morning" is equated with heliacal rising and being bright in the texts herein considered, and not with being stationary (e.g. 8184). The same is meant of Venus in x072, wherein the planet is described as being not yet visible, and so clearly cannot be anywhere near its stationary point. Similarly, in 8157: 7 it states that "a planet (Mercury) stands (gub) in the east" where in line 1 it is made clear that the heliacal rising of the planet is meant. In the late NA period these descriptions did not refer to stationary points, although it is possible they did in earlier times.

it is *ša pî ummâni*. When Mars furthermore retrogrades (*tûra*) from the Head of Leo and touches Cancer and Gemini its interpretation is" (x008: 21f cf. 8386)
"Saturn will push itself (*ramanšu ida'īp*) this very month. There is definitely not a word about it anywhere." (x008: r.25f)
"Mars has reached Cancer and entered it. I kept watch: it did not become stationary, it did not stop (*lā innemid lā izziz*)" (8101: 1)
"Mars which stands inside Scorpius, is about to move out; until the 25th of [..] it will move out of Scorpius" (8387: 3)
"If Mars rides Capricorn – it has gone into Capricorn, halted..." (x104: r.2f)

6 Lunar "Opposition"

12th "If the Moon is seen on the 12th/at an inappropriate time (*ina lā simanišu* igi)" (8088: 1f)

13th "If on the 13th the Moon and Sun are seen together" (8306: 1)
"If the Moon moves str[aight] (*ú-še-*[*šir*]) in its proceeding – on the 13th (the Moon) was seen with the Sun." (8458: 1)

14th "If the Moon and Sun are in opposition (*šutātû*)... It means that on the 14th one god is seen with the other, or that Saturn stands with the Moon on the 14th" (8110: 8f)
"If on the 14th the Moon and Sun are seen together" (8015: 6)
"If the Moon and Sun are in balance (*šitqulū*)" (8015: 1)
"If the Moon is seen on the 14th" (8293: 8)
"If the Moon is proper (*né-eḫ*) in its course – it is seen on the 14th" (8110: 7 cf. 8411)
"The Moon will complete the day in IV – on the 14th it will be seen with the Sun" (8046: 1)

15th "If on the 15th the Moon and Sun are seen together" (x094: r.1)
"If the Moon does not wait for the Sun but sets – it is seen on the 15th with the Sun" (8481: 1)
"If the Moon [keeps setting (on the 15th)] while the Sun rises" (x105: 18)
"If the Moon is seen at an inappropriate time – it is seen on the 15th with the Sun" (8091: 4)
"If the Moon is hasty (*ezî*) in its course" (opp. on 15th) (8173: 7, 8295: 7)

16th "If on the 16th the Moon and Sun are seen together" (8025: 6)
"If the Moon becomes late at an inappropriate time – it sets on the 15th and is seen with the Sun on the 16th" (8082: 1)
"The Moon was seen on the 16th (with the Sun)" (8177: r.3)

7 Occultation (For stars see 17)

"If Jupiter stands inside/enters/comes out (of) the back of the Moon" (8100)
"If the Moon covers Jupiter" (8438: 4)
"If a planet (Saturn) *comes close* (si$_4$) to the top of the Moon, stops and enters the Moon" (8166: 4)
"If the Sun (Saturn) enters the Moon" (8166: 1)

"If the Sun-disk (Saturn) stands above/below/in the position (ki.gub[224]) of the Moon" (8431)
"If Mars come close to the front of the Moon" (8311: 1)
"Mercury is seen at the appearance of the Moon" (8259: r.3)

8 By the Moon's horns (si = qarnu*)*

"If a planet (Saturn?/Mars) stands by the left horn of the Moon" (8350: 1/8311: 3)
"If a star stands at the left horn of the Moon" (8311: r.1)
"If the Moon reaches the Sun and follows it closely and one horn meets the other" (8294: 1)

9 Haloes (tùr = tarbāṣu[225]*)*

"If the Moon is surrounded by a halo and Jupiter/Saturn/Mars/planet (Mars) stands in it" (8006: 1/8118: 1/8168: 10/8412: 5)
"If the Sun (Saturn) stands in the halo of the Moon" (8317: 3)
"If the Moon is surrounded by a black halo – Saturn stands in the halo..." (8040: 5f)
"If the Moon is surrounded by a halo and two stars (Ma+S) stand in the halo" (8383: 1)
"If the Moon is surrounded by a river" (x113: 8)
"If Scorpius/the Bow Star stands in the halo of the Moon" (8377: 7/8378: 1)
"If the Sun is surrounded by a halo" (8413: 1, see also 8210)

10 The Moon's crown (aga = agû[226]*)*

"If the Moon at its appearance wears a crown – it means (the Moon) will complete the day" (8188: 1)
"Regarding I and VII; if the Moon wears a crown" (8189: 3)

11 Eclipses[227]

The Month, Day, Watch, quadrants in which it *started* and *finished*, and the *wind* (8103) were significant:
"If there is an eclipse in month III on the 14th, and the (Moon) god during his eclipse becomes dark on the upper east side, and clears to the lower west side, the north wind rises during the evening watch..." (8004: 1, also 8316)

The *presence of the planets* was important:
"If in the eclipse Jupiter stood there" (8316: r3, x090: r.10')
"The planets Jupiter, Venus, and [Sa]turn were present during this eclipse" (x057: 7)
"The planets Jupiter and Venus were present during the eclipse" (x075: 12)

---

224 See n194, above.
225 A ring around the Moon of radius 22° – see Kugler *SSB* II 103.
226 *Agû* is a name given to the crowns of deities (*CAD agû* A.1). It is also perhaps the circle of the new Moon revealed by earthshine (ibid. 2.1'), but is also present when the Moon is full (ibid.2.2' & x059: 12). It also describes the rainbow coloured Frauenhofer diffraction pattern caused by the droplets in clouds in *ACh.*1Supp 1: 11 & 8513: 4. Planets can have crowns too (ibid. 2.b & 8051 – see also *BPO*3 p12). Venus wears a black crown due to Mercury's proximity in 8051: r.1.
227 See Parpola *LAS* II App.F.4 and Rochberg-Halton *ABCD* Ch.4.

If the Moon *set eclipsed*, this was ominous:
"If the Moon makes an eclipse and sets with unwashed feet" (8103: 6),
"That it set darkly means" (8487: r.3).

Similarly, if it *rose eclipsed*, this was noted:
"If the Moon comes out darkly" (8336: 5f).

The *direction* in which the shadow *moved* and *cleared* was relevant:
"If the eclipse begins in the east and *goes* to the west." (8103: 5)
"(The Moon) pulled the amount of its eclipse to the south and west. That it became clear from the east and north is good" (8316: 8).

Which *part of the Moon* was eclipsed needed to be known:
"Its right side was eclipsed" (x149: 8')
"That (the eclipse) covered all of (the Moon) is a sign for all lands" (8316: 10f).

Where the Moon was eclipsed could also be interpreted:
"[If the Moon] becomes dark [in the region of Sa]gittarius" (8300: r.11).

The *colour* of solar eclipses were ominous:
"[If an eclipse] *is red*" (8384: 3).

Omens were derived from the situation when the shadow cast by the Moon onto the Sun was smaller in diameter than the Sun:
"If the Sun at its rising is like a crescent, and wears a crown like the Moon... If there is a solar eclipse...its left horn is pointed, its right horn long" (8384: 7f)

The *magnitude* and *time* of eclipses were measured, *but were not ominous:*
"The Sun made an eclipse of two fingers (šu.si) in magnitude at Sunrise" (x148: 4).
"On the 28th at 21/2 double hours...in the west...2 fingers" (8104: 1f)
"It was eclipsed in the area of Scorpius. The *kumaru* of the Panther was culminating (*ziqpu* - this gave the time, see App.1 §33). An eclipse of 2 fingers took place" (x149: r.1)

12 Radiance of the planets

"If Jupiter carries radiance (še.er.zi *naši*)" (8004: 16)
"If Jupiter is bright (*ba'īl*)" (8254: 5)
"If Jupiter has awesome radiance (*melammu*) (8489: r.1)
"Mars...is bright and carries radiance" (8491: r7, x048: 15)
"If the Anzu star (Mars) is bright" (8064: r2)
"Mercury is shining [ver]y brightly" (x074: r.4)

13 Faintness of the planets

"If the yoke star (Mercury) at its rising is low and dark" (8073: 1)
"Saturn...is faint" (8491: r9)

"If Mars is faint" (8114: 3)
"The radiance of the Sun diminished in the path of the Anu stars" (x079: 8)
"If the Sun rises and its light does not get stronger" (x104: 9)
"If the Sun is dark" (x104: 11)

14 Colour of the planets

"If Jupiter is red (sa_5) at its appearance" (8326: 5)
"If Venus is red" (8541: 5')
"If its (Mars's) glow is yellowish (sig_7)" (x100: 18)
"If...Nergal is ...whitish (*pūṣu*)" (x100: 20)
"The black (gi_6) planet is [...]" (8180: 4)
"If the Sun rises and is red" (8308: 2)

15 Planets and weather/Adad

"If a planet [...] in the middle of a gust of wind" (8101: 11)
"In the morning, during the Sunrise the south wind blew" (x079: s.1)

16 Planets near to planets (other than the Moon and Sun)

J+ME "J[upiter] and Me[rcury] in the same day came forth together in succession... If the Marduk planet is black[228]...they are at a distance and will keep away from each other" (x067: 10f)

J+V "If Venus reaches and follows (*ikšudamma ireddi*) Jupiter, ditto, approaches and stands (*iqribma* gub)" (8212: 1)
"If Jupiter reaches Venus and passes her (*ikšudamma* dib-*ši*)" (8212: 3)
"If Venus comes close (te) to Jupiter" (8212: 4)
"If Jupiter passes to the right of Venus (8448: 1)
"If Jupiter goes with (*itti illak*) Venus" (8244: r.2)

J+MA "If Mars comes close to Jupiter" (8288: 6).
"If Jupiter stands in front (*ana* igi gub-*iz*) of Mars" (8288: 1)
"If Jupiter and the false star (Mars) meet (uš.meš) (8288: 3)
"If the stars of Jupiter and a planet (Mars) are equal (mul.meš-*šunu mitḫaru*)" (8288: 8)

MA+ME "If Mars goes behind Šulpae (Mercury)" (8114: r.4)

MA+S "Concerning (Mars) who came near to the front of (Saturn)" (8102: r.8, cf. x047)
"If Mars keeps going around a planet (Saturn)" (8082: 4, cf. 8048, 8049)
"Concerning the planets [Satu]rn and [Mars]...There is (still a distance of) about 5 fi[ngers] left; it (the conjunction) is not y[et] certain" (x047: 6f, cf. 8082: 7f)

---

228 Mercury turns constellations, stars, and planets "black" (gi_6-*ma* = *ṣalimma*) by its proximity (x067: r.1, 8051: r.1, 8113: 8, 8371: r.2, 8504: 3, 8545: 1, 8146: 3/5, 8245: 1). It can turn Leo "black" (8146) and "dark" *adir* (8437) with differing prognostications. Saturn can turn Regulus "dark" (8040: r.3), and when it stands inside the halo of the Moon the halo is said to be "black" (8040: r.1). Mars can turn Scorpius "dark" in 8502: 13, where Mercury turns it "black". The apodoses are opposite. Similar protases are found in EAE 51 (*BPO2*). Reiner & Pingree (op.cit. 2.2.6.1) suggest that they originally described an atmospheric phenomenon. By the period of the Letters and Reports being "black" or "dark" nearly always describes the proximity of Mercury, Saturn, or Mars.

MA+V "If Mars carries white radiance[229] (8541: 8')
"If the Pleiades (Mars) flare up and go before Venus" (x063: r.5 cf. 8415)
S+VE "Saturn did not approach (*iqrub*) Venus by (less than) 1 cubit (kùš); there is no sign from it" (8500: 4)
S+ME "If a planet comes close to a planet" (x109: r.14)?
V+ME x064? "If Venus wears a black crown – a planet (Mercury) stands with Venus" (8051: r.1f – see n226)

17 Planets and the constellations (in order of Mul.Apin Iiv33–37)[230]

Mul.mul "If the Marduk planet (Jupiter/Mercury) reaches the Pleiades" (8326: 7/8454: 3)
"If the Pleiades come close to the top of the Moon and stand there" (8296: 1, 8455: 1)
"If the Pleiades enter the Moon" (8351: 1/4, 8443: 1, 8455: 7)
Gud.an.na "If Jupiter comes near (dim_4) to the Bull of Heaven constellation" (8049: 9)
Sipa.zi.an.na "If Jupiter comes close to Orion" (x362: 12)
"If Jupiter enters (*ana* šà...*irrub*) Orion" (x362: 13)
Šu.gi "If the *kurkurru* of Perseus shines (*inambuṭ* – see n229)...[Venu]s stands with the foot of Perseus" (8380: 1f)
"If Mars comes near to Perseus" (8400: 5)
"If Perseus comes close to the top of the Moon" (8408: 1/4)
Gàm "If Auriga carries radiance (see n229)" – Jupiter nearby by calculation (8115: 4, 8170: r.1)
Maš.tab.ba.gal.gal "If Jupiter passed to the back (*ana* egir dib-*iq*) of Gemini" (8084: 5)
"If the strange planet (Mars) comes close to Gemini" (8064: r7)
Al.lul "If Mars/the strange planet comes close to Cancer" (8452: 6f)
"If Mars stands in Cancer" (8080: r3)
"Until the 5th or the 6th (Venus) will reach Cancer. If the Goat constellation comes close to Cancer" (8175: 6)
Ur.gu.la "If Jupiter has awesome radiance – Regulus stands either to the left or to the right of Jupiter. Now it stands three fingers to the left of Jupiter" (8489: r.1f)
"If Leo is dark (*adir* – see n228) – Saturn became visible inside Leo" (8324: 1f)
"If Regulus is dark.(n228)...If Saturn in front of Regulus..." (8041: r.3f)

---

229 Venus causes constellations (8051, 8185, 8370 etc.) to "keep gaining radiance" (*ittananbiṭu* Ntn Pres. of *nabāṭu* "to shine"). The verb is translated as "to shine brightly repeatedly" in the EAE texts in *BPO*2.2.2.3 and thus to have once referred to atmospherically induced scintillation, or possibly to variable stars. By the period of the texts herein considered it mostly refers to the presence of Venus. Perhaps, the translation "to make scintillate" would be better in the Scholars' correspondence also, for then it would better distinguish it from the other phenomenon induced by Venus and other planets. This is the "bearing of radiance" which Venus, Saturn, and Jupiter can cause heavenly bodies to do through being located nearby. In 8218: 4 and 8547: 3' the horns of Scorpius are said to "bear radiance" *šaruri našā* due to the proximity of Saturn (the texts are fragmentary). In 8541: 8 Mars carries white? radiance due to Venus's proximity. Jupiter causes Auriga to bear radiance in 8170: r.1 & 8115: r.4. Jupiter itself can carry radiance in 8004: 16. In 8184: 7 Jupiter carrying radiance is equated with "being bright" *namarma* – in 8254 *ba-'i-il*. The same is no doubt true of Mars in 8274: r.3 and Mercury in x052: r.11. Thus far, it appears that in the Letters and Reports only Venus causes constellations to "scintillate" by being near, but Venus, Jupiter, and Saturn can all cause constellations or planets to "bear radiance", the latter term sometimes being equated with "being bright". The extent to which this represents a late rationalisation of early descriptions of stellar "twinkling" etc. can only be guessed at.

230 Compare the treatment by Pingree in *BPO*3 §5 p6f of the omens in the official series EAE and its commentaries in which Venus is in or near constellations.

"If Leo is black (n228)" – Mercury nearby, probably (8337: 3)
"If Regulus is black – Mercury stands with Regulus" (8245: 1)
"If Regulus come close to the front of the Moon" (8363: 1)
"If Mars enters Leo and stands there" (8081: 3)
Ab.sín "Mercury became visible in the east in the region of Virgo" (8325: 1)
Zi.ba.an.na "If Libra is dark – Mars stands in Scorpius (which is near)" (8502: 15)
"Mars was sighted in V; now it has approached within 2.5 spans (*ūṭu*) of Libra. As soon as it has come close I shall write to the king my lord" (x172: 4')
Gír.tab "If Scorpius is dark... If the sides of Scorpius are very dark – Mars stands in it" (8502: 13f)
"If Scorpius is black" – Mercury in Scorpius (8371: r.2)
"If ^{Mul}apin (Mars) comes close to Scorpius" (8502: r.1)
"If Nergal (Mars) stands in Scorpius" (8502: 11)
"If Scorpius comes close to the front of the Moon" (8430: 1, 8466: 1)
Pab.bil.sag "If Šarur and Šargaz...are made to scintillate (*ittannanbiṭu* – see n229) - Venus stands in Sagittarius" (8502: 7f)
"If Jupiter stands in the *mišḫu* of Sagittarius" (8369: 3)
Suḫur.máš "If the Fish constellation stands close to the Raven – Mercury becomes visible in Capricorn" (8073: r.1f)
Gu.la Aquarius. *Unattested in the texts herein considered, but one omen is listed in the SB material noted in Planetarium No.81 III B 2.*
Kun.meš "If Jupiter stands in Pisces" (x160: 14)
Sim.maḫ "If a planet comes close to the Tigris star – Mercury stands inside the Swallow" (8253: 5f)
Anunītu "If the Worm star is very massive - Venus stands inside Anunitu" (8357: 3f)
^{Lú}ḫun.gá Mars in Aries (8412)

18 General locations of the planets

"If a planet (Saturn/Mars) stands in the north" (8039: 6/8341: 3)
"If a planet (Mercury) stands in the east" (8157: 7)
"If Mars stands in the east" (8081: r.1)

2.2.3 Comments and Comparisons with the Uninterpreted Observational Records

The categorisations of §2.2.2 have indicated that some celestial phenomena appear *not to have been ominous for some planets. This may well be only because no texts so far known to have been composed for the first time between c. 750 and 612 BC attest the existence of the relevant protases. The SB material has been consulted occasionally in order not to argue from silence.*

Most of the references listed above are omen protases, demonstrating explicitly that the phenomena or configurations which fall into one of the above eighteen categories were considered ominous. Those references which do not begin with "If" are found in the Reports and Letters lying between omens. It is sometimes unclear if the information they recorded was considered ominous, but it seems likely. Even if not, they must have been in some way relevant to the work done by the Scholars. It is often the case that the descriptions for which no protases are (thus far) attested are more accurate than those descriptions found in attested protases. Examples include the records of the eclipse (11) in 8104: 1, of Jupiter's invisibility (4) in x362: r.3, of Mercury's date of rising (1) in 8050: r.3, the proximity of Saturn and Venus (17) in 8500: 4, the proximity of Mars and Libra (18) in x172: 4', and Mars's leaving of Scorpius after being stationary (5) in 8387: 3. The presence of these more precise descriptions within the correspondence of the Scholars demonstrates that some at least of their number used (or constructed) more accurate records of the phenomena before they consulted EAE for their interpretation. The references just given show that both Assyrian and Babylonian Scholars had access to these more accurate records.[231] This fact alone draws the authors of the Reports and Letters close to the authors of the early NMAATs. The following comments on what was considered ominous by the Scholars of the late NA period:

1–2 The heliacal rising and setting of all six planets were considered ominous. For all except the Moon, only the month in which they rose was relevant.[232] As for location, protases are attested which indicate that the star path[233] in which they rose was significant. Since the location of the planets within constellations was ominous (17), clearly the constellation into which they rose or set was important. No more accurate a date[234] or location was seemingly required. For the inferior planets evening and morning risings and settings were distinguished, but they appear not to have had any separate relevance. For the Moon the day of heliacal rising ($28^{th[sic]}$-1st) was of great importance, as attested by the large number of Reports that are concerned with this phenomenon. Some relevance was attached to the day of disappearance as well.

3–4 Omens derived from the prolonged presence or absence of Jupiter, Venus, and Mercury are attested. Many other references are made to planetary periods that are not in the form of omens, but which can be traced either to Mul.Apin or to EAE.[235] They are discussed further in Ch.3.1.2 and Ch.4.2.2. The references to the prolonged absences and presences of the planets indicates that the Scholars noted the days on which the planets disappeared and appeared.

5 X008, written by Issar-šumu-ereš, whose reputation could not be higher, indicates that omens derived from the retrograde movement of Jupiter and Mars were in use in the period of interest, and none were known for the retrogradations of the other planets. The Scholar also indicated, however, that he knew when Saturn was going to retrograde. Since retrogradations start and end in stationary points, these were obviously also noted for the superior planets, sometimes separately. There is no evidence that the stationary points or retrogra-

[231] More on the accuracy of the records in Ch.4.2.1.

[232] This is typical of EAE as well – see the discussion of Tablets 59–60, for example, in *BPO*3 21f. Omens arranged by month typify the series *Iqqur īpuš*, tablets 67f of which concern planetary phenomena – see App.1 §35.

[233] Reiner & Pingree *BPO*2.2.1.2.1 (p17) and *BPO*3 §9 (p15) and Koch (1989) 14f & 199f.

[234] However, note EAE 56 in Largement (1957) §III 18–19. Here an omen in the series has a protasis concerning Saturn's or Mercury's rising on the 7th, 14th, 15th, 16th and 17th of the VIth month.

[235] In particular to EAE 56, Largement (1957) §XVIII-XIX, and Mul.Apin IIi44–67. See also EAE 63 §II described here in App.1 §9, and *BPO*3 p15 for references to further EAE omens which describe Venus's presence in the sky for 6 months.

dations of the inferior planets were considered ominous in the period of interest, or to my knowledge in the SB material (also see n223).

6 Planetary opposition can only occur for the superior planets and the Moon. It is not remarked for the superior planets, but since it occurs mid-retrogradation, it was perhaps signified by that. "Opposition" for the Moon was of great importance, however. It was referred to by the Š-stem of *atû* with the meaning "to meet/confront/look at one another" (*CAD*), and also by the phrases "the Sun and the Moon are seen together", "one god is seen with the other", and "the Moon and Sun are in balance (*šitqulu*, Gt of *šaqālu*)". Sometimes the Sun is not referred to at all, and the protasis "If the Moon is seen on day (12–16)" is used. Strictly speaking "opposition", as we define it, was not meant. Rather, acronychal (evening) rising *and* morning setting were watched for.[236]

7–10 It was considered ominous if any of the six planets entered or stood near the Moon, stood in its halo, or by its horns (at new Moon). Since the Sun cannot be occulted by the Moon (except at solar eclipse), all references to it are to Saturn. The Sun can itself have a halo, and this was deemed to be significant. The lunar crown was also ominous and it appears to have described more than one phenomenon (n226).

11 As the list above describes, many aspects of an eclipse were deemed to be ominous. The lunar latitude largely determines the amount by which the Moon is eclipsed, which quadrants are covered, and influences somewhat the direction in which the shadow moves. The solar and lunar anomalies help determine the locations, length and time of the eclipse. The date is also dependent on when the 1st of the month takes place, the calculating of which requires many factors to be taken into account. The anomalies also determine how large the diameter of the shadow of the Moon is with respect to the Sun, which also impacts on an eclipse's magnitude and length and, in the case of solar eclipses, determines whether or not an annular eclipse occurs. It is amongst the more remarkable features of the late lunar MAATs that many of these factors were treated separately, modelled and then brought together in order to permit the calculations of such things as eclipses, month lengths and lunar visibility times to take place – see Ch.4.2.2.

236 In mid-month there is a day after which the Moon will not rise before Sunset. There is also a day before which the Moon will set before morning, and after which it will not. The former is watched for in the evening, the latter in the morning. 8418: 1 states that the Moon had set without waiting for the Sun on the 14th, and was only seen with the Sun on the 15th. In this case, clearly, the morning was the time when the observation had been made. In 8173: 7 and 8295: 7 the Moon is said to be *e-zi* (*AHw* "eilig, hastig sein") "hasty" in its course if seen with the Sun on the 15th. If taken to mean that the Moon itself has moved rapidly, this cannot mean that the morning setting was watched for here, since if the Moon is seen with the Sun on the morning of the 15th it has been moving *slower* than expected. Perhaps the Moon could be said to be "hasty" in setting on the 14th without waiting for the Sun, and thereby being in "opposition" on the 15th, else we must adjust the translation. In 8458: 1 the Moon is seen with the Sun on the 13th, and the Moon is said to have moved "straight (*ušešir* Š-pret. of *ešēru* if the reconstruction is good) in its proceeding". *Ešēru* "to straighten up" also has the additional meaning alluded to in the English of "to set aright", and its use in the context of lunar opposition on the 13th, which bodes ill, is surprising. Perhaps, then, the implication was not "to cause to move along a straight and proper course" (*CAD*) but "to proceed, to march on" as in the Št-stem. This might then imply that the Moon was moving *more quickly* than expected, which would account for the morning setting on the 13th. The recording of "opposition" on the 12th (8088: 1), however, suggests to me that in this case, at least, acronychal (evening) rising – the last day on which the Moon was still visible at Sunset – was also regarded by the Scholars as ominous. The Moon and Sun cannot be seen on opposite horizons in the morning on that day of the month except in exceptional circumstances.

12–15 The faintness and radiance of the planets does vary with respect to their positions relative to the earth. The superior planets are brightest at opposition, the situation is more complicated for the inferior planets. Also, their appearances are affected profoundly by the weather. For a discussion of the colours generated by temperature inversion layers in the atmosphere see Reiner & Pingree *BPO*2.2.2–6. Weather was usually recorded without reference to the planets in the Scholars' correspondence (e.g. 8365, 8385).

16 The conjunction or near-conjunction of the planets was ominous. Their relative latitudes determined whether or not the planets came near enough to each other to bode ill or good, for not every passing by was considered significant. The different terms[237] used in the protases to describe the planets' close approach undoubtedly reflects the many influences at different periods on EAE, but also probably outlined, more or less, degrees of proximity so far as the late NA Scholars were concerned. Certainly, some Scholars used more precise terms of measurement. The kùš = *ammatu* "cubit" is used uniquely in 8500: 4 (one cubit) with regard to Venus and Saturn's conjunction, and šu.si = *ubānu* "finger" is used in x047: r.1 (five fingers) and in 8082: 8 (four fingers) to describe Mars and Saturn's conjunction. In all cases it appears as if the distances were considered too great to constitute something ominous (though in 8082 Balasî sends the relevant omen just in case). One finger equals $1/12^{th}$ of a degree and one cubit either 2° or $2 1/2^\circ$ of arc.[238] Presumably, conjunction was ominous if the planets were separated by less than about $1/3^\circ$. The Moon's diameter subtends an angle of about $1/2^\circ$.

Conjunction between all possible pairs of planets is described in the texts except (perhaps) for that between Mercury and Saturn. Perhaps those omens concerned with Mercury's proximity to the Sun in the SB text *ACh.* Ištar 20: 17/23/32 were meant to apply to their conjunction.

17 Omens are derived from the presence of all the superior and inferior planets in the eighteen ecliptic constellations listed in Mul.Apin Iiv33–37 (except ^{mul}gu.la, but see the appended note). It is noteworthy that very few omens are attested that derive from the location of the Moon in constellations. The references to the location of the Moon near the Pleiades may well be connected to the intercalation scheme known as the "Pleiaden-Schaltregel"[239], which depends on the date of their conjunction for determining whether or not an additional month should be added to the forthcoming year. Other than this the Moon is described in omen protases being located in Scorpius, and being near Regulus and Perseus. Considering how many Reports are attested which describe the Moon's heliacal and acronychal risings and morning settings it is noteworthy that the constellations into or from which these happened were not considered ominous by the Scholars.

It is also clear that to some NA and NB Scholars, at least, the boundaries of the ecliptic constellations were accurately and clearly defined (e.g. x172: 4' for Libra, and 8175: 6 for Cancer). Any prediction of the configurations of the planets and constellations would

---

237 *ana* igi (*pāni*) gub (*uzuzzu*), te (*ṭeḫû/ṭaḫû*), uš (*emēdu*), kur (*kašādu*), *itti alāku, qerēbu*, dib (*etēqu/etāqu*), *sahāru, mitḫūru* etc.

238 See Thureau-Dangin (1931a), Powell (1987–90) 461–3 and Brown *CAJ* forthcoming, where I show that the kuš-system of celestial spatial measure and the UŠ-system of celestial temporal and spatial measure are interlinked, with the two systems being used to record celestial distances *for the first time* in the late NA period. More on this in Ch.4.2.1.

239 References in Ch.3.1.2.

require the accurate locating of the latter, and lead to the invention of something equivalent to the zodiac.

18 The location of the planets by cardinal direction is not precise, but provides yet another description of their positions for which an ominous significance could be determined.

I attempt to show in the following that the observational records and calculations in the earliest NMAATs were influenced fundamentally *by the requirements of the EAE Paradigm (Ch.3), as manifested in the period of concern here. I am* not *attempting to argue that the Diaries, for example, merely provided the raw data from which the Scholars drew prognostications using EAE and related texts, for the earliest NMAATs fulfilled one other important function (at least). They provided what was considered, at that time, the additional data necessary for the accurate prediction of some celestial phenomena. The contents of the very earliest NMAATs reflect this other purpose as well, and I describe this aspect in more detail in Ch.4.2.*

The Diaries were known as *naṣāru ša ginê* "regular watching" to the Akkadians. Short ones included day to day observations. Long ones usually covered half a year and were compiled from the shorter ones, or similar documents. They were carefully inscribed, succinct records of selected celestial events and included not only observational data, but the results of some calculations.[240] The purpose of these (short term) calculations was apparently to ensure that as *continuous* a record of celestial phenomena as was possible was maintained irrespective of weather conditions or other impediments to observation. I believe that this was done in order to facilitate, or at least that it resulted in, the noticing of important periodicities in the records of the phenomena which ultimately led to their (long term) accurate prediction in the later NMAATs and in the MAATs. I develop this theme in Ch.4.1, and it need not detain us further here.

Only one Diary is attested from the period of interest, but it is believed that many more were written in the 7th and 8th centuries BC.[241] Long gaps exist between the first few Diaries, but their form and contents indicate that they are part of the same text group as the more numerous Diaries dating to the 4th and later centuries BC, though some development in terminology can be discerned. There exists late textual evidence that the Marduk temple at Babylon employed certain individuals to observe the sky and compile these records,[242] and it is perhaps this same employment that Marduk-šapik-zeri is referring to in x160: 40 when he states that he has read EAE and "made astral observations" mul.meš an-*e uṣ-ṣab-bi.*[243] This activity no doubt took place all over the empire. Several Eclipse Records, fragmentary records of Mercury observations published in Pingree & Reiner (1975a) – three of which were found in Nineveh and were therefore composed before 612 BC – and records of the observations of Saturn and Mars from Babylon also all date to the period of interest (see App.1 §32).

240 For details, see Hunger in his introduction to *Diaries* Vol.1 and idem (1993b), the Introduction above and App.1 §32 and Ch.4.1.1 below.

241 Sachs (1974) 44.

242 For details see *Diaries* Vol.1 p12

243 *Ṣubbû* with meanings "to look upon from afar/to complete work according to a plan" see *CAD*.

Dates were recorded in years, months, days, and the three watches,[244] each of night and day, are found in the earliest Diaries, Eclipse and Planetary Records. This is exactly as in the Scholarly correspondence.

The time between an eclipse and Sunrise (*ana* zalág) was measured in units of 4 minutes, the UŠ, in the oldest record of an eclipse of this type (*LBAT* 1413: 5) dating to 747 BC. The unit is attested regularly thereafter, though often without UŠ being written explicitly. It is used in the -567 Diary, and thereafter consistently. UŠ are not attested in the Letters and Reports, though the unit of time equivalent to 30 UŠ known as the *bēru* or "double hour" is. UŠ are also attested in EAE 14 Table D: 24, Mul.Apin IIii24f, the NA *ziqpu* texts and in the Babylonian calendar text BM 36731.[245] It is my contention that the unit was also used in OB times.[246] No doubt the absence of UŠ in the Scholars' correspondence reflects the accuracy of time with which *ominous* phenomena were required to be known. The time of the occurrence of all attested ominous phenomena in the texts composed between 750 and 612 BC had to be known only to an accuracy of one day, or possibly to one watch of the night (in the case of eclipses). The accuracy with which the times were recorded in NMAATs reflects a *new* concern in cuneiform astronomy-astrology.

Time was also measured in the letter SAA2249: 12'f in terms of the culminating of the *ziqpu* stars in a non-ominous context, and in x134: 8 and x149: r.1 to mark the times of eclipses. This method is also used in the Diaries to record the time of eclipses, and is attested in Eclipse Records from 194 BC.

Distances between celestial bodies are recorded in kùš "cubits" and (šu).si "fingers" in the -651 Diary: i16, iv15' and in all subsequent Diaries. They appear in the Eclipse Records, marking distance to stars, though attested only from 573 BC on. Fingers are also used to measure eclipse magnitudes, though these appear to subtend half the angle of the other fingers.[247] These units are found a few times in the Letters and Reports – see §2.2.2 (16), above – and *do not* appear to have been used in celestial context before the late NA period.

The UŠ is also attested as a celestial distance unit for the first time in the late NA period, in both the Mercury and Saturn Records, and in the so-called "gu-text", BM 78161.[248] It is not used in this sense in the Letters and Reports, and only rarely in the Eclipse Records and the Diaries, though this is sometimes difficult to determine since an UŠ of right ascension (distance along the celestial equator) is equivalent to an UŠ of time – both are $1/360^{th}$ of a terrestrial revolution, be that 1° or 4 minutes. The UŠ was the unit used when distances along the zodiac were *calculated* in the MAATs of the Hellenistic period. The zodiac is not parallel to the celestial equator, and the UŠ along its length are definitely units of distance and not of time.

---

244 Usan, murub4, and zalág "first, middle, and last parts of the night" are the terms found in the Diaries. The last watch is usually referred to as (en.nun-)ud.zal in the Eclipse Reports since zalág is used for Sunrise and for the end of the eclipse. En.nun-an.usan(2), = *barārītu*, en.nun-murub4 = *qablītu*, and en.nun-ud.zal.(li/la) = *šat urri* "evening, middle, and morning watches of the night" are the terms used in the Reports and Letters. Ud-½-ám and *ina* murub4 an.ne are terms used in the Reports to describe "noon". *Ina šēri* is used for the "morning". *Bādu* is used for the evening. For the terms used in the Diaries to describe these and other parts of the day see *Diaries* I p15 and Koch (1997).

245 For references to these see App.1 §§ 21, 30, 33 & 38.

246 See Brown *CAJ* forthcoming, and App.1 §8

247 Regular celestial fingers are $1/12^{\circ}$, but 12 'eclipse fingers' cover the entire $1/2^{\circ}$ diameter of the Moon or Sun.

248 For details see Brown *CAJ* forthcoming, and for BM 78161 see App.1 §34.

The four relative orientations in space used in the Diaries: *ár* "behind" (approximately = to the east of), *ina* igi "in front of" (to the west of), *šap* "below", and *e* "above",[249] correspond to those used in the Reports and Letters: *ina* egir = *arki* "behind", *ina* igi (*pāni*) "in front", an.ta-*nu* = *šaplānu* "below", and ki.ta-*nu* = *elēnu* "above". The Eclipse Records also use sig for "below". The Saturn Records also use *ina* igi and egir, as well as *ina* dal.ban and šà for "in between" (both fragmentary attestations). The Mercury records use ki.ta-*nu* and gùb "left", presumably for "below".

Hunger in *Diaries* p23 notes that the Diaries describe the close approach of the Moon to the Normal Stars[250] with dim_4 = *sanāqu* if the distance is 2 fingers, and with te = *ṭeḫû* if it is only 1 finger. These terms are used frequently in the Reports and Letters (see n237). The Diaries offer a guide to their meaning in the Scholars' correspondence. The close approach of the planets to each other and to Normal stars were also recorded in both the Diaries and in the Saturn Records (see line 22'). The conjunction of planets was considered ominous, but few omens are attested that concern the approach of planets to Normal stars.[251] However, it is an elementary matter to determine in which constellation a planet was located when it is known which Normal star it was situated near to. What the presence of these Normal stars in the -651 Diary (e.g. line i10) and in the Saturn Records indicates is that an attempt was being made to locate the planets against the background stars *more* accurately than can be done by specifying their location merely in constellations. In the Saturn Records, even when constellations rather than individual bright stars are recorded, the location of the planet is always specified as being "behind the Furrow", and so forth.

It is interesting that although the locations of the "passings by" (dib) of the Normal stars by both the Moon and the other five planets were recorded accurately, their times were not. These were only recorded to the nearest watch of the night. I suspect that this was because the *time* at which planets entered constellations was not ominously significant. It was important to know only that they would come close to certain stars or planets, and the day on which this would happen. Predictions derived from the record of such data would have permitted the celestial diviners to watch for closest approach only on certain days and not all the time. It would have provided the data necessary to predict conjunction, and so warn the king, even if bad weather obscured the sky. In the later Diaries the planets' dates of entrance into zodiacal *signs* were recorded, as well as the dates of their close approaches to the Normal stars. This is again reminiscent of omens attested in the Reports and Letters.

The Diaries record the following inferior and superior planetary phenomena: the first and last appearances of the superior and inferior planets (evening and morning) together with the zodiacal constellation (later zodiacal sign) in which they occurred, and the oppositions and stationary points of the superior planets only. This, again, correlates so completely with the attested ominous planetary phenomena as to make the existence of a divinatory purpose behind the Diaries difficult to avoid. As noted in **(5)** above, no omens are attested for

---

249 Only *ár* and *ina* igi are attested in the -651 Diary (lines 8 &16), but there is every reason to suspect that the terms attested later for "above" and "below" were known and used in the 7th century also. Their precise meaning in terms of a co-ordinate system is not known. See Fatoohi & Stephenson (1997/8).

250 Mulšid.meš, probably *kakkabū minâti* "counted stars" listed in *Diaries* I p17f. They are a set of bright ecliptic stars and "Normal" is a modern designation.

251 They are found explicitly as "counted stars" in EAE Venus omens. Cf. K2226: 13 in *BPO*3 p13 & 93 where they are contrasted with the "uncounted" stars in line 22. Three, at least, of the Normal stars appear in omens in the Reports and Letters: mullugal "King" – α-Leonis; mul.mul (múl.múl in the Diaries) "Bristle" – the Pleiades; , mul$Li_9.si_4$ (si_4 in the Diaries) "the god Lisi" – α-Scorpii.

the retrogradations and stationary points of the inferior planets. Those for the superior planets are apparently few and far between, but those attested do indicate the constellation in which the retrogradation occurred. As noted in **(6)** above, no omens specifically concerned with planetary opposition are known. Only the date of this phenomenon is recorded in the Diaries. The Saturn and Mercury Records note those planets' first and last appearances.

In the -651 Diary: I 7f it is written (between the 14th and the 17th of month I) "Mercury's last appearance in the east behind Pisces, and Saturn's last appearance behind Pisces; I did not watch because the days were overcast". In the Saturn Records, line 7', last appearance was also not observed (nu pap). In line 23' the comment "last appearance, because of cloud, was computed (*muš-šúḫ*)". Clearly, the heliacal settings of these dimmest planets were calculated according to methods as yet unknown.

Lunar and solar haloes were recorded in the Diaries (-651: i 2,5) just as in the divinatory material. So were lunar crowns (Diary -567: 8), and the vicinity of the planets to the Moon's horns (Diary -418: 14). Many types of weather were recorded, some of which had prevented the sighting of planetary phenomena. The brightness of the Moon was noted in the -651 Diary: i 1. A red flare was noted in the -567 Diary: r.10'. The radiance of planets was rarely recorded in the Diaries, but in the -382 Diary line 17' Saturn is said to have been high (nim) and bright (*ba'il*). The same remarks are found in the 7th century BC Saturn Records (lines 8' and 20'). Perhaps, the relationship of the planets' brightness to their orientation, noted above in **(12–15)**, was understood and was not thought necessary to record. The use of "bearing radiance", "causing to scintillate", turning "black" or "dark" (see nn228 & 229) to describe planet locations are not attested in the early NMAATs. This is not surprising. They were formed by the process designed to extend the applicability of celestial omens – the "normal science" of the EAE Paradigm.

In the Diaries and the Eclipse Records eclipses were both recorded and predicted. The first attested predicted eclipse in the Diaries was for -567 (line 17), and in the Eclipse Records in -744. The first attested *successful* prediction is dated to 523 BC, however. Eclipses are also predicted in the Reports and Letters, for which see Ch.4.2.4.3. Very little remains of the eclipses *recorded* in the -651 Diary, so a comparison between what was considered ominous in the Reports and Letters can only be made with that which was recorded in the later Diaries, and with what is found in the Eclipse Records. The comparison[252] shows that *all* the ominous phenomena manifested by eclipses and noted above in §2.2 (11) were recorded in one or more Diaries and Eclipse Records (except for the colour of the solar eclipse). Additional information, including the time between Moonrise and Sunset and between Sunrise and Moonset, the location of the eclipse relative to Normal stars (though this could easily have been translated into an ecliptic constellation), and its accurate length (and perhaps magnitude[253]), was also recorded in these texts. As noted, the time of an eclipse was sometimes measured by a culminating *ziqpu* star in both the NMAATs and in the Letters and Reports. Its presence was apparently not ominous, but indicates, as does the other additional information, that the non-interpreted records

[252] What appears in the Diaries is summarised by Hunger op.cit. Vol. 1 pp23–4, that in the Eclipse Records by Huber (1973) pp5–13.

[253] There is no explicit evidence to my knowledge of the magnitude of the eclipses being ominous, except in so far as the shadow in an eclipse of small magnitude will touch fewer quadrants. This is also stated in Rochberg-Halton *ABCD* Ch.4 III A (p49). The magnitude of lunar and solar eclipses was noted in the Letters x149 and x148 respectively.

contained *more* than the minimum data needed to interpret eclipses and that some, at least, of the Scholars possessed these data. These non-ominous data, in combination with a record of the ominous phenomena they manifest, provided what was considered necessary to make accurate predictions of the times and ultimately locations of the occurrences of eclipses.

On pages 20–22 of *Diaries* Vol.1 Hunger describes what have come to be termed the "lunar six" which are recorded in the Diaries. These are six intervals of time measured in UŠ between Moonset or rise and Sunset or rise at the month beginning, middle, and end. They are amongst the principal objects of prediction in the lunar MAATs. They first appear in the -567 Diary. In the -651 Diary: 6 only the phrase "on the 14th, one god was seen with the other" is attested. This also appears in the -567 Diary: 4, but thereafter is not attested to my knowledge. In the -567 Diary: 4 that phrase, which in the Reports describes the morning setting of the Moon, is immediately followed by 4 NA, "4 (UŠ) Sunrise to Moonset". The identification of one of the lunar six with the ominous occurrence of the date of morning setting is significant in demonstrating the close relationship between the NMAATs and the Scholars' correspondence. It provides a direct link between the omens and the late MAATs as well.[254] I suggested, above **(6)**, that acronychal (evening) rising was also considered ominous by the Scholars of the late NA period. This would correspond with the Diary lunar six term "gi_6". The lunar first visibility was also looked for keenly by the Scholars, as was the Moon's last visibility. These correspond to the lunar six terms "NA" (again) and "kur". These terms designated time intervals, and their lengths were not *directly* ominous. However, they were made significant through the use of "ideal period schemes" in celestial divination, which I describe in the next chapter. I suggest, therefore, that the recording of these lunar six time intervals was undertaken precisely because the *dates* of lunar appearance, disappearance, morning setting and evening rising were ominously significant, and that an effort was made to predict *on which days* those ominous phenomena would occur through the use of the record of their occurrences.

The lunar six are attested in the second oldest surviving Diary, which provides a *terminus ante quem* of -567 for this effort. The similarity of the -651 Diary to the later examples indicates strongly that the purpose of its production was not dissimilar, suggesting that the interest in predicting the dates of these very important ominous lunar phenomena existed in the 7th century BC too.

Similarly, the Diaries from -418: r.3' on record a time interval (yet again called NA) in UŠ between the first observation of the superior and inferior planets after heliacal rising and Sunrise. Were this interval too long (according to rules as yet not understood), the author wrote a date (always earlier) believed to be the date on which he estimated the planet to have *truly* risen.[255] Precisely the reverse was done for the dates of last appearance. If the interval between the last evening visibility of the planet and its setting were too long, a further (later) date was offered which, we presume, was the calculated ideal date of heliacal setting. In the -382 Diary: 17', for example, Saturn is said to have been "high" nim, *elû*. What follows is

---

[254] This is because not only were the lunar six the object of prediction of some lunar ephemerides, but some of the fundamental parameters used in many lunar ephemerides may well have been derived from records of these intervals. Brack-Bernsen (1990) showed how the parameters for column Ø of the system A ephemerides could have been derived from records of four of the lunar six. See further in Ch.4.1.2.

[255] Suggested by Hunger in *Diaries* 1 p25 and exemplified by A 3456, a Seleucid text from Uruk concerned with the dates of Mercury's first and last appearances – see Hunger (1988).

the ideal date of the planet's first appearance. Interestingly, a planet being "high" (nim-a) on first appearance is also noted in the 7th century BC Saturn Records, as mentioned above. In the Reports and Letters references were also made to the planets being "high", using *šaqû*. In x100: 7, for example, Akkullanu says that he had observed Mars when it had (already) risen high. He gives a precise date for this observation before he offers the interpretation. It is also noted for the Moon in x225: 8f. These examples show that in the late NA period there was awareness of a planet's usual altitude at first appearance. Also, the Ninevite Mercury texts, published by Pingree & Reiner (1975a), include measurements in UŠ for what are the distances/times between first visibility and Sunrise, and last visibility and setting, albeit without the explicit mention of the term NA. It seems highly probable, though, that observations of NA for the planets (including the Moon) *were* being made in the period of interest here.

Again, the purpose of the observations of NA and equivalent time intervals was, presumably, so as to provide the accurate dates for the heliacal phenomena of the planets. It was from this *precise* and *continuous* record that accurate values for the periods of time after which the planets repeated their heliacal phenomena were derived.[256] Some of these periods were known by the 7th century (see Ch.4.2.2). The ominous significance of a planet being "high" at heliacal rising (8391: 3, 8093: r.3, x100: 7f, x225: 8f), and of it appearing at a non-ideal time (above §2.2 **3** & **4** and Ch.3.2.2 xix) were all too apparent to the Scholars, and explains why records of the dates of their occurrence were made in the first place, when in many cases only the *month* in which heliacal rising occurred was significant **(1–2)**.

Solstices and equinoxes were recorded in the Diaries, and also in three examples of Scholarly correspondence (8140–2). Some historical information is also present, and data on commodity prices and river levels were included. As with the weather it was perhaps felt by the authors of the Diaries that these phenomena too might be predicted.[257] The presence within the Reports and Letters of weather reports, historical information (e.g. x109f), of information on river levels (x364), and of omens whose apodoses sometimes refer to the prices of goods does nothing but further indicate the intimate relationship of the Diaries to the works of the Scholars.

In summary, I hope to have demonstrated in §2.2 the close connection between the omens and accompanying comments in the Reports and Letters and the data recorded in the earliest NMAATs (as well as, in passing, data in later examples). So close is the connection, that it is probable that the Scholars – the celestial diviners, lamentation chanters and exorcists – used or possibly authored the Diaries, Eclipse, Mercury and Saturn (and Mars) Records. It is not without significance in this context, I suggest, that in *CT* 49 144: 7–10 the individuals hired in the Seleucid period by the Marduk temple to compile Diaries were known as "scribes of EAE". In addition, I have noted that even the earliest Diary included both observational data more accurately recorded than was necessary for their interpretation, and the results of the calculated prediction of phenomena. They included data that were *then* considered to provide the necessary information for the prediction of some celestial phe-

256 Swerdlow (1998) has demonstrated that from a knowledge of the synodic period, the synodic arc can be estimated using the "Sonnenabstandprinzip" (van der Waerden, 1957), and thence the parameters of the planetary ephemerides can be derived. He effectively shows that it is possible to move from observations of NA in UŠ to the planetary MAATs - more in Ch.4.1.2.

257 Hunger's suggestion with regard to the weather in *Diaries* pp27–28 and see idem (1977).

nomena to an accuracy useful for celestial divination. That this was thought possible was the breakthrough of what is here termed the "PCP Paradigm" outlined in Ch.4.1. Again it is relevant in this context, I maintain, that one of the few attested authors of a Diary (-321) is also known to have authored MAATs for Venus and for Mercury, while a descendant of his copied Mul.Apin.[258]

[258] Slotsky (1997) 99–103.

CHAPTER 3

The *Enūma Anu Ellil* (EAE) Paradigm

The NA and NB Scholars interpreted the heavens for their kings. In order to do this they used EAE and other related series, which together I am calling the "EAE Paradigm". The EAE Paradigm had a long history before the late NA period. This chapter will provide a brief study of the cuneiform astrological-astronomical texts thought to have been first composed before this time in an effort to define what the EAE Paradigm constituted so far as the NA and NB Scholars were concerned. The study will be directed towards elucidating the set of premises which underpin celestial divination and many related scholarly texts. This will be in order to argue that this set of beliefs did not *include an awareness of the fact that some celestial phenomena are predictable to a high degree of accuracy. It will show that the methods by which selected planetary phenomena were accurately predicted, or for which attempts at accurate prediction were made, and which are alluded to in certain texts composed in the period c. 750–612 BC, represent* genuine innovations. *The textual evidence for these innovative methods is provided in Ch.4.2.*

I will present evidence that the EAE Paradigm existed by the end of the OB period in a form broadly similar to that being used by many of the Scholars discussed in Ch.1 in the late NA period. The description "EAE Paradigm" has been chosen to indicate that virtually every aspect of the astrological-astronomical texts known to have been composed before c.750 BC can be found in the 70-tablet series known by its opening line as "When Anu and Ellil", *Enūma Anu Ellil*, and to suggest in the first approximation that a transition from one long-lived Paradigm to another took place between c. 750 and 612 BC.

In elucidating the premises underpinning the EAE Paradigm the following approach has been taken. In §3.1 the relevant texts are described. They have been divided up into four categories each treated separately in §§3.1.1–4 leading to a definition of the Paradigm itself.

In the discussions accompanying §3.1, the *premises* underpinning the EAE Paradigm have been grouped into three main classes. These premises permitted the celestial diviners to interpret the heavens and its phenomena, and each class is treated in §§3.2.1–3.

The first class of premises is entitled the "*rules*". By these I mean the methods used by the creators of the omens to invent protases and apodoses. These rules included the possibility of generating a new protasis by analogy with or by contrasting another. For example, Jupiter in one protasis might be replaced with Mars. The resulting apodosis might be created by inverting the apodosis of the first omen, since Mars is sometimes perceived as the opposite of Jupiter. Also, some new apodoses were invented by rules which regulated the "play on the words" of a new protasis. For example in 8219 an omen concerning the proximity of Mars (called by its D-name) and Scorpius is given. Mars is frequently an ill-boding planet, and Scorpius an ill-boding constellation. The proximity of the two would lead one to suspect the prognosis also to bode ill.[259] The omen reads:

[259] *Contra* the supposition in Koch-Westenholz (1995) 11 that a negative sign combined with a negative sign bodes *well*.

> "If Mul.Apin comes close to Scorpius: the ruler will die by a sting of a scorpion; after him his son will not take the throne."

Since constellations cannot approach each other, I believe the protasis to have been derived by analogy with protases that describe the approaching of planets to constellations. Next, the means by which the ruler is predicted to die alludes to the constellation. It is a form of simple word play. The proximity of *two* ill-boding bodies probably accounts for why *both* the ruler and his son fail to continue in charge of the land. These are the ways in which the *rules* can account for the omens. Both "number play" and "textual play" rules are discussed in §3.2.1, where a Structuralist approach has proven useful.

The second class of premises underpinning the EAE Paradigm is discussed in §3.2.2 and these form together what I call the "*code*". The code relates directions or months to countries, for example. North is often related to Akkad, the east to Elam. The connection of royal deaths with eclipses is, I believe, also part of this code. The association of Mars and Scorpius with evil in the example just quoted similarly forms part of it. While we in the western world might associate scorpions with danger, say, in Mesopotamia, or rather amongst the authors of the omen series, the characteristic associated with them may be quite different, and I have been cautious not to assume connections but to try and prove them.[260] Once demonstrated, I believe the code should be understood to be largely *given*. It is mainly the code as revealed in those omens known to have been used between c. 750 and 612 BC that concern us here.

The last premise incorporates the "*categorisation*" of the universe into analysable units. I mean by this the classifying of what make up phenomena so as to make their description possible. For example, the continuous colour spectrum was broken down into four colours which were then used to describe various phenomena. I also include in this class such apparently "obvious" categorisations as days, months, and so forth, but also less obvious examples such as watches, double-hours etc. A special set of categorisations are the "*ideal period schemes*". These include the categorisation of the year into 360 days, or a month into 30 days, for example. I argue that the ideal schemes served a divinatory and not an astronomical purpose. This is a new hypothesis and suggests that what previously has been considered to be "primitive astronomy" (see n267, below) was instead integral to the prevailing divinatory methodology.

This approach to the cuneiform astronomical-astrological texts differs substantially from any previous study, to my knowledge. It has provided new ways of tackling the huge omen corpus, has revealed the importance of celestial divination to other branches of Mesopotamian learning, and has helped make clear the significant influence of the omen collections on the development of "science" in Mesopotamia and elsewhere. It has also required a re-examination of the so-called "canonisation" of EAE, for which see §3.3.

Previous studies of the cuneiform astronomical-astrological texts, aside from their editions listed in App.1, have tended to concentrate on the following issues:

a) The assessment of the historical material recorded in the omen apodoses.[261]

---

260 This note of caution is also sounded by Leichty (1970) 6f.

261 See n35.

b) The reconstruction of Mesopotamian religious beliefs and the perceived characters of the gods.[262]
c) The evaluation of the relationships between the king, the Scholars and the gods.[263]
d) The estimation of the influence of Mesopotamian celestial divination on Greek, Indian, Hebrew, and later astrology.[264]
e) The interpretation of the later NMAATs and MAATs.[265]
f) The study of the transmission of Mesopotamian methods and parameters into Greece, India, and Egypt.[266]
g) The attempt to find precursors to the later MAATs and NMAATs.[267]

3.1 Defining the Paradigm

Enūma Anu Ellil groups together stellar, planetary and weather omina as well as elaborations on the lengths of the year, the month and daylight. By the late NA period variant editions comprised some 70 tablets and about 7000 omina, but its textual history stretches back at least a thousand years earlier. For details of its current state of publication and proposed evolution from an OB "proto-EAE" through its MB redaction see App.1 §§ 7, 15 & 21. Other series concerned with celestial phenomena were in use in the period after c. 750 BC, but are also believed to have been written in the 2nd millennium BC. These include *Iqqur īpuš*, particularly §§ 67f, i.NAM.giš.ḫur.an.ki.a, and Mul.Apin (App.1 §§30 & 35). The astrolabes (App.1 §§ 13, 17 & 26) also bear on celestial divination, particularly with regard to their use of the "ideal calendar" (App.1 §8), and certain literary texts from the OB and MB periods refer to the discipline.

Texts such as *The Babylonian Diviner's Manual* (App.1 §36), the commentaries and other explanatory works (App.1 §§ 28–29), even the *ziqpu* star texts (App.1 §33) and some late NA and NB literary material (App.1 §24) attest to the continued vitality of EAE-style divination during the main period of concern in this work, and also help to show how the

262 E.g. Jastrow (1898, 1912), Jensen (1890), Jeremias (1929), Contenau (1940), Gadd (1948), Lewy (1956), Saggs (1978), Baigent (1994).
263 See Ch.1 here, Parpola *LAS* II and SAAX introductions, Oppenheim *AM* 202f & (1975), Rochberg-Halton (1982).
264 E.g. Pingree (1973b, 1978a, 1978b, 1982, 1987b), Bouché-Leclerq (1899), Jastrow (1912), Ungnad (1944), Neugebauer & Sachs (1952–3), van der Waerden (1952–3, *BA* Ch.8), Weidner *GD*, Kuhrt (1982), Reiner (1985b, 1995), Rochberg-Halton (1987b, 1988a, *ABCD* intro), Ulansey (1989), Greenfield & Sokoloff (1989), Parpola (1993a), Barton (1994), Bobrova & Militarev (1993).
265 The founding works of Epping, Strassmaier, Kugler, and Schaumberger are summarised in Neugebauer *ACT*. Since the publication of that work the works of Aaboe, Brack-Bernsen, Britton, Henderson, Huber, Goldstein, Maeyama, Moesgaard, Neugebauer, Sachs, Schmidt, Stephenson, Swerdlow, Toomer, and van der Waerden cited in the bibliography here or in Walker (1993) are important. The best detailed summary is still Neugebauer *HAMA*, though some important developments have occurred since 1975. The best brief summary is Britton & Walker (1996).
266 Neugebauer *HAMA* 589f summarises the pioneering work of Kugler and others concerning transmissions to the Greek world. See also idem (1988) and Huxley (1964). Pingree (1973a, 1987a, 1989a, 1998) has shown the debt of Indian astronomy to Mesopotamia. Jones, in particular, has continued the study of transmissions.
267 These are most clearly noted by the use of the word "astronomical". E.g. van der Waerden *BA* Chs.2 & 3 "Old-Babylonian Astronomy", idem (1978) 672f, Hunger & Pingree "Mul.Apin An Astronomical Compendium...", Al-Rawi & George (1991/2) "EAE XIV and other early Astronomical Tables", *HAMA* 541f "Early Babylonian Astronomy" and so forth. However, the extent to which any of the texts discussed in these works are more "astronomical" than "astrological" is debated here. See the definitions of these terms offered in I.1.

premises underpinning the Paradigm remained unchanged during the millennium or more of its operation.

Titles of several other series seemingly concerned with celestial happenings are also known from *The Babylonian Diviner's Manual*, from the Ninevite library records published by Parpola (1983b) and Fales & Postgate SAA7: 49f and the *Catalogue of Texts and Authors* published by Lambert (1957 and 1962). Their reconstruction may be possible one day.

Omens in cuneiform are attested from the OB period on in regions neighbouring Mesopotamia. They are in many ways similar to those found in Mesopotamia itself (App.1 §§13 & 14). This spread of the EAE Paradigm is particularly important, for it demonstrates the extent to which Mesopotamian celestial divination was a universally recognised cultural achievement. Omens are also attested until the very end of cuneiform writing. Since parts of EAE appear to have been transmitted to India[268] in the last centuries of the first millennium BC, it would appeaar that the late copies of EAE were not kept simply for antiquarian reasons.

As the chart on the first page of Appendix 1 reveals, I have found it convenient to list all the cuneiform astrological-astronomical texts in four categories:

3.1.1 Omens

Omens are attested in Mesopotamia from the beginning of the second millennium to the end of the first. Omens are known which concern the entrails of lambs and kids, the movement of smoke or of oil on water, malformed births, dreams, happenings in the town, flights of birds, visages, and many others.[269] Omens dealing with the phenomena of heavenly bodies are also known from the OB period until the end of cuneiform writing. They are made up of two parts, a protasis and an apodosis: "If x, then y". The protasis is usually introduced by "If" and describes some celestial event. The apodosis gives the prognostication for the land, the economy, the king, or the people. In celestial divination omens are not attested which concern individuals. This is in contrast to omens derived from other phenomena.[270] As discussed in Ch.1 (around n105) celestial omina are special through dint of being visible to all and in not being able to be discounted on the basis of the impurity of the diviner.

Usually the protasis is in the preterite if a finite verb is being used, and in the stative if a state or condition is meant. The apodosis, of course, is in the future tense (expressed by the present in Akkadian). Sometimes the omens are expressed in a more succinct manner in the form: protasis key word – *ana* "for" – apodosis key word. This is found in EAE 50, for example (*BPO*2 XVI 4 etc.). Often the apodoses are followed by comments on, or explanations of, the protases. They are normally in the present and are frequently marked by the particle *-ma*.[271]

[268] Pingree (1982) & (1987a).

[269] For some general introductions to Mesopotamian divination see Oppenheim *AM*, Bottéro (1974), and the CRRA 14 volume.

[270] For example, the apodoses derived from the movement of oil on water. See Pettinato (1966). See also the *šumma ālu* omens in 8237: 4: "If a falcon hunts in a man's house; there will be deaths in the man's house". This applied to private individuals. The authors of the Reports *did* use non-celestial omens even when writing to the king. See Hunger SAA8 xviii.

[271] See Reiner's analysis in *BPO*2.3.3.

Celestial omens are found in most tablets of EAE, in Tablet 2 of Mul.Apin, and in *Iqqur īpuš*. This evidence alone provides good cause for including those two series within the concept of the EAE Paradigm. It is also apparent that both texts were used by the NA and NB Scholars.[272] I.NAM.giš.ḫur.an.ki.a does not include omens, but calculates the visibilities of the Moon in a way comparable to that found in EAE 14 and in Mul.Apin IIii43f. Also, the means by which learned apodoses in EAE are related to their protases are remarkably similar to the hermeneutic methods manifested in that short series. This will be discussed further in §3.2.1.

The attested OB omens are listed in App.1 §§5–6. None are known to come from the preceding Ur III or OAkk periods, but in §§2–3 I discuss the evidence which points to the existence of celestial divination prior to the OB period. The question of why omens are virtually exclusively attested only in the period of the Old Babylonians and no earlier is of interest, but cannot be considered here except in so far as the *majority* of omens appear to have been generated by techniques which could *only have been used by the literate*.

The oldest attested examples (App.1 §§5–6) include protases that describe phenomena that could never occur, and apodoses that were not observed at a time when the events in the respective protases occurred. Eclipses can never occur on the 21st of a month, for example, and yet protases are attested which describe precisely this eventuality, and the resulting prognosis for the land.[273] These omens were invented, and were not simply the result of observation and recording. My analyses outlined in §3.2.1 demonstrate that *most* of the attested celestial omens were, at least in part, invented. This brings into question what has come to be known as the *empiricism* model of the source of the omens. This exists in two forms, a dilute "non-casuistic" form and a strong "causal" form. Its dilute form is espoused, for example, by Hunger in the SAA8 volume p xiii:

> "It was believed that the gods send messages announcing future events. The messages or "signs," as they were called, could come from very different sources. One looked for them in everyday events like the behaviour of animals, or in the entrails of sacrificed sheep, or in the sky... If a remarkable event occurred shortly after such a sign, people assumed a connection between them: the sign had been sent to announce the event that followed it. Whenever the sign occurred again, it was thought to predict the same event once more... As time went on, the signs and their consequences were collected and organised in a systematic fashion."

The supposition is that the simultaneity of some celestial and terrestrial events *led* to the former being considered to be signs that the latter would occur, without arguing that the former *caused* the latter.[274] The "strong empiricist" position believes that the ultimate source of omens was the false logic of the *post hoc ergo propter hoc* argument – "after this, therefore because of this" – and incorporating an underpinning belief that the celestial happening *caused* the mundane one. This is the supposition of Cicero in *De Divinatione* I xlix, that:

---

272 In x062: 13 Mul.Apin is quoted and in x006: r.12 the *biblāni* of month I are cited, the word referring either to the hemerologies directly or to the handbook which contains them. See *CAD biblu* C. Quotations from the hemerologies appear in a number of the Scholars' missives – e.g. 8231–6.

273 Rochberg-Halton *ABCD* 19f. There are OB omens giving the anticipated terrestrial happening for protases describing eclipses every month, in every watch, for shadows crossing over the right side, the left side, and the middle, for the direction of the shadow's movement in the four cardinal directions, for when the Moon rises or sets while eclipsed, and for eclipses occurring on every day of the third week of the month except day 17.

274 Also Rochberg-Halton *ABCD* 15: "Omens in general functioned as indicators of what would occur...yet did not imply a belief that future events followed inevitably from past events predicted in the apodosis".

> "In every field of enquiry great length of time in continued observation begets an extraordinary fund of knowledge... since repeated observation makes it clear what effect follows any given cause."

Strong empiricist models are held to account for cuneiform omens in some Assyriological circles.[275] Koch-Westenholz (1995) 13–19 argues against both empiricist models by pointing to Popper's argument that there is no such thing as observation without a hypothesis,[276] and noting that the proportion of so-called "historical omens" is extremely small, and that they are closely connected to the "historical tradition" – in other words they are "literary". She also notes that the earliest OB liver omens can be read in largely the same manner as the latest, indicating that the tradition was already established *before* the first omens were written down. She concludes that Mesopotamian divination was not especially empirical, but rather an all-embracing semantic system designed to interpret the universe in a different way to science.

Developing this theme, Denyer (1985) discusses *De Divinatione*, arguing that Cicero has judged divination as a natural science, when in fact Stoic divination, in this case, operated in those fields for which there was no science. In most respects the arguments apply readily to Mesopotamian divination. Denyer points to the assumption of causality that lies behind science, but which is not assumed in (Stoic) divination. That is, science endeavours to produce laws which allow us "to infer the unobserved parts of a causal process from those that have been observed" (ibid. 3). We can know something of the future by knowing the laws whereby things operate.[277] However, many things are not covered by causal laws, and scientific prediction of them is impossible.[278] In these areas divination can operate, without threat of contradiction, to provide knowledge of the future, for unlike science it does not state that the portent which predicts the king's death is the *cause* of the king's death, but only the *sign* for it.[279] From the diviner's point of view those areas where science does not offer an explanation are still places in which the diviner's deity (or equivalent) can leave a message. "And to understand and believe messages about the future is altogether different from coming to know of the future by reasoning from present causes to future effects" (Denyer, loc. cit. 5).

Recognising omens as signs accounts for many of the anomalies of divination which tend to concern the scientist. The job of the diviners is to interpret the signs, much as would grammarians working on a text. There is likely to be controversy over the reading of a sign – several readings may be possible, as with some words. Different prognoses might emerge from the same apodosis. Readings may vary from culture to culture and over time, just as languages do. If it is the entire universe upon which the gods can write their messages, which

275 E.g. Bottéro (1974) 149f, Larsen (1987) 211f, Huber (1987).

276 Popper (1959) 107; there are only "interpretations in the light of theories".

277 There is a type of science which allows us to predict the future movement of Jupiter on the basis of the "laws" of gravity, for example, but there is also a type of science which predicts that about 15 people will die each day on the roads in the UK, based not on laws, but only on statistics and records. In some respects it is this latter type of science which characterises the MAATs, and the like.

278 There are the questions to which science has not *yet* provided answers, but to which it may do one day. Secondly, there are the parts of the universe which science already tells us are *structurally* unpredictable, or knowable only to certain levels of accuracy (e.g. delineated by chaos theory and the uncertainty principle).

279 Sign: *ittum* in Akkadian, giskim in Sumerian – also *ṣaddum* is used, written ^{giš}an.ti.bal in Sumerian.

is generally recognised to be the case in Mesopotamia,[280] then it becomes easily possible to read contradictory messages about the future in rapid succession. These must be reconciled much as must contradictory passages in a text. It is possible even that the gods might wish to deceive the diviners for the good of the people, etc. In order not to be deceived it is probable that one should be god-fearing and as expert as possible. It was precisely these two traits which the Scholars of the NA court celebrated in themselves. See Ch.5.2.

Divination provides (potential) access to knowledge about the universe not provided by science. Denyer writes, loc. cit. p9, of Stoic divination that it was "judged (by Cicero) by the standards of natural science, whereas in fact it was put forward to go alongside lexicography and literary criticism as a branch of applied semiotics, concerned with understanding the utterances of the gods." This accurately describes Mesopotamian celestial divination and the way in which it was used by the Scholars, as we have seen in Ch.2.1.2 and will see further below.

So much for the assumption that an underlying causal link was believed by the compilers of EAE to connect a celestial event and terrestrial prognosis – what of the notion that the connections were made through the observation of *actual* happenings on earth and in the sky? As far as Larsen (1987) 212 is concerned:

> "(in) the omen tradition the relationship between protasis and apodosis is based on empiricism: once it had been observed that when the gall bladder of a sacrificial lamb was sharp like a lance, the king made an incursion in-to the land of the enemy – this relationship is *known* and can be registered for future reference"

I argue, however, that by the time celestial omens appear in the cuneiform record they are already largely dependent on assumptions concerning the *categorisation* of phenomena, on a *code* which associates days with countries, for example, and on *rules* which regulate the possible relationships between omens, apodoses, and protases. I suggest that by the time they are first attested they are already manifestly a *literate* creation in which the passive observation of simultaneous events in the sky and on earth plays only a small part.

An eclipse, for example, might always have inspired awe. Similarly, folklore such as "red sky at night; shepherd's delight" may have been discovered empirically. Divination of this form probably extends back into the mists of pre-literate thinking and (probably) beyond the reach of useful analysis. While there is a clear distinction to be drawn on the level of causality between signs such as "clouds mean rain" and those such as "a road sign means a bridge ahead" (Denyer, loc. cit. 5) since the road sign manifestly does not cause the bridge, there is also a distinction at the level of empiricism. Car drivers did not come to an understanding of the meanings of road signs through years of noticing them and then crossing bridges, say. They, or their representatives, *constructed* the signs in order to make the hazards and pleasures of unknown roads interpretable to new users. Likewise with the celestial diviners – they constructed signs so as to make the behaviour of the heavens *interpretable* to themselves and to other experts. Causal knowledge, such as that about clouds and rain, was only one part of the background information brought to bear on the creation of this system of divination – on this *encoding* of the heavens. Other information included the prevailing theology, literature, geo-political state, and so forth. Indeed, I suggest that only those celestial phenomena that co-occurred with certain terrestrial events that were felt to *corroborate*

[280] E.g. Reiner (1991) 316.

the already existing interpretations of the heavenly events were recorded in the divinatory series.[281] Of course, an infinite number of terrestrial events happen simultaneously with any given celestial occurrence.

While this is not the place to argue against the empirical origins of *non-celestial* omens, I note that the protasis cited by Larsen above: "if a gall-bladder is sharp like a lance" can readily be seen to be linked with the apodosis "the king will make an incursion into the land of the enemy" through the connection of potential battles with a feature shaped like a weapon of war. It is as likely that this omen was invented, as that it recorded an actual incursion and the shape of a real gall-bladder.

To summarise, I propose that cuneiform celestial divination was an invention, a deliberate encoding of the sky, *justified* by the assumption that the powers which manipulated the heavens would be so good as to leave messages there concerning the future. No assumption of causality connected the heavenly and earthly domains and prolonged observation played little or no part in assigning celestial phenomena to terrestrial ones. Instead, I argue that not only the categorisation of celestial phenomena, but the establishment of a simple code and a series of rules, which enabled them to be interpreted, had taken place before the writing down of the first celestial omens took place. Some of these *premises* must, to a large extent, be understood to be *given* – or in other words recognised that they derive from an oral background, or are "traditional". This applies particularly to some of the categories and to some of the code. It is, however, not without relevance that omens in Mesopotamia do not appear until a millennium after the invention of writing. When they do first appear some *already* demonstrate the effects of their *literate* production. That is, even the earliest attested omens use the rules of listing and word-play to be discussed in §3.2.1.

Importantly the scribes themselves also resort to the metaphors of writing, drawing/designing, and measuring when describing the basis of celestial divination. For example, in *KAR* 307 (= SAA3 39):33 it is written:

> "He (Bēl) drew/designed (*e-ṣir*) the constellations of the gods (on the lower heavens)"

and in the incipits to EAE itself Anu, Ellil and Ea are said to have designed the constellations and measured the year thereby establishing the signs.[282] The term *šiṭir burūmê*, "writing on the night sky", is found in SAA3 1: 21[283] and *šiṭir šamê*, "writing on the sky", on a MB entitlement narû.[284] The metaphor of writing is also commonly found in the context of extispicy.[285] More on this in Ch.5.1.2, where I argue that despite such examples spanning the centuries, because of the continuity in the *premises* which underpin the EAE Paradigm, they all describe the same attitude towards celestial divination. If such natural events were thought of by their Mesopotamian interpreters as signs akin to those of writing or designs, is it any wonder that they should have applied the learned techniques of literary interpreta-

281 As Koch-Westenholz notes (op.cit. 19), this would account for the "historical" apodoses, and yet is "the exact reverse of the 'empiricism' hypothesis!"

282 Translated in App.1 §21. Horowitz (1998) 14 also gives some parallels using *eṣēru*.

283 See Horowitz (loc. cit.) 226, also *CAD* under *burūmû*, Gadd (1948) 93.

284 A "kudurru" see App.1 §18. References in *ABCD* n54, and for *lumāšu* writing "constellation script" or "astrolglyphs" see Finkel & Reade (1996). Finkel & Reade loc. cit. 257f suggest reading *lumāšu* as "twin-image", but its association with "constellation" is well founded – see *CAD*.

285 *CAD* Š/2 231e) & 239c)

tion to them? The use of these metaphors alone should warn us to steer clear of seeking an empirical origin of that same divination they purport to account for.[286]

3.1.2 Period Schemes

These are those parts of the cuneiform astrological-astronomical texts that have, for the most part, been quoted out of context in an effort to demonstrate the existence of a predictive astronomy before the late period (see n267). They describe "ideal" periods for celestial phenomena based on the nearest "round" numbers (of years, months, days, or double-hours) to the true periods. However, these did not represent inaccurate assessments of the true periods for recurring celestial events, but had divinatory purposes. I propose that their aim was to provide a means of judging when a celestial body was behaving according to the ideal, and when it was not. The evidence is that the former was considered propitious, the latter not. The ideal period was a category *which permitted the times of events to be interpreted. The interpretations given to correspondence and non-correspondence with the ideal formed part of the* code, *and the means by which the ideal periods were elaborated into schemes which modelled other celestial phenomena were made possible through the application of* rules. *This is a new description of these kinds of text, and is important in the definition of the EAE Paradigm proposed here.*

A period of 360 days, comprising 12 months of 30 days each, was assigned by the Mesopotamians to the year in days and months at least by the third millennium BC. This period was used extensively in administrative circles since it simplified those transactions with temporal components.[287] By the OB period, at least, a ratio of 2: 1 for the longest night to the shortest accompanied the 360-day year. The longest night was located on the 15th of month IX, the shortest on the 15th of month III. The equinoxes were dated to the 15th of months XII and VI. This complete scheme I entitle the "*ideal year*". See App.1 §§8 & 11.

Also in the third millennium the *eššešu* and other lunar "cultic" festivals were celebrated on certain days of the lunar cycle which were related to those times when the Moon was *ideally* new, half-waxed, full, half-waned, about to disappear and absent (App.1 §3). This scheme is known here as the "*ideal month*". It is attested, for example, in the second "divi-

[286] The comparison with Derrida's (1981) discussion of Plato's views on the priority of orality over writing in the *Phaedrus* is deliberate. This search for origins Derrida calls logocentrism, which he argues must always breakdown under textual complications thrown up by the search itself. For example, Plato in looking to justify the priority of speech, is forced to resort to the metaphor of writing in order to explain this stance. Speech is "good writing", truth is "writing on the soul" etc. The warning not to seek origins, is well heeded, I believe, for it relies on the assumption that we can think as did the first diviners, that we can relate to them mind to mind. Based on our suppositions as to the origins of divination, for example, we seek to fit the written evidence into our scheme. This is what Larsen op.cit. has done, but there is no way of knowing, or testing, if his chronological ordering of the liver omens from Mari (op.cit. 212f) is valid. Nevertheless, that the compilers of the Mari models *themselves* considered that liver divination was empirically derived is perhaps suggested by such phrases as: "when (*inūmi*) event X took place, it (the liver) looked like this (*kīam iššakin*)" (op.cit. n30). I have located no evidence for such logocentrism in celestial divination.

[287] For details of the earliest attestations of the 360-day, 12-month "administrative year" see Englund (1988). For a study of the rôle the 360-day year played in the development of units used for measuring celestial distances and times see Brown *CAJ* forthcoming. An actual Mesopotamian year, sometimes called the "cultic" year, was made up of 12 lunations, each lasting either 29 or 30 days each. Since this fell short of a solar year, every three years or so a 13-month year was required to keep the lunar year and solar year more or less synchronised. See below.

sion" *pirsu* of i.NAM.giš.ḫur.an.ki.a. Lines IIi1–9 (K2164+[288]) describe critical days in the month including the 1st, the 7th, the 14th, 15th, 21st, 27th, and 28th. In *Iqqur īpuš* §67: 1 and §68: 1 omens concerned with the Moon's visibility on the 14th, 15th and the 30th are found.

In EAE 14, a scheme describing the ideal periods of the Moon's visibility and invisibility at new Moon and at mid-month are presented. It is based on the "ideal year" and on a premise of the "ideal month" that on the 15th the Moon is visible all night. For nights n: 1–14 the length of the Moon's visibility after Sunset is thus calculated to be n/15 multiplied by the length of the night on the 15th derived from the "ideal year" scheme. For nights n: 16–30, the length of the Moon's invisibility after Sunset is (n-15)/15 multiplied by the length of the night of the 15th. The constants "40" and "3 45" used in this EAE 14 scheme[289] are also attested in OB coefficient lists (see App.1 §8) indicating that this, the "*ideal lunar visibility/invisibility interval*" scheme, is also of OB origin.

In Mul.Apin IIii43–iii 15, and in i.NAM.giš.ḫur.an.ki.a II r.1–24, related schemes are found. They are identical to each other and the same as the EAE 14 scheme except for three subtle changes. Firstly, the dates of the equinoxes and solstices have moved forward by one month to the 15th of *nisannu* (I).[290] Secondly, *ideal* acronychal rising (the last day when the Moon is still visible before Sunset) of the Moon has been located on the 14th, and not on the 15th. The 15th is thus *ideally* the first day the Moon is visible in the morning before Sunrise, hence the reference to both these days in i.NAM.giš.ḫur.an.ki.a IIi1–9, noted above. Thirdly, the length of the lunar visibility period at the beginning of the month has been made a proportion of the length of the night at the beginning of the month, and not of the length of night in the middle of the month, as in EAE 14. However, the Mul.Apin and i.NAM.giš.ḫur.an.ki.a scheme is *not* a more accurate reproduction of actual lunar behaviour. It was no more empirically based than was the EAE 14 scheme. Both were *elaborations* based on "ideal years" and "ideal months". It is inconceivable that they were the best estimates of the periods of these celestial phenomena made by the Mesopotamians. It would not have taken long to recognise that about half of all months lasted only 29 days, for example! The numbers 360, 30 and 2: 1 are simple, or "round", particularly in the prevailing number-base used for calculations – base 60. Both lunar visibility/invisibility schemes were simply extensions of these simple numbers – ideal *hypothetical deductions* whose purpose was not to reproduce reality, but to widen the applicability of the EAE Paradigm. More on the evidence for this in §3.2.2.

EAE 63, the "Venus tablet of Ammiṣaduqa", which apparently includes observational material from the OB period, also includes in section II a scheme for the periods of time for

---

288 Livingstone *MMEW* 22f.

289 In the Nippur tradition of EAE 14, 3;45 UŠ is the length of first lunar visibility derived from a slight modification of the simple rule outlined above. See Al-Rawi & George (1991) Table A p55f.

290 This is not particularly significant. It does appear from limited evidence that in the OB period the vernal equinox was conventionally placed by the Scholars in month XII, and that by the NA/NB periods it was commonly located in month I. It is possible that this change was a consequence of the gradual slipping apart of the seasons and the stars known as the "precession of the equinoxes", but it is equally possible that in the later period it was only a new convention which moved the date. Because 12 lunar months are 11 days shorter than a solar year, even if in the OB period the calendar was organised such that the vernal equinox *mostly* occurred in month XII, approximately one year in three it would still occur in month I. It is only in the "ideal year" that the date of the equinox is fixed to one month. It is entirely probable that both conventions existed side by side in the OB period. Certainly, they existed simultaneously in the later period given the Scholars' use of both EAE 14 and Mul.Apin in the late NA period. The use of a vernal equinox in month XII might suggest an OB origin, but the use of the vernal equinox in month I does not preclude one.

which the planet was visible and invisible (see App.1 §9). This scheme assigned 8 months and 5 days to the planet's visibility in the east and the west, 7 days to inferior conjunction, and 3 months to superior conjunction. This ideal scheme should be compared with the more accurate description in Ch.2.2.1, here. It remains unclear whether this period scheme for Venus dates from the same time as the observations in EAE 63 section I, though the name "Ninsianna" *is* used for Venus in both section II and in the section believed to contain observations from the reign of the OB king Ammiṣaduqa. It is known here as the "*ideal Venus*" scheme.

The evidence concerning the so-called "astrolabe" is outlined in detail in App.1 §§13, 16 and 17. I suggest there that it was an OB (or early MB) creation. Essentially the astrolabe was a scheme in which three stars, one lying in each star-path,[291] were meant *ideally* to rise heliacally in each of the 12 months of the "ideal year". In some examples of the genre, numbers corresponding to the lengths of the watches of the night accompanied the star names, and these numbers indicated that a 2: 1 ratio between the longest and the shortest night was used in the scheme. The "*ideal astrolabe*" was thus underpinned by the "ideal year". Significantly, the order of constellations and the dates of their ideal first appearances reappear in EAE 51, in commentaries on EAE 50, and in Mul.Apin Iii36 – iii12, once more attesting the interconnectedness of these texts. Horowitz (1998) 162–5 argues that the astrolabes fulfilled the rôle of a sidereal calendar, with the heliacal rising of some star marked by a *date* in the lunar calendar.[292] No days, only months, are noted in the astrolabes, however, and even in Mul.Apin Iii36–iii12 the days of heliacal rising are either the first of the month or multiples of five, more suggestive of an invented scheme than the record of particular observations. While certain seasonal events, harvesting and so forth, may have been marked by certain stars (App. 1 §1), I think it highly unlikely that the astrolabes served this "astronomical" purpose, though the residue of certain traditional seasonal-stellar associations may have filtered into them.[293] They were, instead, learned elaborations based only very loosely on observational reality with regard to the heavens, whose purpose was *not* to regulate the calendar, but to permit celestial diviners to *interpret* the occasion of a star's first appearance as good-boding if it corresponded with the scheme and ill-boding if it did not. More on this in §3.2.2.

An "*ideal seasonal hour*" scheme appears for the first time in late NA period texts (App.1 §31). It was very simply related to the ideal year and did not represent an improvement in the accuracy with which times were recorded. Seasonal hours are not attested in EAE to my knowledge, but the published fragment has recently been joined to another tablet which contains part of EAE 14. Seasonal hours were presumably not thought to have been particularly different in purpose to that of EAE 14. They are perhaps an example of those post-OB innovations, such as moving the vernal equinox to I 15, which still relied on the same basic premises found in EAE itself.

Mul.Apin (App.1 §30) is the best known cuneiform series in which period schemes are attested. Many of these appear at first sight to be substantial improvements on the schemes

---

291 The paths of Anu, Ellil and Ea – see *BPO*2 17–18 and Horowitz (1998) 252f.

292 Op. cit. 164 "it is probable that the earliest "Astrolabes" were intended, in part at least, to help farmers determine the optimum dates for farming activities." I imagine that the farmers did not turn to the literate temple and palace employees to determine when and when not to sow, since clearly they had managed for millennia without such help.

293 E.g. in *Astrolabe B* month II is said to be the month of the "turning of the soil" and was marked by the Pleiades. Was this a traditional association between the stars and an agricultural event?

just described, and the text has usually been considered "astronomical" – that is, different from EAE, and with a primary aim of regulating the luni-solar year (e.g. Chadwick, 1992, 18). I indicate below, however, that most of the period schemes attested in Mul.Apin are not substantially different from those attested in EAE, and that Mul.Apin contains material which is substantially older than the date assigned to its composition by Hunger & Pingree (1989). I suggest that it does *not* represent a significant improvement on the schemes of known OB date, and that Mul.Apin's aims are in no way distinct from those of EAE. Mul.Apin, as with all the pre-750 BC period schemes, falls well within the remit of the EAE Paradigm.

In its very first section (Ii1–ii35) Mul.Apin lists many more stars in each of the three star paths than are found in the astrolabes, and the planets are more clearly distinguished. However, mere "star ordering" is not "astronomy", so far as the modern usage of the term implies, regardless of the word's etymology (see n15). For that, the prediction of celestial phenomena must be intended.[294]

Many of the stars appear to be out of order in this list. This could have come about because the late copies of Mul.Apin, the only ones we have, are in some way "corrupt".[295] However, it may also be because the star lists were never intended accurately to reflect reality. Pingree suggests (*Mul.Apin* 139) that Jupiter was located at the end of the Ellil stars because of its "association" with Marduk and Nēbiru – a divinatory reason. On the other hand he argues that the four other planets were located at the end of the group of Anu-stars because the latter "lies more or less in the middle between the northern and southern extremes of the ecliptic". This observational reason is plausible, but does Jupiter's different location not demonstrate the precedence of divinatory thinking over "astronomical"? In general, Pingree's explanations for the schemes in Mul.Apin emphasise that which is closest to observational reality, and dismiss as corrupt those aspects which do not correspond with reality.[296] He wishes to interpret the text as an "astronomical compendium". Here it is interpreted as a text of the EAE Paradigm.

The dates of the heliacal rising of some stars are given in the following section, Iii36–iii12. I commented on these lines above in regard of the "ideal astrolabe". The dates given in Mul.Apin for these first appearances are not accurate. They were not the record of a series of observations, but were produced "artificially" while corresponding very broadly to reality. They could *not* have been used for precise prediction.

In Iiii13–33 a list of simultaneously rising and setting stars is given. Undoubtedly, an observational component was involved in the construction of this scheme. However, at the same time Mul.Apin includes the ideal proposition that a star rising heliacally in month n will rise acronychally in month n+6 (Iii42 – Iiii8). I call this the "*ideal acronychal rising*

---

294 I do not wish to press unduly for this definition of "astronomy", but I maintain that a difference in the understanding of the purpose of the material *was* intended by those modern students who described Mul.Apin and the like as "astronomical" and yet referred to EAE as "astrological" or "divinatory".

295 By which is meant that the original texts *did* correspond to reality, but that as a consequence of copying errors, misreadings, and lacunae within texts, the later versions no longer do. Arguing for corruption as an explanation for the cruxes within these texts is tantamount to arguing for a loss of wisdom from the time when the text matched observation to the later period when it did not. This will not be argued here. To argue for a "loss of wisdom" is to prejudge the intention of these texts, I believe, and reflects the desire of the contemporary student to have the ancient scribes share his or her particular interests as to the purpose of texts concerned with the sky – namely the accurate mapping of the heavens in order to make possible the accurate prediction of phenomena , or "astronomy".

296 Repeated in *BPO*3 p29 2.

date" scheme. It also equates a sidereal year with 12 months, much as do the astrolabes. The list in Iiii34–48 is entirely derived from the list of dates of heliacal rising (Iii36–iii12).

In Iiii49–50 Mul.Apin states that the stars move by 1 UŠ (relative to the Sun) each day. Given that an UŠ is 1/360th of a revolution, this scheme implies that one full sidereal rotation, or year, would take 360 days. It is the "ideal year", once again.

In Iiv1–30 the *ziqpu* stars and their dates of culminating are mentioned. The use of culminating stars may represent an innovation of Mul.Apin itself or of the late MB period,[297] but once more the list appears not fully to correspond with reality. Pingree remarks (loc. cit. 141–2): "Assuming our identifications are correct, one would conclude that in *ziqpu* star lists, as elsewhere, tradition often determines content rather than a strict adherence to observed fact." It is precisely the existence of this "tradition" in texts formerly considered "astronomical" that is of great interest. The fact that so many anomalies and astronomical errors occur, errors which could so easily have been remedied by observing the sky, suggests to me that the purpose of these texts was not accurately to record celestial phenomena, and certainly not to make them accurately predictable.

In Iiv31–IIi8 the ecliptic constellations are listed, and the seven planets that move through them are named. While the ecliptic constellations are not named together in texts thus far attested from the OB period, many of them are noted individually and it could hardly be considered "astronomical" that they were so listed in Mul.Apin. Since many EAE omens pertain to the locations of the planets in constellations, listing them in Mul.Apin clearly had a divinatory purpose *as well*, perhaps *alone*.

In IIi9–21 a model concerning the movement of the Sun is presented. According to this "*ideal solar movement*" scheme the Sun rises at its most northerly point at the ideal summer solstice of IV/15. 90 days later the Sun appears directly in the east. 90 days after this it rises at its most southerly point, and the scheme repeats after 360 days. The OB "ideal year" is invoked yet again, for 40 ninda are said to be the daily change in the Sun. This is 1/180th of the change in the length of the day between solstices (120 UŠ, 8 hours, or 4 mina by water clock units[298]) if and only if the longest day is *twice* the length of the shortest. It is possible that in this section of Mul.Apin an equation was being made between the solstices and the most northerly or southerly positions of the Sun in the morning. However, it is the ideal *dates* for these events[299] (a date which can vary by a month or more either side of the 15th of months IV or X) which are actually correlated to the rising position of the Sun. This is neither accurate with respect to the lunar year nor to the solar year, and could not have been used as a basis for accurate predictions. However, the scheme was perfect from the point of view of permitting Scholars to await the solstice, and then to derive significance from its date in relation to the ideal date of its occurrence.

In IIi22–24 of Mul.Apin it is remarked that the above scheme can be used to assess "how many days are in excess", by which is meant how many days over 12 lunar months the year(s) has lasted. This is an intercalation scheme. It suggests that the extent to which the

---

297 See Brown *CAJ* forthcoming on the antiquity of celestial timing and the possible rôle of *ziqpu* stars in the development of celestial distance units.

298 See Brown, Fermor & Walker (1999).

299 The text describes the rising of ^{mul}kak.si.sá at the solstice, so in theory the sidereal and equinoctial years were also being correlated. In fact in Iii42 several stars are said to rise on the ideal date of the summer solstice, which in reality rise on different dates. It is most likely that this constellation was related to the summer solstice simply because the ideal date for the appearance of *both* was IV/15.

ideal year (as exemplified by the "ideal solar movement" scheme) exceeded a real year was judged to an accuracy of a day. Apparently the author(s) of Mul.Apin were quite aware that the ideal schemes did not correspond with reality. These lines make it absolutely clear that the exponents of the wisdom of Mul.Apin were not making primitive, inaccurate, or naïve estimates of the periods of celestial phenomena, but were fully aware that a real year was not 360 days long, and so forth. Why they persisted in designating the length of the year by this number suggests to me that what was attractively "round" from an administrative point of had remained useful from a divinatory one.

In IIi44–67 schemes concerning the lengths of the appearances and disappearances of the superior and inferior planets are presented. Pingree loc. cit. 149 calls them "crude...when compared with the values found in *ACT*". This misrepresents their purpose, I suggest. A brief scrutiny of the numbers will reveal that they are unusually "round" (e.g. *Ju*: 1 year, 1y+20d, 1y+1m; *Sa*: 1y, 1y+20d; *Ma*: 1y, 1y+6m, 1y+10m, 2y, 2m, 3m+10d, 6m+20d; *Ve*: 9m, 1m, 1m+15d, 2m, 1d, 3d, 7d, 14d; *Me*: 7d, 14d, 21d, 1m, 1m+15d). Relationships exist between the various numbers – the first three periods for Mercury are multiples of 7. The Mars invisibility period of 6m+20d is double the period 3m+10d, just as the visibility period of 2y is double 1y, for example. Clearly, many of these numbers were not derived from observation, but from mathematical manipulation. Some, at best, are distantly related to observation. There is nothing in this section or elsewhere that suggests that these figures would ever have been used to predict the occurrence of celestial phenomena. Comparing these figures with those derived from later schemes whose authors knew that some phenomena could be predicted accurately, does not compare texts the purposes of which are the same. I suggest that these figures for planetary periods provided useful *divinatory* material, and were derived with techniques considered *legitimate* by diviners – learned techniques of number play (see §3.2.1). They were schemes of "*ideal planetary visibility/invisibility*" which, as with the other schemes, only broadly corresponded to reality. Their purpose was, I suggest, to provide the Scholars with figures against which real phenomena could be judged. It is worth remarking that the visibility and invisibility periods of Venus are "cruder" than those in EAE 63 (see above), and are partially repeated in EAE 64.[300] The same invisibility period for Saturn, and the same periods for Mercury are found in EAE 56[301].

In Mul.Apin IIGapA1–7 a scheme relating the star path into which the Sun rises with the months of the year and the seasons is provided. Yet again it depends on the ideal year and the ideal solar movement scheme.

IIGapA8–9 presents the "*Pleiaden-Schaltregel*"[302] for intercalation:

> "If the Moon is in conjunction with the Pleiades on the 1st of *nisannu* (I), there is no need to intercalate a month. If the Moon is in conjunction with the Pleiades on the 3rd of *nisannu*, this year is a leap year."

The Moon moves about 13° per day, so in essence this scheme implies that if the lunar calendar has fallen behind the stellar by some 26° it is time to add in an extra month. It is immensely imprecise. Pingree loc. cit. 152 notes that in order to make sense of it, "to be in

300 *BPO*3 p15 and 244f, K2346: 21–2, group F (approx. EAE 64).

301 Largement (1957) §XVIII–XIX.

302 So named by Schaumberger *SSB* Erg.III 340f.

conjunction with" (*šitqulu*) must really mean "to be the closest to". This is unsatisfactory, considering that several other terms are used to describe the mere proximity of the Moon to constellations, and *šitqulu* (Gt *šaqālu*) literally means "to be equally weighted" and is regularly used in the Reports and Letters to mean "opposition" or "conjunction". I suggest, therefore, that the scheme was not derived from observation, but was derived directly from the simple proposition found in Mul.Apin IIGapA10–ii17 that the luni-solar year was kept synchronised by adding one month every three years – which I term the "*ideal intercalation*" scheme. This is equivalent to saying that a month should be added when the lunar calendar has fallen behind the sidereal year by 30°, since one month is 1/12th of the (ideal) year during which time the stars' rising points move (ideally) 1/12th of 360°, or 30°. Given the Moon's daily movement of about 13°,[303] the best estimate for the day of the month after the 1st *after which* it was perceived to have moved about 30° is the 3rd. Fewer than two days and the Moon would have moved less then c. 26°, more than three and it would have moved too far. Only on the 3rd day after the 1st will the Moon have moved c.30°, according to this line of thinking, and hence the "Pleiaden-Schaltregel", I argue. We should not try to see in the "Pleiaden-Schaltregel" any more precise an attempt to regulate the calendar than the *addition of one month in every three years*. It is an ideal scheme derived by learned methods fully comparable to those which elaborated other aspects of the EAE Paradigm.[304] It is not a model of planetary movement based on empirical evidence, it is a divinatory device derived from a well known rule of thumb. Its divinatory purpose will be mentioned shortly.

In Mul.Apin IIii11–17 (where this rule of thumb is noted explicitly) it is remarked that the "correction for the year" is 10 days (i.e. one-third of an ideal month) making the ideal intercalated year 370 days long. Again this is wildly inaccurate, and could not possibly

303 Derived from $^{1}/_{30}{}^{th}$ of [360° (one month's revolution) + 30° (the additional movement of the earth in that same month)].

304 Hunger & Reiner (1975) published another intercalation scheme attested in texts from Nineveh. This scheme only stipulates the celestial configuration that shows that no intercalation is necessary, stating that if the configuration does not occur on the stipulated date "(the year) is left behind" *ezbet* (op.cit. p24). The configuration in question is the conjunction of the Moon and the Pleiades (igi-*šunūtima ištaqlu* "they are seen balanced with each other") on day 27–2n of month n, for n: 1–12. It implies that, according to the ideal, the Moon and Pleiades are in conjunction on day 3 of month XII, and on day 25 of month I, for example. Their analysis (op.cit. 26–28) shows that each line of this scheme implies a different date for the beginning of the year, *and* that the vernal equinox fell before the 1st of *nisannu* (I) (OB?). Also, the authors note the existence of text *ACh.* 2Supp.19: 22f where yet another intercalation rule is found. This latter states that the year does not require intercalation if the Moon and Pleiades are in conjunction on the 3rd of *nisannu*. This is in complete contradiction to the "Pleiaden-Schaltregel". Hunger & Reiner write: "The differences between the texts suggest that no conclusion about the time of origin of any one of these rules can be drawn from their contents. It seems as if there were different attempts to solve the problem of intercalation rules."

I suggest that the discrepancy between these attempts shows that their intention was never to regulate the luni-solar calendar accurately, but simply to provide ideal scenarios against which observed situations could be compared for ominous significance. I have shown how the "Pleiaden-Schaltregel" could have been derived from the most simple intercalation scheme of all, the adding of one month every three years. The Hunger & Reiner (1975) scheme could have been similarly derived. According to the "ideal year" scheme the location of the Moon on day m of month n is 30° behind the location of the Moon on day m of month n+1. On day m+2 (48 hrs later) it will have ideally advanced by c.26° and will continue to advance a little since it is visible for longer than it was on day m. By day m+3 it will have moved by c.39°. The scheme supposes, then, that the Moon will be at the same location in month n+1 on day m, as it was on day m+2 in month n. This is the entire basis of the scheme. It is not really an intercalation rule at all, but a neat pattern derived from equating the monthly movement of the Moon with twice its daily movement. If an equation between monthly and daily movement had to be made, the factor of two would be the best to choose, assuming that one had only whole numbers to choose from. This is the essential point. The scheme is based on simple "round" numbers, just as is the case in all the other ideal schemes.

have been believed to describe the length of a solar year accurately. It was, instead, the consequence of a mathematical elaboration on what had long since been recognised in Mesopotamia, that one additional month every three years kept the lunar calendar and the seasons pretty much aligned.[305] (The evidence is simply that because some at least of the oldest month names described seasonal activities, the lunar and solar calendars *must have been* kept synchronised – see App.1 §4.) This well-known astronomical fact was simply worked into a scheme in Mul.Apin which had a divinatory purpose – the "ideal year".

The final period schemes in Mul.Apin pertain to the time intervals for changes in the lengths of shadows at the solstices and equinoxes (IIii21–42) and the above mentioned "ideal lunar visibility/invisibility" scheme (IIii43–iii15). As Neugebauer *HAMA* 544–5 showed, the time intervals in the shadows-section were computed according to tables of reciprocals. They bear little relationship to observation and could not have been used to time anything even remotely accurately. The 360 days of the "ideal year" are implicit in all the accompanying calculations. Pingree, in *Mul.Apin* p153, suggests that since the ratios of the times for the occurrence of a shadow of one cubit in length at the solstices is 3: 2, this provides evidence that this, more accurate, ratio between the lengths of the longest and shortest nights was known to Mul.Apin's authors. This is unlikely, however, as the 2: 1 ratio is otherwise used throughout the series, and because it is also now apparent that what was believed to have been the earliest attestation of this ratio in i.NAM.giš.ḫur.an.ki.a II26–7, in fact turns out to be no more than the "correction for a day" quoted in Mul.Apin IIii14.[306] The 2: 1 ratio is thus the *only* ratio of maximum: minimum night lengths known in texts composed in the period before c. 750 BC. Despite certain attempts to account for the inaccuracies in this 2: 1 ratio in terms of the timing apparatus used,[307] it is apparent that in many, if not all the known examples, the ratio was indeed understood to be in terms of *time*.[308] The use of this very inaccurate ratio in Mul.Apin indicates that the series was not an "astronomical" compendium, but a divinatory collection of ideal schemes. It is no surprise, then, that from IIii16 to its end[309] (IIiv12) Mul.Apin was concerned only with celestial omens. Its purpose was not distinct from that of EAE.

The final text we will look at in this section is known as *The Babylonian Diviner's Manual* (App.1 §36). It includes a "period scheme" involving the length of the year and various indications of when intercalation should take place, but its main purpose was the averting of portended evil and the creating of new omen apodoses. Its editor, Oppenheim, dated its composition to the late NA period (idem, 1974, 209), though it is attested in many copies in

---

305 Since the average length of a lunation is 29 1/2 days, 12 months is about 354 days. One additional month every 36 means that 12 1/3 months last 364 days, close to the real length of a year, especially considering that in any given year the number of 29-day and 30-day months may not be the same. The figure of 364 UŠ in the *ziqpu* list AO6478 (App.1 §33) perhaps derives from this reasoning – see Horowitz (1994) 94, but also Koch (1996).

306 In i.NAM.giš.ḫur.an.ki.a II26 it states that 1,40 ud.da.zal-*e* u_4-*mu*, which is translated by Livingstone *MMEW* p25 as "1 2/3 of a longest day is a day" based on the apparent meaning of the Sumerian. This would have implied a ratio of the longest to the shortest day of 3: 2. However, line Iiii14 in Mul.Apin makes it clear that the text should now be read as "1,40, the correction for a day" which derives from the 10 "ideal intercalary days per year – the ideal *epact*" divided by 360. The earliest attestation of the 3: 2 ratio is in the late NA period in BM 36731 – see App.1 §38.

307 Most famously by Neugebauer (1947a). This desire to find amongst the Mesopotamians the same interests as we might have is typical of much that is said about the so-called "astronomical" texts written before the late NA period.

308 For details see Brown, Fermor & Walker (1999).

309 Assuming it to be only two tablets long.

both Babylonian and Assyrian scripts from Nineveh. In line 59 it states that at the start of the year the location of the constellation ^{mul}aš.gán is important for intercalation purposes. In some of the (OB-style) astrolabes ^{mul}aš.gán's ideal date of rising is in month I. Also, in line 60, the appearances of the Moon and Sun in months XII and VI are considered important. They are reminiscent of the equinoctial dates of the OB ideal year. I suggest, then, that the text might be substantially older than Oppenheim suggests and certainly alludes to much older schemes for determining when intercalation should take place. These schemes are those found in Mul.Apin, though in this latter text ^{mul}aš.gán is said to rise in month XI (Iiii10) and, as noted, the equinoxes are in months I and VII. The method for averting evil in *The Diviner's Manual* is perhaps older than the proposed date of Mul.Apin's redaction.[310]

The text describes an omen series derived from signs seen on earth, and one derived from signs in the sky, and goes on to say that the good and evil prognoses of the two series should correspond, because "the sky and earth are related" (*itḫuzu*). If evil has been portended on earth, and if on the same date an evil portent has occurred in the sky then the portent is confirmed. The text asks "when they ask you to save the city...how will you make (the evil consequences) bypass (them)?" The text continues:

> "This is the Namburbu (the means to avert evil):
>
> 12 are the months of the year, 360 are its days. Take the count of the new year in your hand, and look repeatedly for the days of lunar disappearance (*bibli*), the corrections (*uddazallê*) for the first appearances of the stars and the correspondence (*mitḫurti*) of the beginning of the year with ^{mul}aš.gán, the first appearances of the Sun and Moon in months XII and VI, the risings and first appearances of the Moon as observed each month. Observe the conjunction (*šitqulta)* of the Pleiades and the Moon, and this will give you the answer... establish the (length of) the year and complete its intercalation."[311]

A scheme very similar to that in Mul.Apin GapA10–ii6 for determining when a year should be intercalated is described. It is proposed as a way to avert the consequences of portended evil, assuming it was not used when the omens boded well. The text finishes with a hemerological table, listing the months beneath each of which are the words "favourable" še, or "not favourable" nu, corresponding to a variety of prognoses – "for the army to enter camp"; "for the army to give battle"; "to take booty", and so on. Different versions of the text distribute the "favourables" and "not favourables" differently, and in one version the three watches of the night are listed with še and nu listed beneath them.

Oppenheim writes (loc. cit. 206f): "a different divinatory message is introduced: the validity of an omen...depends on the month and the watch" arguing that the "systematic transfer of the time criterion to ominous signs...represents an original contribution by the author."

It appears to me that this text is describing how, through intercalation, it may be possible to change the date upon which a phenomenon happened, and thus avert the evil of that sign. That the prognostications of omens changed dramatically for identical phenomena observed in different months is well attested in the hemerologies, menologies and, of

310 Parpola *LAS* II n565 favours a late second millennium date and points to x100: r.8f, a Letter written by Akkullanu in which he states: "If a sign occurs in the sky and cannot be cancelled..." which is perhaps referring to *The Babylonian Diviner's Manual*, in which case it would show that the text was employed by the Scholars in their work.

311 My translation based on Oppenheim (loc. cit.) and Horowitz (1998) 151–2.

course, in EAE itself. Oppenheim provides a few examples from other divinatory genres in his n46. These methods for deciding on the occasion for intercalation, I have demonstrated, form part of the EAE Paradigm. Despite Oppenheim's: "the *namburbû* prescribed in our text consists in establishing the exact date of the event observed by means of sound astronomical observations and calculations," there is absolutely no way that the exact date, let alone the time, could have been calculated using the methods described in this text. What could have been done, however, is that an additional month could have been considered justified. This would presumably have been done retrospectively, allowing the diviner to say: "you may have thought it happened in month IV; in fact it happened in month V" and thereby changing the prognostication. In the Report 8338: 4 Ašaredu writes:

> "The lord of the kings will say: "the month is not (yet) finished, why do you write to me good or bad (boding omens)"."

The implication is that not until the month was over were omens sent to the king, and that at the month's end the prognostications could vary from good to bad. I suggest that this was precisely because of the possibilities afforded by intercalation for changing the prognostications in the manner just stated.

To summarise, *The Babylonian Diviner's Manual* shows one purpose to which those period schemes concerned with intercalation outlined above could have been put by the diviner in pursuance of his art. As I will show in §3.2.2 xix there is good evidence that all of the so-called period schemes described above served a divinatory and not an astronomical purpose. The reverse is true for all the remaining entries in App.1 under the column headed "period schemes". They all formed part of what I term the Prediction of Celestial Phenomena or PCP Paradigm and will be discussed in the next chapter.

3.1.3 Observational Texts

These are texts which appear only to record observations of celestial bodies or phenomena without *any direct evidence that their purpose was divinatory. As I will demonstrate, however, in many cases their use was probably close to that of the texts of the EAE Paradigm, as well.*

In App.1 §10 I note the surviving OB Sumerian star lists. I have been unable to establish any particular rationale behind the order in which the constellations and stars are listed therein. Their order is neither observational nor divinatory. I include them in this section simply because they provide clear evidence that naming certain heavenly bodies and grouping particular stars into constellations[312] was an activity undertaken by the "Sumerians", no doubt long before their efforts were recorded in the lexical tradition. "They" categorised the heavens and obviously they had to observe them to do this. This was the first step in making them interpretable. In the OB forerunner *OECT* 4 161[313] col.X 22 mul.udu.idim is found. There is no reason not to interpret this as "planet" and since $^{mul/d}$Ninsianna (Venus) is listed in line 19, it was doubtless referring to one of the other planets. I suggested in Ch.2.1.3 that the

[312] Constellations are not "obvious" constructions, as a look at the night sky will demonstrate. Particularly important are the "limbs" of the creatures perceived in the constellations, which sometimes include relatively faint stars.

[313] Civil & Reiner *MSL* XI p143.

evidence of the name associations is that Saturn and Mercury were the last two planets to be discovered.

The dub.mul.an "star tablet" mentioned in *Gudea's Cylinder A*, a text composed towards the end of the third millennium, is nowhere explicitly described as having a divinatory purpose (App.1 §3). However, that it is being consulted by Nisaba with regard to Gudea's rebuilding of Ningirsu's Eninnu temple suggests that it may have been used to find out whether or not the heavens foretold that the time for this endeavour was appropriate. This was part of the aim of celestial divination as can be seen from the many examples in which the Scholars inform their kings of the propitiousness or otherwise of the timing of certain events.[314] Equally, the "star tablet" may have been thought to have contained within it the plans or "designs" for the temple. When in Sennacherib's annals (*OIP* 2 94: 64) it is written:

> "Nineveh... whose plan had been designed since the beginning in the writing of the night sky",[315]

a similar idea is probably being referred to (see above §3.1.1). Either way, I suggest that the dub.mul.an in *Gudea Cylinder A* and other texts[316] attests a Sumerian perception of the heavens and its phenomena in decodable categories – the same encoding which underpins EAE.

In EAE 63 §I a series of observations of Venus's dates of rising and setting are recorded. They are noteworthy in that despite recording the days on which these events occurred they are structured as omens with apodoses based entirely on the *month* of heliacal rising (see App.3). They do provide evidence that these events were recorded *to the day* as early as the OB period. It is, however, not without importance that unlike in the case of the systematic recording of the dates of heliacal phenomena in the LB Diaries and other NMAATs, they constitute only a short record and were presented with prognostications. There is no evidence that any special use was made of the dates of heliacal setting or rising – these did not add to the interpretation put on Venus's behaviour, nor were they used in order to make the date of the planet's first and last appearances predictable to the day. As noted above, the scheme which accompanies these omens in EAE 63 §II was "ideal" and incapable of providing accurate predictions. Section I of EAE 63, in being unique amongst texts composed before c. 750 BC, shows only that despite the possibility of accurate observations being made, the scribes of EAE, Mul.Apin and so forth were *in general* not interested in recording them. They recorded the results of observations only in so far as they provided useful material for divination.

This applies equally to those several lists of stars in the astrolabes and in Mul.Apin described in §3.1.2. While the recorded order and dates of heliacal rising, the locating of the stars in the three star paths, and the noting of which stars culminate and the dates between such events were based on original observations, little effort was apparently made to eliminate the inconsistencies in them. "Traditional" associations were allowed to take precedence over observed reality in many cases and the yearly and monthly schemes into which these observational records were imbedded preventing any possibility of their being used to make accurate predictions. This in no way prevented them from being useful to diviners wishing to interpret the temporal component of a phenomenon, however, and it is for this reason that

---

[314] E.g. x044 & x048.

[315] *Ninua...ša ultu ullâ itti šiṭir burūmê iṣrassu eṣretma.*

[316] Horowitz (1998) 166–8.

despite these star lists etc. not being formulated in omen form, I still consider them to have been "divinatory" in aim and not "astronomical".

Although in the late NA period *ziqpu* or culminating stars (App.1 §33) were used to record the times of events with a certain degree of precision, it is probably not without significance that in Mul.Apin they are *not* accompanied by values for the time between their culminations. Only in the period after c.750 BC are they listed in this way, which suggests perhaps that only *then* were they adapted to an "astronomical" use.

HS 245 (App.1 §17) probably describes an actual series of stellar risings around mid-month near the autumnal equinox. Rochberg-Halton (1983) successfully demonstrated that the numbers in the text were derived from a mathematical exercise rather than from observation, but op.cit. 216–7 mentions that the *order* of the Moon and stars in the text does correspond with the order in which they appear mid-month, near the autumnal equinox. It seems probable then that an order originally derived from observation was transformed into a mathematical exercise. This is precisely how I believe the astrolabes and Mul.Apin-type star lists were formed – broadly realistic observational data squeezed into ideal schemes of divinatory purpose by certain learned numerical methods.

For my comments on entitlement narûs (kudurrus) providing records of celestial observations, see App.1§18.

A variety of observational records from the 8th and 7th centuries BC are attested for which no interpretation is given of the observations' significance – see App.1 §§22 & 32. Some of these texts are found in the Scholars' Reports. In some cases it has to be assumed that only the record of the observation was sent because its interpretation was obvious or well known. In other cases the recording of more information than was considered ominous (see §2.2.2;16) was, I suggest, indicative of the emergence of the PCP Paradigm.

3.1.4 Other Related Material

I discuss in Ch.5.1.2 and 5.1.3 the background against which EAE developed. It is well-known that Greek astronomies adopted in varying amounts notions as to what constituted the universe. The same is true of Mesopotamian astronomy-astrology, some of the evidence for which appears in literary and other texts. However, while some literary and religious compositions suggest ideas as to the design of the heavens that underpin the EAE Paradigm, it is also the case that EAE itself influenced literary material from the OB period on.

The Prayer to the Gods of the Night lists ten stars or constellations that anticipate the 10-star astrolabe tradition (App.1 §26) and refers to Ištar, Sîn, Šamaš, and Adad bringing judgement, which reflects the four divisions of EAE itself. The ideal year and ideal month appear in a number of OB literary texts (see App.1 §11) which reveal the widespread acceptance of those celestial categorisations that underpin EAE.

Enūma Eliš, whose composition in the late OB or MB period is believed to have marked the rise to prominence of Marduk in Babylonia, strongly reflects the concerns of celestial divination. It remarks on the design of the heavens and its yearly patterns, which are identical to those of the ideal schemes of EAE and Mul.Apin. It comments on the signs inscribed in the sky and indulges in the kind of word play found in celestial omen creation (see §3.2.1) and in the explanations of omens attested in the so-called commentary texts (App.1 §28). For example, Bottéro (1977) shows how the signs used to make up some of the names of

Marduk in the final sections of the epic are reflected in the epithets attached to those names. E.g. in Tablet VII: 1, the following is found:

> ^{d}Asare.(ri) *šārik mērešti ša eṣrāta ukinnū*
> "Asare, giver of farmland, who fixes the designs."

Bottéro (loc. cit. 5f) notices that the Sumerograms RU can be read in Akkadian as *šarāku*, SAR as *mēreštu*, A as *eṣrātu*, and RA_2 as *kânu*. The epithet can be derived from breaking down the name into its syllables and treating them as Sumerograms, just as we saw in the case of some of the planet names in Ch.2.1.1. In the case of RA_2 a homophone has been used to derive a suitable epithet in Akkadian. Many other examples are listed by Bottéro, including cases where two or more words in the Akkadian epithet are derived from one Sumerogram. For further details on the celestial interests of *Enūma Eliš* see App.1 §19. For those SB texts which share similar concerns see App.1 §20, and for the influence of celestial divination on late NA and NB literary compositions see App.1 §24. Just as the presence of copies and translations of EAE in areas neighbouring Mesopotamia indicates, so does the existence of EAE-type material in literary texts composed within Mesopotamia demonstrate – that celestial divination was a profoundly important activity in which the literate were engaged throughout the second and early first millennia BC.

To summarise, I hope to have demonstrated in §§3.1.1–4 the longevity and pervasiveness of the EAE Paradigm. As such, its definition has not yet been attempted, though this is a relatively simple matter now:

All of the cuneiform astrological-astronomical texts listed in App.1 and authored before c.750 BC fall either partially or fully within the EAE Paradigm, I argue. In many cases their contents are actually found repeated in certain parts of the great celestial divination series. This applies to some omens in Mul.Apin, the Diviner's Manual, and *Iqqur îpuš*.[317] The astrolabes are reflected in EAE 50[318] and the Mul.Apin and i.NAM.giš.ḫur.an.ki.a schemes in EAE 14, 56 and 63. None of these schemes was presented in a form in which it was possible to model the behaviour of the celestial bodies with any precision. It is also the case that the ideal schemes did *not* incorporate the "Mesopotamians"' best estimates for the periods between particular celestial phenomena. All used whole, round numbers which made their adaptation to the modelling of other phenomena simple, as was the case with the "ideal lunar visibility/invisibility scheme" which was derived from the "ideal year" and "ideal month", or the "Pleiaden-Schaltregel" which was derived from the "ideal intercalation" scheme. Calculating the *resultant* ideal lunar visibility times, say, was mathematically straightforward and despite the almost total lack of observational content in their creation they were obviously considered to have been important. That these schemes are found in many different text types suggests that they may have been thought to reflect something fundamental about the underlying or original state of the universe (more on this in Ch.5.1.3), but their repeated use in EAE shows that they were also or primarily useful to the diviners. In §3.2.2 xix I outline the evidence which indicates that they were useful because they permitted the

[317] Hunger+Pingree op.cit. 9, Labat Calendrier 170 n6.
[318] E.g. *BPO2* III: 28 "In month IV, the Arrow, the Twins, and the Heroic rise heliacally". This is a quote from *Astrolabe B* C 19, as noted by Reiner & Pingree op.cit. 43.

diviners to compare the time or interval of a celestial event with that predicted by the ideal models and interpret it accordingly.

The presence in other works of the early second millennium BC of the categories and notion of encoding which underpin the EAE Paradigm indicates that by the time celestial omens appeared in cuneiform a long history already lay behind them. Certain celestial phenomena had long been considered signs, and being such they were *read*. The EAE Paradigm exemplified that decoding *par excellence*, and involved the use of those techniques of exegesis which accompanied other literate activities – play on both text and numbers. This play had begun at least by the OB period. As I indicate in App.1 §§7 & 21 a proto-EAE existed in the OB period, by which I mean a precursor to the final series, but still entitled EAE.

Incidentally, the EAE Paradigm suits well those aspects of a Paradigm defined in Kuhn (1962) and refined in Masterman (1970). It was a "metaphysical Paradigm", for it assumed an underlying set of beliefs – namely that the universe was designed and full of signs sent by the gods, signs which could be read using the underlying premises of the categorisations, the code, and the rules. It was a "sociological Paradigm", in that it was used by an identifiable group of Scholars, in that its underlying premises pervaded other forms of writing, and in that it was successfully transferred abroad wholesale. Finally, it was an "artefact Paradigm" in that near-identical copies of the series were used in different cities, in that it assumed the usage of one type of water clock, one type of gnomon, one set of units etc., and in that the rules by which the arena of applicability of the Paradigm was widened were universally applied. This widening of applicability was undertaken in what might be termed the "*normal science*" of the EAE Paradigm, to borrow again from Kuhn that which was used to describe the activity in which the premises of a Paradigm are applied to more and more situations without any effort being expended to try and falsify the Paradigm itself.

While the truth or otherwise of Kuhn's model that science develops in uneven stages from Paradigm to Paradigm is not within my capabilities to judge, I make two comments that are perhaps of importance in the history of science: Firstly, it so happens that Mesopotamian celestial divination can be described as a Paradigm in much the same way as can, say, Newtonian mechanics, and this provides at least a start to a means of useful cross-cultural comparison. Secondly, as I will show in the next two chapters, a dramatic transition in the concerns of those writing on the heavenly bodies and their phenomena took place in the period immediately following c. 750 BC, which could also be described as a "revolution" in Paradigms in the manner of Kuhn. Again this provides for comparison with other so-called revolutions and will ensure, I hope, that Mesopotamian celestial achievements take their place in any future discussions as to the nature and purpose of science, which is after all one of the agendas lying behind this work.

3.2 Making the Heavens Interpretable with the EAE Paradigm

3.2.1 The Rules of Omen Invention

The following analyses the rules used by the NA and NB Scholars to create ideal schemes and omens. These rules fall into two main types, those concerned with numbers and those concerned with text.

In K4364+ = Zimmern *BBR* 24: 18 it is written:

> "... which are evaluated according to the explanatory word lists (*ṣâtu*) of Enūma Anu Ellil and the mathematical table (*arû*)[319]"

The text as a whole concerns the acquisition by the king Enmeduranki of various texts and arts of divination, and his subsequent teaching of them to the Scholars. Amongst those texts taught are EAE, *ṣâtu* and *arû*.

A *ṣâtu* is a text which contains readings of particular cuneiform signs used in the elaboration of the meaning of passages, particularly omen passages.[320] Texts which comment on parts of these omen series by analysing the signs used in the omens were also entitled *ṣâtu*s and they were preserved along with the official series in the late NA and NB libraries. The existence of *ṣâtu*s on EAE[321] shows that this form of "word-play" explanation was used by celestial diviners in the late NA period. As the passage quoted suggests, however, and as I will show below, this "word play" formed part of a wider "textual play" which was used from the very first time celestial divination was committed to writing in the form of EAE and the proto-series of the Paradigm.

An *arû* is also a term for a type of text, and is attested from the OB period on to mean a table of numbers produced as the result of some mathematical endeavour, often multiplication since *arû* also meant "product" in that sense. One at least of the MAATs (*ACT* No.135) was described as an *arû*, as were the LB water clock text BM 29371[322] and i.NAM.giš.ḫur.an.ki.a.[323] Lieberman argues, however, (1987) 188 that *arû* also described the kind of gematriac speculation noted above in Ch.2.1.2 (n205) and found in i.NAM.giš.ḫur.an.ki.a itself, as the following example shows:

Lines IIi1–9 describe the critical days in the "ideal month". In lines IIi11f these are justified by using different readings of the signs and multiplying:

11) ud.7.[kam *agâ ma-á*]*š-la* bar bà bà *za-a-zu* bà *pa-r*[*a-su*]
12) bà [*ba-an-t*]*u* bà *mi-šil* bà (30) *Sîn* (30) *mi-šil meš-*[*li*]
13) 30 [a.rá 30] 15 15 a.rá 4 60 60 ^{d}*a-nù im-bi inbu* (gurun)
14) [ud.13.kam)..] ud.12.kam 12 130 an.ta ki.ta ki.ta an.t[a]
15) *ta-*[*nam-b*]*i* ka.inim.ma *nēmeq* (nam.kù.zu) *a-re-e šu-a-tu*

Translated as follows (after Livingstone *MMEW*):

11) "The 7th day (the Moon is) a ha[lf crown], bar (half) bà (the sign 30), bà: (can mean) 'to div[ide]', bà: (can mean) 'to cut'.
12) bà: (can mean) 'to sh[are]' bà: (can mean) 'half', bà Sîn (is therefore) 'ha[lf] of half'.

319 *Ša* ki *ṣâti* ud an ^{d}en.líl *u* a.rá-*a šutābulu*. In his edition and publication of the duplicate K3357+ Lambert (1967) 132–3 reads the line as "'that with commentary'; *When Anu Enlil*; and how to make mathematical calculations". Lines 16–18 are an intrusion into the original text, and may have been added no earlier than the late NA period.

320 *Šumma ina ṣâti šumšu ana panīka* BAL *enû* BAL *nabalkutu* "if you have at your disposal (a reference to) its name in a *ṣâtu* (you will note that) BAL = *enû* and BAL = *nabalkutu*" *CT* 31 40 r.iii 12 quoted in *CAD* Ṣ 119.

321 "EAE, the official series of the gods, together with its *ṣâtu*," Rm.150: 11' quoted in *CAD* Ṣ 119.

322 Published in Brown, Fermor & Walker (1999).

323 *MMEW* 29 – the second *pirsu* "division" of the series is called a long *arû* of Nabû-zuqup-kēna and this is the section which contains both the table of "ideal lunar visibility and invisibility times" and the elaborations on the days of the Moon.

13) 'Half [multiplied by half]' is 15 (sixtieths). 15 (multiplied by) 4 is 60. 60 is Anu. He called the 'fruit'.
14) [The 13th day]: the 12th day. You call 12 130 reversed.
15) Words of wisdom of that mathematical tablet (*arû*)[324]"

In these lines the statement that the Moon on the 7th is a "half crown" permits a connection to be made between the sign for "half", which is 'bar', and the Sumerogram 'bà', which is the same sign as that used for *Sîn*, the Moon. The Sumerogram 'bà' can be read in Akkadian as "to divide, to cut, to share", or "half". The signs used to write a half Moon (on the 7th day) are therefore equivalent to the multiplication of a half by a half, which in base 60 is 0;30 times 0;30 which is 0;15. Hence, another of the Moon's significant days has been derived. Similarly, half times a half is a quarter, and 15 is a quarter of 60. 60 is a cuneiform sign for the god Anu,[325] and in myth it was he who called out the Moon naming it the "fruit" when it first appeared on day 1. Hence, by associating logograms through near homophony (bar and bà), through graphic play on bà and *Sîn* which share the same sign (cf. the example from *Enūma Eliš* quoted in the previous section), and through a reference to a myth, the author of this text has justified the significant lunar month days 1,7, and 15. In lines 14–15, days 12 and 13 account for each other, for 12 is written with a sign for 10 followed by two signs for the unit, where as 130 is written with two signs for the unit (now equal to 60) followed by the sign for 10. This particular bit of number metathesis[326] is described as *nēmeq arê*. The text continues in this vein, justifying the other significant days of the month.

However, does the *arû* in the colophon to this section of i.NAM.giš.ḫur.an.ki.a (n323) refer to such numerological-philological elaboration as line 15 suggests, or to the table of lunar visibility and invisibility times on the reverse, which does more closely parallel the tables of numbers in the MAAT and in the LB water clock text just cited, both also described as *arû*s? It perhaps refers to both, and should be understood to refer to "number play" more generally – that is to the mathematical elaborations exemplified by EAE 14 as well as to the doubling and inverting of numbers exemplified by the section on planet periods in Mul.Apin (Iiii44–67) and repeated in EAE 56 (see above §3.1.2). K4364+ quoted above indicates that for the compilers of that work, *arû*, whether it described a type of text or a particular skill, was closely connected to EAE and, since both types of "number play" are present in the series and typify the Paradigm, I will use the term to describe both. Like *ṣâtu*, *arû* was thought by these late NA Scholars (see n319) to have been part of the original "wisdom" of celestial divination.

I have described the close numerical manipulation present in i.NAM.giš.ḫur.an.ki and in parts of EAE 56 that utilise rules such as multiplying, gematria and metathesis. What of the more "mathematical" techniques employed in the ideal period schemes? Many of them model variations in celestial phenomena using piece-wise linear techniques – either step functions as in the OB daylight scheme in BM 17175+, or zigzag functions as in EAE 14.

324 Lieberman loc. cit. n305 translates this line as "it (is) a dictum of the skill of calculation".
325 This is an example of gematria – see n205.
326 Another well-known example of metathesis in the texts herein considered is found in Esarhaddon's inscriptions (Borger, 1956, 15, Ep.10), which accounts for the reduction in the period of Babylon's abandonment from 70 (sign for 60 followed by the sign for 10) years to 11 (sign for 10 followed by the sign for 60=1).

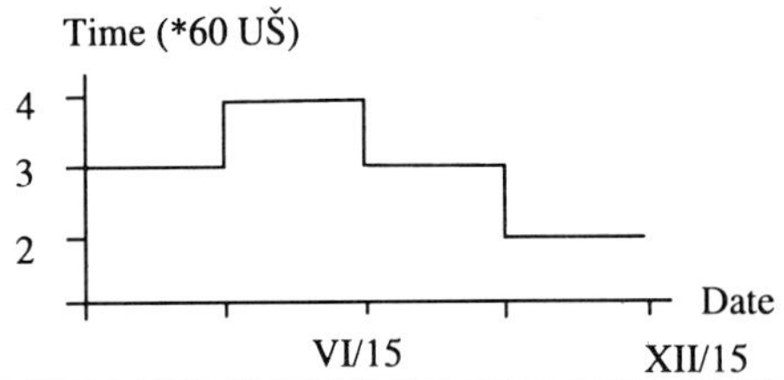

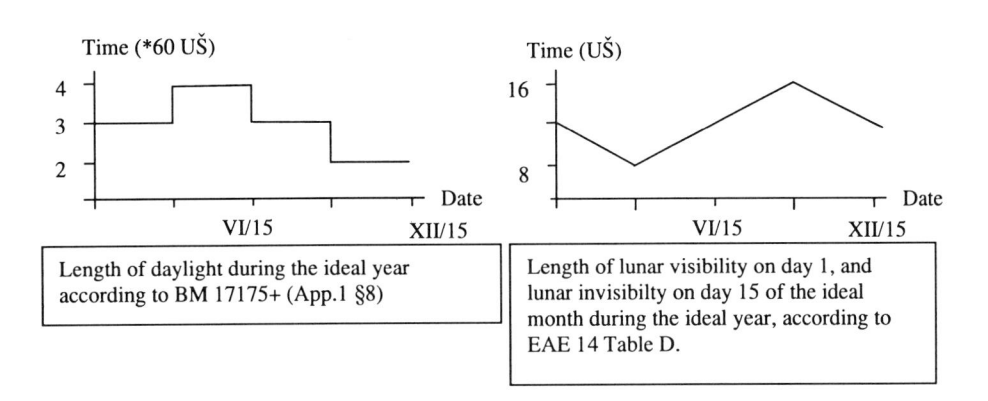

Length of daylight during the ideal year according to BM 17175+ (App.1 §8)

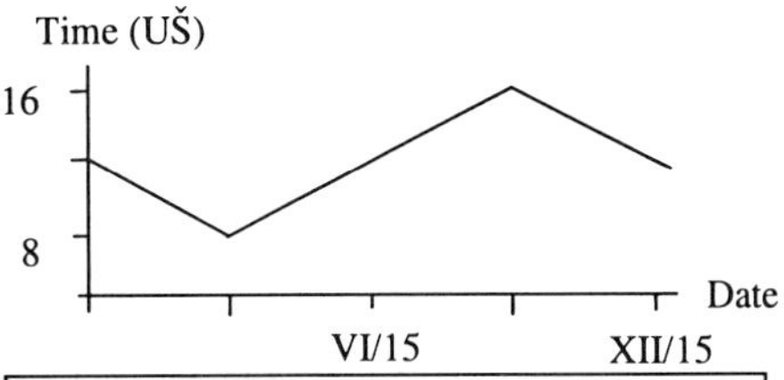

Length of lunar visibility on day 1, and lunar invisibilty on day 15 of the ideal month during the ideal year, according to EAE 14 Table D.

Tables A and B[327] of EAE 14 describe the durations of lunar visibility and invisibility during an ideal equinoctial month. In the so-called "Babylon variant" this is expressed as a straight-line function, such that on day 15 the Moon is visible all night – namely for 3 times 60 or 180 UŠ – and on day 1 it is visible for $1/15^{th}$ of that time – namely for 12 UŠ. On day 2 it is visible for 24 UŠ and by day 5 it is ideally visible for 60 UŠ, etc. However, in the "Nippur variant" of EAE 14, a "geometric gloss" is added to this straight-line scheme, such that the lunar visibility period on day 4 is said to be 30 UŠ (half that on day 5), on day 3 it equals 15 UŠ (half again), on day 2 it equals $7\frac{1}{2}$ UŠ and on day 1 it is ideally $3\frac{3}{4}$ UŠ, or 3;45 in sexagesimal. This is the value which appears in the OB coefficient lists (see App.1 §8 and n289) attesting to the antiquity of this mathematical embellishment. I stress that neither the additive (of 12 UŠ per day) nor the multiplicative (by a factor 2) variant correspond even remotely accurately to reality.

In addition to these piecewise linear and geometric elaborations, I showed above how the ideal "Pleiaden-Schaltregel" scheme in Mul.Apin and elsewhere derived from round number approximations to the relationship between the daily movement of the Moon and the monthly movement of its point of first appearance, in combination with the "ideal intercalation" scheme. I also noted how the shadow clock section of Mul.Apin was based on a table of reciprocals. *We* may perceive such "number play" to be different to the techniques of metathesis and gematria just outlined, but it is not clear that they were thought to be different by the compilers of EAE and its related texts. In HS 245, for example, what is little more than a mathematical exercise in "remarkable" irreducible numbers[328] (those whose division into 60 cannot be expressed as a terminating sexagesimal) was appended to a list of stars as if it *could* have described the distances between them without any apparent explanation as to why this might be. Even such a "reasonable" assumption as the Moon being visible all night on the 15th as is found in EAE 14 and elsewhere is missing. It seems most likely, then, that the term "*arû*" should have described the "number-play" techniques on *both* the obverse and reverse of the second division of i.NAM.giš.ḫur.an.ki.a, for both types of "number play" were legitimate means by which such texts could be elaborated.

The use of linear and geometric functions and reciprocal tables is characteristic of OB mathematics, and no doubt a strong relationship existed between the early mathematical

[327] Defined according to Al-Rawi and George (1991/2).

[328] See above §3.1.3, App.1 §17 and Høyrup (1993).

and divinatory texts – a relationship which I cannot explore further here. These mathematical tools along with numerological-philological techniques formed together what I term here the "number play" rules of the EAE Paradigm. These rules permitted the ideal models of celestial phenomena such as the "ideal year", the "ideal month" and the "ideal intercalation" scheme to be elaborated into such things as the "ideal lunar visibility/invisibility" scheme, the "Pleiaden-Schaltregel", parts of the "ideal astrolabe", the "ideal seasonal hour" scheme, and the "ideal solar movement" scheme. They thus made more and more phenomena amenable to interpretation – not just the length of the month, but the length of time for which the Moon was visible every day of the year, say. These rules in effect "created protases". Turning now to the textual part of divination – the omens – it will be seen that *both* protasis and apodosis invention was also undertaken through the rules that governed "textual play".

I have approached the study of the celestial omen corpus on Structuralist lines. This is not because I subscribe to the belief in fundamental underlying structures implied by this "philosophy",[329] for I accept that there will be inconsistencies in my description, but in order to make sense of the thousands of cuneiform omens. To this end I have found some of Lévi-Strauss's proposals very useful.[330] I will employ terminology used by the Structuralists, which is to be compared with that I developed in Ch.2.1.2. I take as the object of analysis the table of all possible permutations,[331] and only then compare them with the empirical evidence – the texts:

Both the protasis and the apodosis of an omen *could* have recorded real, that is observed,[332] phenomena, or have been invented. Omens would thus fall into four possible types:

(I) Observed (protasis): Observed (apodosis)
(II) Observed: Invented
(III) Invented: Invented
(IV) Invented: Observed

In verbal language a *metaphoric* relationship exists between similar parts of speech (nouns to nouns, verbs to verbs etc.), and a *syntagmatic* "sentence" relationship is found between different parts of speech (verbs to nouns etc.). Visualised as two axes, vertical relationships between elements classified in this scheme are metaphoric. Horizontal relationships are syntagmatic. With regard to the language of divination, to the omens which are partly, or fully invented, the relationships of the protases to apodoses are (potentially) syntagmatic, the re-

329 Leach (1970) 21 "Lévi-Strauss is distinguished among the intellectuals of his own country as the leading exponent of 'Structuralism', a word which has come to be used as if it denoted a whole new philosophy of life..."

330 Structuralism classifies thought, but particularly "primitive thought". The extent to which Mesopotamian celestial divination is the product of primitive thought is not of importance here. The classificatory method used by Lévi-Strauss, for example, has proved rewarding when applied to the "early literacy" which exemplifies the EAE Paradigm.

331 Lévi-Strauss (1964) 16.

332 By observed celestial phenomena, I mean those that were described in terms of the underlying categories, and by observed terrestrial phenomena, I mean those that in general *corroborated* the existing reading of those heavenly signs as determined by the code.

lationships of protases (or elements within those protases) to the collection of other protases, and of apodoses to other apodoses are (potentially) metaphoric.[333]

protasis — *syntagmatic* —	apodosis
|	|
metaphoric	*metaphoric*
|	|
protasis	apodosis

Invented apodoses and protases that have come about as a consequence of their mutual relationship are called here syntagmatic (A), those invented as a consequence of their relationships to other protases and apodoses are referred to here as metaphoric (B). The following omen types can therefore be postulated:

(I) Observed protasis: Observed apodosis

(IIA) Obs: Inv – apodosis is syntagmatically related to the observed protasis.
(IIB) Obs: Inv – apodosis is metaphorically related to the collection of other apodoses.

(IIIBA) Inv: Inv – protasis is related to the collection of protases, apodosis is related to it.
(IIIAB) Inv: Inv – apodosis is related to the collection of apodoses, the protasis is related to it.
(IIIBB) Inv: Inv – both protasis and apodosis are related to the collection of protases and apodoses.

(IVA) Inv: Obs – an observed apodosis is appended to a syntagmatically related invented protasis.
(IVB) Inv: Obs – an observed apodosis is appended to a protasis derived from the protases collection.

There are, of course, those omens for which the reasons why the protasis or apodosis or both were invented are not yet understood, or were entirely arbitrary. Any analysis of this kind must be aware of the difficulties caused by the fact that some allusions (syntagmatic connections) between apodoses and protases may be lost on modern translators, and that the collection of protases or apodoses which suggested the creation by analogy (metaphor) of the attested protasis or apodosis may be lost.

Some of the invented apodoses are used repeatedly in many celestial omens, and sometimes in other contexts. They are rather like epithets or formulae, and are known as "stock

333 Lévi-Strauss (1966) 149 describes this classificatory system with two axes which is so fundamental to Structuralism (also Leach, 1970, 42f, and Jakobson, 1990, 115f). It classifies "languages" in the broadest sense of the word, including non-verbal forms of communication such as the clothes people wear, the language of architecture etc. Some elements of these languages can be replaced by others that perform the same function. For example, in the language of clothing shoes can be replaced by boots, or by sandals, and so forth. The relationship between boots and sandals is said to be "metaphoric" in Structuralist terminology. Equally boots possess relationships to the other garments worn – to the trousers, the overcoat, to the bowler hat. These relationships depend on the rules of the language (e.g. boots are not normally worn with bowler hats). They depend on the rules of syntax – they are "syntagmatic".

apodoses".[334] Other apodoses were apparently invented specifically for an invented or observed protasis.

In Chapter 2.1.2 I described a syntagmatic connection (A) between an apodosis and its protasis as one deriving "from the words themselves (whether this be etymologically, ideographically, semantically, or graphically)" and a metaphoric connection (B) as "drawn from what I term "Listenwissenschaft" or "the technology of listing"". This Structuralist analysis corresponds to what were there described as "learned" associations between names, and it is entirely appropriate also to refer to invented apodoses and protases as "learned", for they too could only have been made by experts in the field. Using both terminologies a more descriptive account of the possible omen types is now possible:

(I) The protasis describes an actual celestial event and the apodosis records an actual, though not necessarily simultaneous (see §3.1.1), terrestrial occurrence. An eclipse, say, occurs and this was *known* to signal a royal death. Its details, and those of the death of a famous king are noted, adding colour and a sense of authenticity to the resultant omen.[335]
(II) Observed celestial events, recorded in protases, have appended to them invented, "learned" apodoses, generating new omens. These apodoses do *not* relate events on earth which occurred at any special time, and certainly not when the events in the protases were first observed. They have either been drawn from the stock of existing apodoses, or invented anew. Each may bear a "syntagmatic" relationship to its protasis (IIA), or a "metaphorical" relationship to the apodoses above and below it in the omen list (IIB).
(III) Alternatively, learned protases are generated as a consequence of Listenwissenschaft relationships with other protases (IIIB), or as a result of word play with an invented apodosis (IIIA). They can sometimes be spotted because they describe celestial events which do not ever occur. The apodoses to which they are then attached are either syntagmatically related to the protases (IIIBA and IIIAB), or only to the apodoses of the collection of omens (IIIBB).
(IV) Finally, an observed terrestrial event recorded in an apodosis has added to it an invented protasis constructed either through the technology of listing (IVB), or through a play on the words of the apodosis (IVA).

It is noteworthy that this last type of omen is based on the assumption that an important event in the terrestrial or human milieu *ought* to have a celestial marker and it was perhaps for such reasons that the heavens were encoded in the first place. This celestial marker might have been invented (IVA-B) or observed, producing a type (I) omen. Now that the structure of all possible omens has been outlined, the empirical evidence can be approached. All examples derive from the texts known to have been composed during the period c. 750–612 BC, though others will be referred to in passing:

---

[334] The definition is somewhat arbitrary. Here, an apodosis is regarded as "stock" if it is attested with a different protasis, *or* if it is entirely general in its prediction. For a further discussion of stock apodoses see the references cited in Leichty (1970) p4.

[335] This is how I would interpret the supposed Old Akkadian eclipse omens analysed in Huber (1987), though I accept that the death of a king nearest in time to an observed eclipse *may* have been the one recorded in the final omen.

1) Obs: Obs – Type I:[336]

> "If there is an eclipse in *simānu* (III) on the 14th, and the god in his eclipse becomes dark on the east side above, and clears to the west side below, the north wind rises during the evening watch and touches the middle watch; you observe his eclipse and keep the north wind in mind; thereby a decision is given for Ur: the king of Ur will experience famine; deaths will become many; as for the king of Ur, his son will wrong him, but Šamaš will catch the son who wronged his father, and he will die in the mourning place of his father, a son of the king who has not been named for kingship will seize the throne" (8004: 1f).

The protasis of this omen is so explicit it probably referred to an actual eclipse. It occurs in EAE 20 §III (*ABCD* 189), a tablet which records a series of lunar eclipses in great detail, each with a long accompanying apodosis – see also App.3 below. The apodosis of the omen appears to record a real historical event concerning the demise of the king of Ur. The detail with which it describes his death and his disloyal son suggests this.[337] Incidentally, 8004 can be dated to June 11th, 669 on the basis of other data contained in the Report. The eclipse in that year in that month actually took place in the morning (Parpola *LAS* II App.F, Steele & Stephenson, 1997/8). Presumably Issar-šumu-ereš felt that the protasis description in EAE was close enough to the eclipse he saw to merit sending the omen to Esarhaddon. Other examples of type I omens are found in 8115: 1 and 8158: 9.

I sought type II omens amongst those whose protases were similar to those of type I omens, but whose apodoses lacked detail. Many omens include protases that were once derived from observation, but which have become so stylised that often they appear to have been invented. For example many omens are attested which describe the presence of a planet or constellation lying within the halo of the Moon. They are usually of the form:

> "If the Moon is surrounded by a halo and Jupiter/Šulpae/planet/Mars/Saturn stands in it" (8147: 3/8147: r6/8049: 6/8168: 10/8181: 1).

It is impossible to decide which, if any, of these omens records an original observation, and which were constructed by analogy. That is, even if the events described in the protases *were* observed before the omens were fully constructed, only those which corresponded to an existing pattern were looked for. In other words, they were already encoded and the resultant interpretations reflected the *decoding*, which may or may not have been described in terms of an actual terrestrial event. To some extent, then, the distinction between many observed and invented omens is artificial.

2) Obs: Inv – Type IIA:

> "If the Moon's horns at its appearance are very dark: disbanding of the fortified outposts, retiring of the guards; there will be reconciliation and peace (*salīmu*) in the land" (8107: 6/8304: 3).

---

336 I have sought those that provide a large amount of detail in both protasis and apodosis, or that record unusual, but possible, celestial phenomena, and have a detailed apodosis.

337 It is of no great concern here whether the omens *faithfully* record actual historical events. I accept that this short tale of family tragedy is what Reiner (1973) 261 would call "a historiette" of little more than "anecdotal" value, but I am also prepared to believe that some real, observed event underpinned it.

I believe that this omen may well record an observed phenomenon, since it was presumably an atmospheric effect, and few analogous protases are known to me. The apodosis is, at least in part, syntagmatically related to the protasis. This is made clear by the following lines in both Reports which state: "GI (means) 'to be dark' (as in the protasis). GI (means) 'to be well' (*šalāmu*). GI (means) 'to be stable', its horns are stable". The connection between *šalāmu* and *salāmu* is being alluded to here. As ever, it is unclear if this *ṣâtu*-style syntagmatic link between protasis and apodosis was made *after* the omen was formed, or *when* it was being formed. I suggest that the apodosis was invented, however, because it is a stock apodosis, and was chosen because of the link between "to be dark" and "to have peace" via the sign GI.

3) Obs: Inv – Type IIB:

> "If a star is darkened in the area of Sagittarius: a decision for Muttabal and Babylon" (8004: 8).

This omen appears to record a relatively unusual event, but its apodosis is derived from a well known code which relates constellations to cities, particular concerning the darkening of the Moon (and not a star) in each ecliptic constellation. See §3.2.2 xviii.

The following omen is also type II, for the description of what are presumably meteoric phenomena appears to be a faithful recording of a real event. However, the apodosis seems both unrelated and yet stock, so much so that I do not believe it to have recorded a real expedition:

4) "If a flash appears and appears again in the south, makes a circle and again makes a circle, stands there and again stands there, flickers, and flickers again, and is scattered: the ruler will capture property and possessions in his expedition" (x111: 5).

Many type III and IV omens are easily identifiable, for they describe celestial phenomena that cannot occur.[338] I believe that these protases were invented by analogy with the other omens which were, at least in part, observed. Similarly, many omens are attested in the omen series which are manifestly extensions or parallels of other omens, and yet do describe possible celestial phenomena. For example in *ACh.*Išt.19: 2–3 we find:

> "If Jupiter stands by the right horn of the Moon; the king of Akkad (and) the king of Amurru will die."
> "If Jupiter stands by the left horn of the Moon; the king of Amurru (and) the king of Akkad will die."

At least one of these omens is a IIIBB omen! Examples from the Letters and Reports are:

5) Inv: Inv – Type IIIBA:

> "If the Fish constellation stands close to the Raven constellation: fish (and) birds will become abundant" (8073: r.1).

[338] See n273. See also *ABCD* 38f and 52. Other examples of *impossible protases* are those describing the relative movement of constellations and those describing Venus entering the Moon on the 15th, or staying occulted for more than 1 watch in K3601: r34 and K3111+10672: no.13, quoted in Pingree (1993).

Two constellations cannot stand close to each other where they formally did not. It appears as if this protasis was first derived by analogy with other protases describing the close approach of planets to constellations. The apodosis is clearly semantically linked to the protasis with "birds" and "fish" referring to the constellation names.

6) Inv: Inv – Type IIIBB:

> "If the Moon makes an eclipse in month VII on the 21st day, and sets eclipsed; they will take the crowned king from his palace in fetters" (8103: 12).[339]

No eclipse can occur on the 21st, so the protasis must have been invented. This undoubtedly took place by analogy with other protases. The stock apodosis appears to be unrelated to the protasis, but is in the same semantic area as many other eclipse apodoses which portend the demise of the king, and was no doubt chosen accordingly.

One at least of the following two omens is a **Inv: Inv – Type IIIAB**:

7) "If the Moon is surrounded by a halo, and Regulus stands in it: in that year women will give birth to male children" (8278: 1).

8) "If the Moon is surrounded by a halo, and the Pleiades stand in it: in that year women will give birth to male children" (8005: r.2).

The apodosis is very general, and hardly the product of observation. It was attached to at least one of these two protases because it was attached to the other, since the protases are of the same form. I suggest that the reason it was accompanied by these particular protases was because a celestial body lying within a lunar halo was thought to represent a foetus in the womb – a syntagmatic connection. Another example is perhaps:

9) "If the Moon rides a chariot in the month of Sililiti; the dominion of the king of Akkad will prosper, and he will capture his enemies" (8112: r.3).

This protasis appears unrelated to those of other lunar omens, and yet is so general as to suggest that it was invented, rather than observed. The apodosis is stock, but is probably syntagmatically linked to the protasis through the connection of the (war) chariot in the protasis to the portent concerning the capture of enemies (in war). See also x104: r.5. In the following however:

10) "If the Moon is surrounded by a halo and two stars stand in it; a reign of long days" (8020: 1),

the apodosis is clearly stock, but does not appear to be related syntagmatically to the invented protasis.

Type IV omens are infrequent since the vast majority of apodoses are stock, or lack the detail to suggest that they record actual events. Also, for the few for whom this is not true, the

---

[339] This omen is from EAE 19 §3 13) *ABCD* p171. Many other lunar eclipse omens occurring on the 21st are found in EAE 17, 18, 19, 21 & 22.

attached protases also tend to record phenomena with great detail, leading to Type I omens. However, the following example was found:

11) Inv: Obs – Type IVA/B:

> "If a planet comes close to a planet, the son of the king who lives in a city on my border will make a rebellion against his father, but will not seize the throne; a son of a nobody will come out and seize the throne; he will restore the temples and establish sacrifices of the gods; he will provide jointly for all temples" (x109: r.14).

This omen's apodosis seems to describe an actual event, and while it is possible that at that time the only significant celestial happening was the close approach of two planets, I suggest that the protasis was invented on the basis that the two planets represent the king and his son, and that ^{mul}udu.idim is a name more commonly used for the generally ill-boding planets (see Ch.2.1.1).

These examples from the Letters and Reports demonstrate how useful this type of analysis can be in revealing the several ways in which the omens in EAE came about. My study of the omens used in the Letters and Reports indicates that the number of type I and type IV omens is small. By far the most numerous are types II, IIIBB and IIIBA. I mean by this that the majority of celestial omens used in the Letters and Reports were either fully or partly invented. Many were made up of what appear to be genuine records of celestial phenomena, but for which the prognoses were created rather then watched for. This indicates that the interpretations of these observed phenomena were *already known*, and it was for this reason that they were watched for in the first place. Many others were wholly created, their protases having been invented by analogy with other protases, and since through the application of the *code* (§3.2.2) these unseen scenarios were already understood to bode good or ill and for whom, it was unproblematic to attach to them stock or newly invented apodoses.

This is a significant result, and to the best of my knowledge is borne out by EAE itself (see App.3, where I analyse the published EAE material). It reveals that Mesopotamian celestial divination did not involve the collecting together of a mass of observations of the sky combined with observations of happenings in the human sphere. Instead it involved the mass production of both protases and apodoses, the *majority* of which involved little or no empirical input.[340]

The huge number of invented omens, particular those with impossible, non-occurring protases, demonstrates that their creators were not interested in accurately recording observations of the heavens. It reveals instead that once the basic *categories* of directions, constellations, planets, watches, heliacal risings, occultations, eclipses, colours etc., had been made, there was little need felt to observe the sky again before writing new protases. It was possible to invent the protasis: "If the Moon is surrounded by a halo, and X stands in it", and use the name of any celestial body for X. Often X would correspond to a heavenly body that could be found in a lunar halo. Sometimes, however, X would not, as was the case for ^{mul}ban (8378: 1), Canis Maioris, which is located too far from the ecliptic. In Report 8378, how-

340 Compare Parpola (1993b) 53 who calls EAE "primarily a scientific collection of signals". For Rochberg-Halton *ABCD* 8 EAE is "a text that included not only omens based on observed occurrences, but also omens based on non-occurring phenomena."

ever, the impossible protasis was re-interpreted by "associating" ^{mul}ban with a planet (it is not clear which), since the lunar halo is actually said to have been located in Virgo. This is an example of the ongoing and complex development of the EAE Paradigm.

The inventing of new protases and the attachment of appropriate apodoses based on an interpretation derived from the *code* was a form of *hypothetico-deductive* method. For example, some of the code – §3.2.2 (xi) & (xiii) – indicates that Jupiter was a benefic planet and that its brightness boded well for the home country:

> "If Jupiter is bright; the king of Akkad will reach the highest rank" (8254: 5).

We would expect that were Jupiter to be dim, it would bode ill:

> "If the Marduk planet is very black; deaths" (*ACh.* 2Supp.69: r.3, Mul.Apin IIGapB6).

The apodoses correspond with our interpretation of the code. We can go further. In *ACh.*2Supp.58: 1–4 we find:

"If Jupiter is darkened from its front;	the weapons of	Subartu	will [...]	
id.	behind	id.	Elam	id.
id.	right	id.	Akkad	id.
id.	left	id.	Amurru	id. "

The first two omens correspond with the code again, since what remains of the apodoses suggests success in war for the enemy countries – a dim Jupiter bodes ill. However, the third omen probably bodes *well* for Akkad. What has happened here is that two schemata of the code have come into conflict. That concerning the assumption that the brightness of Jupiter bodes well, and its dimness bodes ill, and the one relating directions to countries. In this way the code can account for these and many other apodoses, at least in terms of whether or not they bode well and for whom. It does not account for which apodoses were used, however. Nor can it account for the creation of many protases. For these we need the *rules* of "text play" and "number play".

For example, how do we account for the protases in the omens just quoted? Three at least of them have been derived by analogy with the other. They are in a *metaphoric* relationship to each other. They are IIIB type. Why was this relationship established? At least in part because the omens were written, I argue. This, I believe, imposed particular pressures on the creators of the omens, which included the tendency to seek fixed, linear sets of associations.[341] Many of the metaphoric connections between protases came about precisely because they were written down one after another in lists. A particularly good example is found in *ACh.* 2Supp.57: 26f:

> "If Jupiter twinkles in the *1*st watch to the *north*; the *head* of the land of *Akkad* will be seized by illness"

which is repeated 12 times, with only the italicised words being replaced. 1st, 2nd, and 3rd correspond to head, middle, and base. North, south, east, and west correspond to Akkad, Elam, Subartu, and Amurru. Although this kind of elaboration is not impossible in an oral

[341] See n203, above.

environment, it is highly likely that these omens were only invented after they had begun to be written down.

In the Type IIIBA omen in (5) above, the apodosis had been chosen because of the *syntagmatic* link with the protasis. There the link was SEMANTIC. See also (9). In (7), (8) and (11) above, the protases and apodoses were related, I suggested, through a semantic link based on the idea that the heavenly bodies PICTORIALLY represented the descriptions in the apodoses. In 8029: 1 the omen:

šumma (mul) *šarru šarūri naši šar Akkadî gamirūtu ippuš*
"If Regulus carries radiance; the king of Akkad will exercise complete dominion,"

demonstrates a play on the SOUNDS (alliteration and assonance) and meaning, which syntagmatically account for the apodosis. In (2) above, the apodosis and protasis are linked through MULTI-VALENT play on the two-readings of one sign. In EAE 50 (*BPO*2 III7a) the (short-form) omen is given thus:

^{mul}lul.la ^{mul}al.lul
"The False Planet is Cancer,"

in which the two parts are related through RETROPHONY.[342] For the use of the techniques of GEMATRIA and NOTARIKON, and of ETYMOLOGICAL and GRAPHIC derivations in astrological-astronomical texts see Ch.2.1.2, and the example from *Enūma Eliš* in §3.1.4.

To conclude, I have outlined the "number play" rules known as *arû* that included both numerological speculation and a variety arithmetic techniques. I have also explained the rules of "textual play" present in the omens, which include those utilising the "technologies of listing" as well as the "word play" which sometimes connected the two sides of an omen. This last set of rules was most closely described by the term *ṣâtu*. Together these are the hermeneutic rules employed by the Scholars who compiled and authored EAE and the other series and texts of the Paradigm. They were used to extend the applicability of celestial divination, creating new protases whether by elaborating on the "ideal year", or by substituting one heavenly body for another in an existing omen. The Scholars were able to formulate the appropriate apodoses because, as we shall see in the next section, the interpretation of many heavenly events could *in large part* be *read* directly from the sky through the use of what I term the underlying code. This, in its basic form, assigned the prognostication of each event or configuration to one of four geographical entities and determined whether it boded well or ill.

This *decoding* is itself suggestive of writing and it is even plausible to argue that the concept of an ideogram, a sign indicating the meaning of something, must have preceded the invention of this form of divination. I do not assert, however, that the non-literate could not have prognosticated from heavenly happenings.[343] However, some of the "textual play"

[342] Beaulieu (1995) 6f.

[343] Bottéro (1992) 97f argues that with ideograms the relationship between the sign and the thing signified is direct and immediate, and that giving something a name was to give it existence, which was similar to giving it a destiny. I note x030: r.3f: "you will speak [a wo]rd [that] is as perfect as that of a sage; a word that has been spoken just as it is meant by its nature (*šiknišа*)...is that not the very acme of the scribal craft?". The concept of immutable fates inscribed on the "Tablet of Destinies" is well known. See George (1986). Beaulieu (1995) 10 writes: "even

rules present in the omens cannot be older than the invention of writing, and those based on bilingual allusions no earlier than the adoption of the Sumerian script by Akkadian speakers. It is clearly the case that by the time divination appears in the written record writing has influenced and altered it to such an extent that the discussion of its pre-literate origins cannot hope to be derived from the texts themselves. Such a discussion remains speculative.

Structuralist analysis was originally designed to reveal the organisation of the "savage mind", but I make no such claims for a transcendent Mesopotamian mind, for this analysis has restricted itself to the products not of orality, but of early literacy. I have also limited myself to the works of a select group[344] within Mesopotamian society, a group that for the period of the last NA kings can be reasonably well described, as has been attempted in Ch.1. As Leach (1970) 50 points out, Lévi-Strauss's theories cannot be rigorously tested. The same is not true for the analysis here. More of the material upon which it is based is gradually appearing in modern editions, providing an opportunity to *test* whether the categorisations, the code, and the hermeneutic rules have changed over time, and thus whether the omen types posited really do persist over time and whether or not my estimation of their relative frequencies is in fact also the case prior to the NA and NB period. See my provisional study on this in App.3.

3.2.2 The EAE Paradigm Code

This has been discussed before many times in the secondary literature on cuneiform celestial divination. Almost as soon as one begins to read the omens it becomes apparent that certain words in the protases are repeatedly found with certain words in the apodoses. For example "north" in the protases is frequently found with "Akkad" in the apodoses. The particular *choice of countries, and the directions with which they were associated, for example, will be referred to as the schema. Some texts are preserved which give the schemata alone. They effectively spell out the means by which apodoses can be deduced from protases* from first principles, *and are examples of "Mesopotamian abstraction".*

The oldest known celestial omen texts demonstrate the use of schemata, and they are attested both in other omen texts and in non-divinatory texts. Rochberg-Halton *ABCD* calls them "traditionally accepted schemata", and perhaps very little more than this can be said about them. They are preceded by the categorisation of the universe. I mean by this, that in order to describe a phenomenon the number of variables of which it is constituted have to be reduced. Each celestial event has to be assigned a location ("north", "in front" etc.), a date

Marduk 'whose decrees cannot be altered' found himself compelled to reinterpret his own decrees by reversing the order of the numerals in order to shorten Babylon's period of abandonment" (see n326). I tentatively propose that this form of thinking in many cases *required* the use of the syntagmatic rules such as those used to extend the range of applicability of the EAE Paradigm. Bottéro goes on to suggest that writing which reproduces speech is more distant from this concept of "giving existence and destiny", but that even in later periods it still remained as an idea, as is indicated by the evidence of continued *scrutiny* of the ideograms in order to elicit more meaning from texts. It is also possible to argue, however, that the close study of the cuneiform signs served only to *justify* meaning already considered present in the material. All this is closely related to Goody's (1977) arguments concerning the influence of writing on cognitive patterns, and more study would be useful.

344 Leach (1970) 113 writes "Although many of us may be willing to concede that the structures which he (Lévi-Strauss) displays are manifestations of an unconscious mental process, I for one must part company when he insists on treating this unconscious as an attribute common to all humanity rather than as an attribute of particular individuals or of a particular cultural group."

(implying a calendar), a time (a "watch", say), a colour (one of only four normally used in EAE) and so forth. The code indicates *in general* that the members of these categories were related to countries, cities, and so forth. Particular schemata are attested in which the *actual* choice between the possibilities has been made. Many of the attested schemata have been described before. Where this has been done the examples will be summarily presented and the appropriate references given:

A) Code Relating Four-Fold Divisions: Countries[345]

i) Location in one of the Cardinal Directions

N	Akkad	
S	Elam	
E	Subartu	
W	Amurru	All – *ACh.* 2Supp.57: 26.

ii) Orientation or Movement in one of the Cardinal Directions

N	Akkad	
S	Elam	
E	Subartu	
W	Amurru	All – *ACh.* 2Supp.19: 17', Parpola *LAS* II p407.

E.g. the direction of shadow movement – this encoding appears in the attested OB eclipse omen texts – see *ABCD* 20–21.

iii) Months

I,V,IX	Akkad	
II,VI,X	Elam	
III,VII,XI	Amurru	All – *ABCD* EAE 20 excerpt text f r.3f
IV,VIII,XII	Subartu + Guti	Dates of eclipses.

GSL 274–7 and 2Supp.19: 13' omit Guti. See *ABCD* p37 n11, and *LAS* II p407. It is contradicted in EAE 16 §III.3 where a month III eclipse is connected with Akkad. See also EAE 20 §V and 21 §X. Variant schemes are found in Rm 2,38: r.6, *ACh.* Sîn 22: 25 and IM 62257: 32'f (van Dijk, 1976).

iv) Mid-Month Days

13th	Akkad	
14th	Elam	8300: r.15

[345] Related to ub.da.límmu.ba = *kibrāt arbā'i, kibrāt erbetti* "the four regions". See *CAD kibrātu* and Horowitz (1998) 298.

15th	Amurru	8174: r.2, 8275: 5
16th	Subartu	8177: r.5
		All – *ACh.* 2Supp.19: 12' *LAS* II p407.

The 8174, 8177 examples apply to days of ordinary months, the others to eclipse-month days.

v) Eclipse Quadrants

upper (north) quadrant	Amurru	8316: 12
lower (south)	Subartu	8316: 12
right (west)	Akkad	8316: 11
left (east)	Elam	8316: 11
		All – Ref. in *ABCD* p53 n101.

An OB version has n = Akk, s = Elam, e = Sub , w = Amurru, *ABCD* fig. 4.6. Yet another schema is found, described *ABCD* fig. 4.3. For the association of the quadrants in which the eclipses start with colours see *ABCD* p57. Basically it is e = sa_5 (red), w = sig_7 (green), n = babbar (white), and s = gi_6 (black). It suggests that part of the code may have included the association of these four main omen colours with countries.

vi) Winds During Eclipses

North wind	Akkad	
South wind	Elam	
East wind	Subartu	
West wind	Amurru	All – *ABCD* p59, derived from EAE 16.

"This schema does not operate with regularity," Rochberg-Halton loc. cit. 59. The four winds are related to sheep, cattle, horses and asses respectively in *GSL*:295f.

B) Code Relating to the Three-Fold Division of Watches

vii) Watches: Countries

Evening watch	Akkad	8300: r.15
Middle watch	Subartu	
Morning watch	Elam	All – *ACh.* 2Supp.19: 12' *LAS* II p407.

Variant found in EAE 20 text f r.11–12 where Elam is replaced by Subartu and Guti, and Subartu by Amurru. This last is also found in op.cit. r.5. For code relating the watch in which an eclipse starts to the watch in which it finishes (and consequently determining the prognosis in part), see *ABCD* p47 table 4–3.

viii) Watch in which an Eclipse Occurs: the Period (*adannu*) for which the Evil Portended by the Eclipse was considered to last

Evening watch	100 days	8336: r.10
Middle watch	200 days	
Morning watch	300 days	8004: r.9 All – *ABCD* p43 n52.

ix) Watches: Illness/Business

Evening watch	plague	8336: r.9
Middle watch	diminishing business	
Morning watch	healing of the sick	8004: 9 *ABCD* p20 OB texts A and B and in EAE 19 §1 A iv 2'-4'.

C) Binary Code

Many of the following binary distinctions found in the omens also appear as part of the four-fold schemata described above, or as part of the planetary code described below:

Left: Right (see v) the *pars hostilis* and the *pars familiaris* discussed by Jeyes (1991–2) 35.
Up: Down (see v).
In Front Of: Behind "If Jupiter stands in front of Mars: *there will be* barley; animals will fall; variant: a large army will fall" (8288: 1). "If Mars goes behind Šulpae: this year is good" (8114: r.4).
Bright: Dim (see xiii).
On time: Not On Time (see xx).
Have: Not Have "If Venus has a *ṣirḫu*, not favourable. If Venus does not have a *ṣirḫu*: favourable" (K35: 1–2 = *BPO*3 p101).

Some of this apparent part of the code may represent no more than the metaphoric *rule* which replaces one word in the protasis with its opposite, and does the same in the apodosis. These binary oppositions form part of the code *only* when it is known that, *in general*, left = bad and right = good, say. Sometimes, this is hard to ascertain because the code and/or rule can often come into conflict with other parts of the code – see for example n348 below.

D) Code Related to Planets

x) Presence of Planets in Eclipses: Fate of the King/Cattle

Mars	destruction of cattle	*ABCD* p62
Jupiter	the king will not die, a famous person will in his stead	
Jupiter + Venus	the king will be well	All – *LAS* II pp407–8

xi) General Planet Values

The order in which the planets are "most frequently"[346] enumerated in EAE is:

Ju,Ve,Sa,Me,Ma.

In the late period the order has changed slightly to:

Ju,Ve,Me,Sa,Ma.

For details see Neugebauer *HAMA* 690, and Hunger & Pingree *Mul.Apin* 147. There is good reason to believe that the order corresponds to a code which arranged the planets according to their good-boding or ill-boding qualities, from benefic to malefic.[347]

xii) Planet Colours

Due to horizon effects any planet can appear reddish, greenish, or violet/black, as explained in *BPO*2 p19. Nevertheless, the planets were assigned ideal colours.

Ju	white	Ch.2.1.1. J
Ve	green	In 8114: 6 & x100: 18 Mars is green
Sa/Me	black	Ch.2.1 B-names & n228
Ma	red	Ch.2.1 D-names
		All – K2346+, mentioned in Mul.Apin 150.

This four-fold division perhaps reflects the association of the colours with the four countries noted in A (v) and the colours probably mimic the general planet values, with red being the *most* ill-boding. Note that it is not black that is the most malefic – see n195. Mars, of course, is most often perceived of as red due to properties intrinsic to the planet, rather than to any atmospheric effects. See the discussions in *Mul.Apin* 149–50, *BPO*3 19 and 23, and below.

xiii) Planetary Brightness

Broadly, it appears as if a bright benefic planet bodes well, and a dim one bodes ill. A dim malefic planet bodes well, and a bright one ill.

Ju	See the evidence in the previous section – §3.2.1.
Ve	"If the worm star (Ve) is massive; there will be mercy and peace in the land" (8538: 3).
	"If Venus is dimmed on her right side: women will have difficulty giving birth. If Venus is dimmed on her left side: women will have easy childbirth." K2226: 19–20, *BPO*3 93.[348]

346 Hardly a testable assumption at present.

347 As argued by Rochberg-Halton (1988b).

348 The metaphoric rule has come into play here, just as described in §3.2.1, which explains the good boding nature of the second of these two omens. Usually right bodes well and left ill, so *without* the code implying that a dim Venus bodes ill, one would have expected the first of these two omens to be the ill-boding.

Sa "Saturn...is faint; this is bad" (8491: r.9).
Me "Mercury is shining very brightly. This is [propitious] for Subartu" (x074: r.5). (Subartu is Akkad's enemy so far as EAE is concerned.)
"If Centaurus (Me) *flickers* when it comes out; prospering of the harvest, business will be steady" (8158: r.7).
Ma "If Mars becomes faint; it is good; if bright; misfortune" (8114: 8).

xiv) Planets: Countries

Ju	Akkad star *GSL*:209
Ve	Elam star 8302: r.2, *GSL*:201
Sa	Amurru star 8491: r.9, of Akkad 8383: r.7
Me	Probably a Subartu star x074: r.7
Ma	Subartu star 8491: r.7, of Amurru 8412: r.2 & *GSL*:219, of Elam *Planetarium* 318.

Again, a four-fold division of the five planets into the four countries was perhaps sometimes intended, and it is probably no coincidence that the most propitious planet, Jupiter, was an Akkad star.

xv) Planets: The Royal Family

Me	Crown Prince	Ch.2.1 A-names
Sa/Su	King Ch.2.1	B-names

xvi) Planet-Planet Interaction

If the Moon occults or closely approaches a benefic planet this bodes ill, if it occults a malefic planet this is good.

Ju "If Jupiter stands inside the Moon; in this year the king will die" (8100: 1). See also 8438.
Ve "If Venus stands inside the Moon: the king's son will rise to (make) a revolt, upon divine order Elam will perish,[349] there will be rains in the land, (and) upon divine order the land will diminish" (VAT 10218: 35 = *BPO*3 43f – for omens describing Venus's proximity to the Moon, all of which portend ill except where modified by the rules, see VAT 10218: 25–47).
Sa "If the Sun (Saturn) enters the Moon; universal peace" (8166: 1).
"Tonight Saturn approached the Moon... it is good for the king" (8095: r.1).
Me
Ma "If Mars comes close to the front of the Moon and stands there; the Moon god will resettle a bad land" (8311: 1).

[349] Venus is an Elam star – see xiv.

A code undoubtedly existed which served to interpret the conjunctions of the superior and inferior planets with the Sun and with each other, determining in part the interpretation to be put on the proximity of benefic and malefic planets, for example.

Venus's disappearance into the Sun and its proximity to Jupiter boded ill according to VAT 10218: 48–50 and 51–59 (*BPO*3 45 and 8212 & 8214). Its "going" with Jupiter bodes well in 8244, however. Mars's approach to Venus appears to bode well in VAT 10218: 63 and in 8541, but its entering Venus at the latter's rising bodes ill in *BPO*2 IV: 5a and VI: 5–5a. Mercury standing with Venus bodes ill in 8051: r.7f, and Mars and Saturn's close approach bodes ill in 8049 and 8125, for example. Too little is preserved to decipher the code as yet, which may have depended on which planet was perceived to be doing the approaching.

Similarly, a code probably existed which accounted for the interpretation put on planets located inside lunar haloes or constellations. Certainly, some constellations were considered particularly malefic. Scorpius was one such case. The location of Mars within Scorpius was considered to be of particularly evil portent – see 8502: 11f & x008: 24. Other constellations were perhaps reckoned to be benefic. Unfortunately, insufficient material is thus far available to check these suspicions. The so-called *ašar* or *bīt niṣirti* "secret place or house", attested for Venus in EAE text K2346: 21–22 (*BPO*3 245, and see also pp14–15), appear to be constellations within which the particular planets in question bode well.[350]

Finally, codes may underpin the readings put on planets located in one of the three star paths, and on the part played by such phenomena as "crowns", "*ṣirḫu*" (luminescence or some form), "*mešḫu*" (mirage?) and "the cross". For these terms in the context of Venus omens in EAE see *BPO*3 14f.

xvii) Eclipse: Royal Death

Many omens pertaining to this particular luni-terrestrial-solar interaction directly predict the king's death:[351]

> "If there is an eclipse in month III, the king of the universe will die" (8004: r.14).

Some omens, however, apparently do not:

> "If there is an eclipse, and the north wind blows: the gods will have mercy upon the land" (8004: r.5).

I believe, though, that these latter omens *assumed* that the eclipse portended the death of the king, providing the relevant quadrant had been darkened, and Jupiter was not visible, and that they were simply describing *additional* prognostications. The best evidence for this view is offered by the Letters and Reports data concerning the substitute king ritual. Parpola's table in *LAS* II pxxiii shows that *every* eclipse between 679 and 666 BC, for which the relevant quadrants were obscured, triggered the ritual, except those during which Jupiter was visible – cf. (x) above. This was done regardless of the movement of the shadow, the month or the day, regardless of the watch, or the winds. The substitute king ritual was needed because the king's death had been portended.

350 So argues Rochberg (1988a) 53–57.

351 An equation attested from the earliest known eclipse omens (App.1 §5) to the latest eclipse records (Walker, 1997, 22).

Evidence from earlier periods is not available. I suggest, however, that at first eclipses were read to portend the king's death. Eclipses were then "encoded" according to (v) and (x), so that not every eclipse would portend the death of the king of Akkad. The further encoding that pertained to the watches, days, duration, meteorological factors, direction of shadow movement etc., was used only to provide *additional* prognostications. Some of these simply confirmed that the king was predicted to die, but others were concerned with different eventualities. For full details on the schemata employed in the lunar eclipse omen section of EAE see *ABCD* Ch.4.

E) Miscellaneous Code

xviii) Constellations: Cities

Hired Man	Uruk and Kullaba
Bull of Heaven	Aratta, Ur, Der, or Duranki (Nippur)
Pleiades	Der, Ur, or Duranki
Twins	Kutha
Crab	Sippar, Tigris, or Euphrates
Leo	Enamḫe, Paše, Egalmaḫ, Eridu, Kumar?, and Kullaba
Furrow/Virgo	
Scales	Sippar, Larsa, Girsu, Lagaš and Paše
Scorpius	Seeland, Dilmun or Borsippa, Hursagkalama, Borsippa
Sagittarius	Muttabal and Babylon, Kiš, Dilbat, and Girsu
Capricorn	Eridu or ^{d}íd
Aquarius	Eridu or ^{d}íd
Northern Pisces	Tigris and Akkad
Southern Pisces	Euphrates and Seeland, Dilmun
Orion	Sippar and Larsa
Perseus	Nippur
Wagon	Nippur
Field	Babylon
Wolf	Hursagkalama
Fox	Enamtila
Šu.pa	Babylon
Rooster	Kullaba

A list reconstructed from a number of NA and LB texts by Weidner (1963).

(xix) Phenomena Behaving According to/ Not According to the Ideal Schemes

If the phenomenon occurs in the manner or at the time anticipated by the ideal period schemes, this bodes well, if it does not, this bodes ill.

For example, in 8290: 3f we find:

"– The 30th day completes the measure of the month.
If the Moon becomes visible on the 1st; reliable speech, the land will become happy."

The Moon is said to appear on the 1st, meaning the previous month lasted 30 days. This is also the length of the "ideal month". The prognostication is positive. This omen is one of the most common of all those attested in the Reports. A variant apodosis is attested in 8086: r.1:

> "If the Moon becomes visible on the 1st; good for Akkad, bad for Elam and Amurru. On the 14th day it will be seen with the Sun."

The prognosis is still positive since the underlying schema is that 30-day months bode well.[352]

If the Moon does not appear on the 1st, but earlier, the following omens are attested:

> "If the Moon becomes visible on the 30th; there will be frost; variant: rumour of the enemy" (8011: 1).
> "If the Moon becomes visible on the 29th; Adad will devastate" (8457: 1).
> "If the Moon becomes visible on the 28th; good for Akkad, bad for Amurru" (8014: 1 variant in 8063: 1).

The first of these omens is often quoted in the Reports. The prognoses are not good if the month does not last 30 days. The one exception is the omen derived from a month of only 27 days. A 27-day month is impossible, unless at the beginning of the month, the Moon was so obscured by poor weather that it was not seen for days. It appears as if a process of encoding the days at the end of the month has occurred, assigning day 28 to Amurru, and thus lifting the essentially ill-boding prognoses from Akkad and heaping them on to another nation. It does not alter the general notion that if a month lasts for fewer days than the ideal, this bodes ill. As it says in 8391: 5:

> "If the Moon at its appearance is visible early: the month will bring worry."

If the Moon appears on the 1st in the 1st month, thereby starting off the year according to the ideal, this appears to bode particularly well:

> "The mo[on] will complete [the day] on the eve[ning] of the beginning of the year; it is favourable fo[r the king], my lord" (8083: 1).

During the middle of the month, the same situation pertains. If the Moon and the Sun are seen together[353] on the 14th, this bodes well; if not, this bodes ill.

> "If on the 14th the Moon and Sun are seen together; reliable speech, the land will become happy..." (8015: 6).
> "If the Moon and Sun are in balance; the land will become stable; reliable speech..." (8015: 1 also 8015: 4, 8110: 7/r.4, 8293: 8).
> "If on the 13th the Moon and the Sun are seen together; unreliable speech; the ways of the land will not be straight... (8306: 1 also 8458: 1f).

---

352 Noted before anecdotally from Weidner (1912b), Schaumberger *SSB*III Erg. 251f, to Beaulieu (1993). See e.g. Parpola LASII 45r.8 & Chadwick (1992) 11f. Despite this, Oppenheim writes in (1978) 634 "after experience had taught (Mesopotamian man) to recognise a pattern in the sequence of certain events and in the predictable features of specific phenomena, he considered any *deviations and irregularities* to be endowed with meaning... " (my italics). Clearly non-deviations and regularities were also imbued with meaning by some Mesopotamian men.
353 Ch.2.2.2 (6).

"If on the 15th the Moon and the Sun are seen together; a strong enemy will raise his weapons against the country. . . " (x094: r.1 also op.cit. r.2, x105: 18, 8091: 4, 8295: 7).
"If on the 16th the Moon and the Sun are seen together; the king of Subartu will become strong. . . " (8082: 11 also 8177: 4f).

The 14th day was the "ideal" day for the Moon's evening rising according to the EAE Paradigm. In *ABCD* 39–40 Rochberg-Halton argues that the 14th was the "normal day of opposition" of the Moon, and was the day upon which eclipses were expected to fall. This is perhaps meant in the omen:

"If there is an eclipse in Month III on an unappointed date (*ina lā minātišu*): the king of the universe will die" (8004: r.14),

though *ina lā minātišu* could refer to an unanticipated month, or even watch.[354] *Manû*, the root, does suggest something countable, and not just "unappointed time", however. Note the similarity of the apodosis quoted from 8015: 6 above with that used for the Moon rising on the 1st. The omens were obviously invented together. The variant omen for the Moon rising on the 1st, quoted in 8086 and above, also states that the Moon rising on that date implies that it will be seen (rising at Sunset) on the 14th. Such is the assumption of the "ideal lunar visibility/invisibility scheme" described in §3.1.2. One correspondence with the ideal deserves another, it appears. A correspondence with the ideal bodes well, a failure to do so bodes ill. This part of the code is summarised by the omens:

"If the Moon is seen not at its normal time/number (*lā simanišu* / *lā minātišu*); business will diminish" (8088: 1, 8474: 1).
"If the Moon and Sun are seen together not at their normal (day) number (*lā minātišunu*); a strong enemy will oppress the land" (8088: 4).

It is also implicit in Esarhaddon' inscription dating to 680 BC:

"The twin gods Sîn and Šamaš, in order to bestow a righteous and just judgement upon the land and the people maintained monthly a path of righteousness and justice, appearing regularly on the 1st and the 14th days."[355]

The length of the day also appears to have been used to generate omens:

"If the day reaches its normal length (*ana minātišu ērik*): a reign of long days" (8007: 3).
"If the day is short compared to its normal length; a reign of short days. . . " (8457: 4).[356]

354 *ABCD* p42, where Rochberg suggest that an "unanticipated eclipse" may have been one which has occurred after fewer or more than 6 months, and points to a statement of this kind in *ACh.* Sîn 3: 26f and to 82–5-22, 501 edited in her App. 2.3. This would imply that the 6-month interval between eclipses was another of the EAE Paradigm "ideal schemes". However, since both *ACh.* Sîn 3 and 82–5-22, 501 are commentaries which may only date from the late NA period I have not included this "scheme" in my discussion in §3.1.2 but have treated the 6 month eclipse period as a potential NA innovation indicative of the PCP Paradigm.

355 Borger (1956) Ass. A. i31f and quoted in Parpola SAA9 lxxiv.

356 Note also the verbal correspondences protasis to apodosis – *ērik* to "long", and lúgud.da to "short", as a secondary, justifying device using word-play rules.

Once again, observation matching the ideal bodes well, where the opposite bodes ill. It must be mentioned that these two omens are only ever used in the Reports and Letters in conjunction with the Moon rising on the 1st, or not on the 1st, respectively. This is spelt out specifically in 8251 and 8457. It is probable, however, that these omens (once and perhaps still) meant that the length of the day was also measured, and compared to that calculated using linear interpolation in combination with the "ideal year".

In 8391: 3 the following omen is given for an occasion when the Moon rose on the 30th:

> "If the Moon at its appearance is high and becomes visible; the enemy will plunder the land,"

which suggests that not only was the Moon seen a day earlier than the ideal, but that it was higher than expected when first spotted. This would indicate that an ideal height, and perhaps a time, for the Moon's first visibility was known. This was, I suggest, something similar to the $^1/_{15}{}^{th}$ of the night derived in the ideal lunar visibility/invisibility schemes in EAE 14, for example. In x225, Adad-šumu-uṣur notes that the Moon was too high for the 30th. "Its position was like that of the 2nd," he writes, but he gives no interpretation. The negative prognosis in the omen above suggests that not corresponding with the usual ("ideal") height at first visibility did not bode well.

Lunar disappearance could also occur at a non-ideal time:

> "If the day of the disappearance of the Moon is not on its normal date; the ruin of the Gutians will take place" (8346: 1).

Planets, similarly, could be said to be too high for their day of first appearance:

> "Mars has appeared... I saw it on the 26th of month II when it had (already) risen high" (x100: 5f),

but Akkullanu quotes no omens pertaining to its excessive height. Mercury's movement is noted in 8093: r.3:

> "Mercury, is going beyond its (normal) position and ascends."

Hunger adds the "normal", which is not fully justified. Mercury's *manzāzu* "position, station" could apply to the location into which it rose or its horizon position,[357] and not to its ongoing position. Nevertheless, it demonstrates that the Scholars carefully watched the planets' risings. This is because, I believe, it was considered portentous if the planets' dates of appearance, and periods of visibility and invisibility corresponded, or failed to correspond with the ideals set out for them in texts such as Mul.Apin and in EAE itself.

> "Venus made her position perfect [...] she became visible quickly – a good sign for the king and his reign.
> If Venus stays in her position for long: the days of the king will become long.
> If the rising of Venus is seen early: the king will extend the life" (8027: 6f).
> "If Venus gets a *flare* this is not good... she does not complete her days, but sets" (8145: 2).
> "Jupiter... additional days... If Jupiter [xx] in the sky; the days of the king will become long" (8167: r.1f).

---

357 *BPO*3 18.

> "If the Sun rises in a *nīdu*;[358] the king will become furious and *raise* weapons – Jupiter stands in the sky for excessive days (ud.me diri.meš)(8329: r.5) / Jupiter stood there one month over its period (*ana muḫḫi adannišu*)" (8456: 3) / "Jupiter remained steady in the sky for a month of days. May the king of the lands be everlasting" (8339: r.3) / Jupiter retained its position; it was present for 15 more days. That is propitious" (x100: 30f).

These omens seem to indicate that if the two good-boding planets were present in the sky for longer than expected, this was considered to be propitious. If however, they (only attested for Venus) disappeared earlier than expected this did not bode well. This variation on the encoding of (xix) is due to the encoding of (xi), I suggest. The periods against which these observations were being compared were those ideal ones found in Mul.Apin Iii44–67 and EAE 63 §2 and 56 §XVIII–XIX. The invisibility periods can be analysed similarly:

> "If the rising of Venus is seen early: the king of the land will extend the life" (8247: 6).
> "If Nēbiru drags; the gods will get angry, righteousness will be put to shame. . . Jupiter [*may* remain invisible] from 20 to 30 days. Now it has kept itself back from the sky for 35 days. . . exceeding its period (*edānišu*) by 5 days. . . furthermore when it had moved on (by a further) 5 days it completed 40 days" (x362: 5'f).
> "If the planet Mercury becomes visible *within* a month; flood and rain" (8281: 3).

Similarly, the prolonged invisibility of a benefic planet boded ill, and the early rising of one was good. In x362, although fragmentary, the reference to Jupiter and to 20 or 30 days undoubtedly derived from Mul.Apin IIi49–50 or similar:

> "Jupiter disappears in the west and remains (invisible) in the sky for 20 days or remains for a month, and rises and becomes visible in the east in the path of the Sun."

Omens were derived for the superior and inferior planets directly from their ideal period schemes.

With regard to the "ideal intercalation schemes", the following omens:

> "If the Pleiades enter the Moon and come out to the north; Akkad will become happy. . . " (8443: 1),
> "If the Pleiades come close to the front of the Moon and stand there: the great gods will. . . for the better understanding of the land" (8072: 1),

bode well either because those schemes were based on the proximity of the Moon and Pleiades on certain dates in the year (the "Pleiaden-Schaltregel"), or on the basis of the code in (xvi), and the identification of the Pleiades with the malefic Mars.

In 8098: 18 Balasî writes:

> "Let them intercalate a month; all the stars of the sky have fallen behind. Month XII must not pass unfavourably. Let them intercalate."

He has observed that the stars were not rising in their ideal months according to the "ideal astrolabe", or similar. The implication of this was unfavourable, since non-correspondence with the ideal boded ill. In *BPO2* pp56–61 a series of omens from EAE 51 describe the ominous significance of stars rising early (nim-*ma* = *iḫ̮rupma* igi), late (zal-*ma* = *uḫ̮ḫ̮iramma*),

[358] See n221.

at, or not "according to its period" (*adannišu*).[359] Those that rose heliacally according to their periods boded well, and those that did not boded ill. For example:

> "In month II, the Stars, the seven (great) gods; if it rises heliacally according to its period; the great gods will assemble and give good counsel to the land, good winds will blow; if it rises heliacally not according to its period; (they) will give bad counsel to the land; evil winds will blow, there will be grief for the people" (*BPO2* IX: 13) –

omens which obey code (xix) perfectly, mirror each other metaphorically and in which the apodoses in question have been construed in such a way that syntagmatic links with the protases are made through the "great gods". In general the omens in which the stars rise "early" bode well, and those in which they rise "late" bode ill. This appears to be an extension of code (xix), but can be explained by the equation that was apparently sometimes made between rising early and the expected date of rising. For example in *BPO2* IX: 8 we find:

> "Venus rises heliacally in its month – if this planet rises early; the king will have a long life; if this planet rises late; the king of the land will die soon."

Balasî in 8098 (above) also pointed towards the divinatory use offered by intercalation, in addition to its rôle in regulating the luni-stellar year. This was, I noted in §3.1.2, the ability to avert the evil in forthcoming or recent ill-boding celestial events by adding in an extra month and thereby changing the dates upon which the phenomena occurred or will occur, and so reversing the prognostications. I suggested that this was the purpose of the apotropaic method in *The Babylonian Diviner's Manual*, and in this light I interpret the comment made by Marduk-šakin-šumi in x253: 15f:

> "[Concern]ing the intercalation [of] the year a[bout which the k]ing said thus: 'Let us add an intercalary month VI.' – the matter is (now) settled. [May the kin]g, my lord, be everlasting on account of that (*ina muḫḫi lū da-a-ir*)"[360]

To sum up, this study has demonstrated how extensive the code underlying celestial divination was. Undoubtedly it was substantially larger than that outlined above (some suggestions are included in my analysis in App.3). The most striking fact about the code is, however, that most of it relates members of the categories of date, direction, colour etc., either to *countries*, or to *good and bad*. I call this the "***simple code***". In other words, on the one hand the cardinal locations and orientations, months, watches, mid-month days, quadrants, winds, the planets, and probably the colours were all associated with one of the four countries – Akkad, Amurru, Elam, and Subartu (sometimes including or replaced by Guti). On the other hand the watches,[361] planet presence during eclipses, the planets themselves, their brightness and their occultations, the binary opposites and correspondence or non-correspondence with an ideal boded either well or ill. This simple code is perhaps alluded to in the late commentary text to EAE 61, K.148, published in *BPO3* 57f line 1 in which it is written:

---

359 See also *BPO2* 2.2.1.1.

360 Note also the word play between *da-a-ir* < *darû* A and "intercalate" *darû* B.

361 It seems reasonable to assume that behind the particular schema quoted in (ix) lay the code: evening watch = bad, middle watch = partly bad, morning watch = good.

> ḫul *u* sig$_5$ *ṭup-pi* ki.meš *u ṭup-pi* 15 *u* 2,30 sum-*in*
> "(whether) it is evil or good, the tablet of regions and the tablet of right and left will give (the answer),"

where left and right refer to the *pars hostilis* and the *pars familiaris* noted in (C) above.

The remainder of the reconstructed code was somewhat more "complex", relating as it did watches to periods (viii), planets to members of the royal family (xv), and constellations to cities (xviii). The relating of planets to colours (xii) probably also reflected an underpinning binary division into good or bad, and the association of eclipses to royal deaths clearly portended ill for the main subject of concern to the celestial diviners.

It appears, then, that in most cases the code did little more than determine whether or not a particular (observable or hypothesised) celestial happening boded well, and to which country the prognosis applied. Since Akkad was "home", and the other three countries were enemies (until the use of EAE by the Assyrians), in most cases the code *simply* determined whether the event boded ill or well. This is an important result, for it shows how the code underpinning EAE could have been drawn from ancient, simple equations made between celestial phenomena and whether they portended ill or good for the nation.

Two more observations derive from this. Firstly, the existence of a simple code running through EAE makes the transition from an oral to a written discipline easier to understand. A simple code, which assigned good boding or ill-boding qualities to the categories of the universe, and determined whether the phenomena applied to home or abroad, probably did not need to be written down. This could explain the absence of "Sumerian" celestial omens dating to the third millennium BC, and their appearance in the OB period when, as with much else recorded during that time, the disappearance of Sumerian as a spoken language lead to their preservation in writing. Secondly, that such a simple code could date to a time before the first written omens suggests what was asserted in §3.1.1, that the omens were already being read *before* being written, and that consequently any discussion of their "empirical origin" reveals little more than a prejudice for oral rather then written wisdom (see n286). No doubt the "encoding" of the sky was a partly theological exercise, assigning Jupiter and Venus to benefic gods and goddesses, and Mars to malefic ones and so forth. The later discovery of Mercury and Saturn, for which some evidence exists in the record of their name associations (see Ch.2.1.2), may also account for their intermediate, good and bad, qualities found in the code. Similarly, the encoding of the months, watches and directions etc., probably happened long before the first omens were written down, as a consequence of the need on the part of diviners to avert the evil portended by their reading of other phenomena. It could hardly have been discovered from the passive observation of phenomena that month V pertained to Akkad, or that Mars boded ill, say! These were *a priori* encodings of the heavens.

The EAE omens may reveal the workings of a basic underlying code, but they are usually much richer than simple bland statements such as 'X bodes ill for Akkad'. This richness came from two sources, one much more dominant than the other. Firstly, real phenomena in the sky and phenomena in the human arena which *verified* the simple decoding, were written down (reversing the so-called empiricist model described in §3.1.1). Secondly, and much more frequently, omens were elaborated, embellished and indeed invented according to rules described in §3.2.1.

The code that was not simple was, I suggest, a later development and part of the ongoing expansion of the EAE Paradigm. This is particularly likely for the code relating constellations to cities, and Saturn and Mercury to the king and crown prince. Similarly, while variants in the *particular* schemata preserved attest the idiosyncrasies of the various versions of EAE used in different temples and cities, the underlying code remains consistent throughout. This premise, and those of the rules and the categorisations were what underpinned the EAE Paradigm, and they did not change throughout its millennia of use.

3.2.3 Categorising the Universe – Variable Reducing and Anomaly Producing

The final premise of the EAE Paradigm, which permitted the heavens to be made interpretable by celestial diviners, concerns the manner in which space, time and the phenomena described in omen protases were categorised.

I propose that the categories underlying the EAE omens and ideal schemes were devised in order to make the sky above interpretable in the sense of being amenable to divinatory inspection rather than necessarily to inquiry as to its origins and purpose. That is, I argue it was categorised in this manner *in order* that it could be encoded with signs, while other categorisations, such as those which proposed a series of heavenly levels,[362] say, were perhaps aimed at answering cosmogonical queries. Since we have no schemata relating countries to intermediate directions (e.g. to north-east), or to other colours, for example, this suggests that the celestial code and the celestial categorisations were developed hand-in-hand. The ideal period schemes, discussed in §3.1.2, were part of this categorisation of the universe, and they too were in possession of a particular part of the code which made them similarly interpretable (§3.2.1 – xix). In Ch.5.1 I will further indicate that the particular categorisations employed in celestial divination had their background in ideas of "design" attested in Sumerian literature. I also argue in Ch.4 that they laid emphasis on particular heavenly phenomena, the record of which formed the basis of the predictive MAATs of the last centuries BC. Finally, I discuss them in Ch.5.1.3 in the light of the extensive literature on so-called "primitive classifications".

The "Mesopotamians" classified the stars into constellations, many of which are still used today in the West (see n312 above). Almost always only four colours were used to describe celestial phenomena, although other names for colours and tones are known in both Akkadian and Sumerian.[363] The brightness and dimness of the heavenly bodies were noted without any reference to a scale or even to intermediate levels of luminescence, though the precise meanings of a number of terms still elude us.[364] The times of phenomena in EAE

---

362 See now Horowitz (1998) Ch.1.

363 E.g. gùn = *burrum* "multi-coloured" and si_4 = *pelu* "light red" are attested, though the four main colours of babbar = *peṣû* "white", gi_g = *ṣalmu* "black", sa_5 = *samu* "red – perhaps brown/red", and sig_7 (*w*)*arqu* = "green – perhaps yellow/green" are far and away the most commonly used terms in the lexical and omen material. Many names for colour tones or qualities are known (*da'mu* = "dark"; *ukkul* = "dark"; *namir* = "shiny", *tarku* = "livid" etc.), as well as particular similes (*eddetu* = "boxthorn" – yellowish; *ṣurru* "obsidian" – black or dark brown; *uqnû* "lapiz lazuli" – blue etc.). For the very occasional use of some of these terms in EAE see *ABCD* p56.

364 E.g. in *BPO*2 18 Reiner has opted for translations such as "obscured" for *adir* (though in *BPO*3 19 this has become "dimmed"), "dark" for *da'mu*, "dim" for *ekil*, and "to be faint" for *unnutu* without being able to determine which, if any, describes the least bright object. The translation "very faint" for *lummunat* in *BPO*3 does not improve the situation.

were recorded by the month, the day, or by the watch.[365] The locations were given by one of the four cardinal directions, the relative orientations similarly, although the alternatives "above", "below", "in front of", and "behind"[366] were used as well.

Other categories included the planets – known as muludu.idim, which reflected their wandering nature (see Ch.2.1 B-names) – the ecliptic constellations, the culminating *ziqpu* stars, and the three star paths into which all the heavenly bodies were categorised. Finally, certain regularly recurring celestial phenomena were classified as ominous. These included the heliacal phenomena of the planets, including eclipses, their proximity to each other and to the constellations. To this end, the times and locations in space of these occurrences were made significant, as were their concomitant and non-repeating colours, brightness, haloes, crowns, and other luminous phenomena. The ominous planetary phenomena noted in the Reports and Letters included most of those deemed ominous in the reconstructed copies of EAE, and are listed and discussed in Ch.2.2.2–3.

Why were the locations, directions, times and meteorological effects not described more precisely? Because the intention, so far as the diviners were concerned, was not to generate an accurate record, but to describe the phenomena *only* with enough detail to make interpretation possible. The categories effectively *reduced the number of variables* to a point whereby their decipherment was made possible. The heliacal rising of Venus was interpreted simply according to the month in which it occurred (EAE 63), and not according to the precise times of these events. An eclipse of the Moon was (mostly) decoded according to the month, one of four mid-month days, one of three watches, one of four shadow directions with which it occurred or moved and so forth. The *precise* direction of movement, or the precise time and location of the event was not important. To have generated omens for each heliacal phenomenon for each minute of each day, or for each degree of orientation would have made EAE unfeasibly large. These few categories made the universe amenable to interpretation by being so broad, and the presence of so few omens with protases that contain more than the broad detail afforded by them shows how unimportant more accurate observation was in the creation of the divinatory series.

The recurring phenomena were made amenable to decoding by means of a particular form of classification which used "round numbers" to describe their periods. The year was described by a period of 360 days, the month by 30, for example. Intercalation was modelled on a one-in-every-three years basis. Venus was assigned approximate values for its repeating periods of invisibility and visibility, as were the other superior and inferior planets. These "ideal" periods were then elaborated into "ideal schemes".[367] The solstices were

365 Note in *LKA* 29d ii 3: en.nun.meš *ša mūši lidbubanikki*, "let the watches of the night speak to you", quoted in Reiner (1995) 16 with further references in n42. This prayer indicates that the watches were thought by some to provide relevant messages. This undoubtedly refers to that part of the code which indicates that a phenomenon's significance can change depending on the watch in which it occurred – e.g. §3.2.2 ix. For the occasional use of *bēru* in Venus omens in EAE and commentaries see *BPO*3 index. It does not appear to add anything meaningful so far as the interpretation of the relevant omens goes.

366 See n249, above.

367 Despite no one word being used in the texts discussed to describe the term "ideal", a number of them adopted this sense on particular occasions. In Mul.Apin IIA8,10 & IIii1,3,5 the year was deemed to be gi.na-*ta* (*át*) = *kīnāta*, as opposed to "in need of intercalation". Hunger translated this as "normal", though clearly something along the lines of "ideal" would be equally suitable, implying that in these cases the phenomena would be corresponding with those implied by the ideal year. Similarly, in x363: 13 the appearance of the Moon on the 1st is said to "fix" the month, also using *kânu*. The expressions outlined in §3.2.2 xix using *minātišu* "its count", *adannišu* "its period" and *simanišu* "its interval" all alluded to the expected period between two phenomena, and the translations "ideal

separated by 6 months, the middle of the month was placed on the 15th, and using piece-wise linear arithmetic techniques ideal values for the length of the night, and for the visibility of the Moon throughout the year were obtained. Further ideal values were extrapolated using number play techniques such as "metathesis" and "doubling", as was the case in the "ideal planetary visibility/invisibility schemes", and for the significant days of the "ideal month". And, much as occultations were classified as "signs" and were encoded with the notion that the Moon inverted the evil of the planet it obscured (§3.2.2 – xvi), so the ideal schemes were encoded with the notion that a correspondence with the ideal boded well, non-correspondence did not, as demonstrated in §3.2.2 xix.

In other words the categorising of the temporal component of a phenomenon into an "ideal period" and its subsequent elaboration into "ideal schemes" made not just the infrequent or exceptional events in the sky open to interpretation, it ensured that the regular running of the universe could be deciphered (*contra* Oppenheim, 1978, 634 – quoted in n352, above). The ideal schemes made possible a comparison between observed reality and an anticipated "ideal". They *produced anomalies and coherences* from the universe's regular and repeating behaviour, since on occasions the stars did *not* rise in their ideal months, the month was *not* 30 days long, and so forth. Both eventualities were interpretable.

To summarise, the categorisations attested in EAE and related texts performed two main functions, both of which ensured that the heavens could be interpreted. Firstly, for those unusual celestial events, such as heliacal phenomena, the categories reduced the number of their variables in such a way that a finite and none-too-large number of data could be encoded for the purposes of future decoding. Secondly, the unexceptional, continuously varying circumstances of the heavens were made amenable to interpretation by establishing ideal schemes against which reality appeared either anomalous or coherent, the two being encoded as quintessentially malefic and benefic respectively. It is, to push a functionalist interpretation, possible to suggest that the benefits of this two-fold system of categorisation to diviners who might be called upon both to offer decipherments of unusual happenings and interpretations on a *daily* basis are clear. Certainly, the late NA kings demanded of their Scholars interpretations as to the propitiousness or otherwise of particular scenarios on a daily basis.[368]

These two main functions performed by these particular categorisations of the heavens had a direct impact on the kinds of celestial records taken when predictive astronomy became important during the 8th and 7th centuries BC, and thus ultimately on the methods and parameters used in the MAATs to predict the recurring heavenly phenomena.

Certain categorisations of the heavens appear in the written record only in later centuries. The "Normal stars" (n250), although attested as a group in EAE text K2226: 13 (*BPO*3 p13 and 93), were perhaps not differentiated until after the late NA period. The classifying of the ecliptic into twelve signs of the zodiac probably only occurs in the 5th century BC (App.1 §42). Some categorisations of time and extension also appear for the first time only in the late NA period.[369] These categorisations were employed in the NMAATs and MAATs, and

count/period/interval" would have been just as appropriate as "normal".

368 E.g. In x044 and x053, Balasî explains to the king which days and months are suitable for a trip or for visitors. Although he may be referring to good and bad days determined by hemerologies, it still demonstrates the need on the part of the celestial diviners to be able to produce prognostications *on demand*, much as their colleagues, the extispicers, could.

369 Brown, *CAJ* forthcoming.

were the result of demands for accuracy in the records of observations brought about in the wake of the new predictive astronomy.

So, it is on the one hand *not* surprising to us that some learned Mesopotamians designated the bright stars which form a curved tail by ^{mul}gír.tab, "scorpion", nor is it odd that they consequently considered it to be malefic. (Of course, of all the animals with curved tails, why they chose a scorpion is beyond knowing.) On the other hand, the categorisation of selected recurring phenomena into ideal periods made up of the nearest whole-number estimates of the real period, and schemes derived from that, *is* perhaps surprising if this was *not* done in order to make them predictable, but was done in order to make them amenable to interpretation. Surprising, I suggest, only because *we* are steeped in the desire to predict phenomena. However, just as a constellation makes it possible to locate a planet by reducing the number of spatial variables, and its naming means that its interaction with the planet can be interpreted, so an ideal period reduces all the variations in the observed periods to one number – a number that can be remembered, that can be *justified* by being round and thereby appearing significant – *and* which through its existence means that periodic phenomena can be interpreted. When the observed phenomenon occurs again after an interval different from the ideal, this provides an anomaly which bodes ill. When its appearance coheres with the ideal, this bodes well. The smooth running of the universe bodes well. I therefore suggest that all previous studies have misunderstood the ideal period scheme texts written before c.750 BC by confusing their *intention* with their *potential* to predict phenomena. They were never intended to predict celestial phenomena to any useful level of accuracy (an accuracy that would permit the diviner to know when at night to watch, or whether an event obscured by bad weather had actually occurred, say) as texts from the period after c. 750 BC were capable of doing. They were intended to make the regularly recurring phenomena of the heavens predictable. This is why Mul.Apin, which includes so many of the ideal schemes, also includes many omens, and why the main collection of celestial omens, EAE, includes so many period schemes. Both formed part of the same Paradigm. Both were *variable reducing*, as found in the omens, and *anomaly producing*, as found in the ideal schemes, in order that the heavens be made interpretable.

3.3 Reflections on the EAE Paradigm – Canonisation

I have posited the existence of a set of three fundamental and unchanging premises lying behind cuneiform celestial divination. Firstly, I argued, the universe was categorised in such a way that the number of variables by which one celestial event could differ from another was small. Next, I proposed that each variable was encoded with either a good or ill-boding meaning, and with a sense which meant that its reading pertained either to home or to others' homes. Finally, I suggested that this simple code was elaborated through a series of number-based and/or text-based rules into the omens and ideal schemes found in EAE. I do not suggest that celestial divination in Mesopotamia evolved in a simple way from categorisation, through encoding to elaboration, for it is possible to imagine that some encoding preceded certain categories, for example, or at least emerged simultaneously. Any discussion of this earliest pre-literate evolution of the EAE Paradigm would be speculative. The premises, once outlined, must be understood as given, I argued. However, the evolution of celestial divination thereafter can be determined to some extent, and I have already made a number of comments to this end, which I will now bring together.

The Scholars who used EAE in the 7th century BC, and those who first wrote down celestial omens in the OB period were practitioners of the same Paradigm. They employed most of the same names for the heavenly bodies and their phenomena, interpreted the phenomena similarly, and used the same set of texts – EAE, Mul.Apin, the Astrolabes and so forth. They shared the same way of seeing the universe, and used the same code and rules to read meaning into the configurations of the bodies and phenomena of the heavens. To borrow terminology designed by Lakatos (1978) 48 to describe sciences, the "irrefutable hard core" of the hypotheses underpinning EAE and the related texts was unchanging throughout this period. The same, however, could not be said of the "progressive protective belt" which surrounds the core.

The vast majority of omens found in EAE, and used by the NA and NB Scholars were the result of learned, rational thought, and were in no way a mere collection of observations and examples. The EAE Paradigm was very much *the* Paradigm for those interested in the sky throughout a large area of the Near East, and for at least a millennium up to the late NA period. While the 70-tablet EAE was its greatest exposition, it was not sufficient that a celestial diviner of the level of a Scholar or *ummânu* should merely be familiar with the text. The methods, premises, hypotheses, which led to its creation were also known to them as the following indicates:

When Balasî writes in x060r.1f about omens concerned with malformed births, his remarks undoubtedly applied to celestial omens as well:

> "*Šumma izbu* is difficult to interpret (*ana parāsi*)... really [the one] who has [not] had (the meaning) pointed out to him cannot possibly understand it."

He is saying that the correct understanding of the written texts can only be made by an expert who knows *more* about the discipline than is apparent from simply reading the omens. This was a knowledge that was probably only transmitted orally, and its acquisition probably enabled a Scholar to "master" (*gummuru*) a series rather than simply being able to "read" it, as discussed in Ch.1.4. The categories, the code and the rules outlined above formed, I suggest, a small part of that additional orally transmitted knowledge. This would account in part for the Scholars' long training and also for the presence of only a few texts in which the code and rules are written out explicitly. Those concerned with word-play, such as the *ṣâtu* material (§3.2.1), relied particularly on the equations between certain signs found in the lexical material, and were consequently best suited to being transmitted in a written format alongside any oral explanation as to why and when this kind of elaboration was justified.

I suggest, then, that far from being the mere *users* of the EAE series, the Scholars of the NA and NB period were fully aware of the underlying premises that made up the Paradigm they used. This is made most apparent by their continual adaptation of the omens to suit the prevailing socio-political situation. Some examples of this behaviour were provided in Ch.2.1.2, where I showed that the Assyrian Scholars sometimes reinterpreted omens that boded ill for Akkad on the basis that Subartu could be equated with Assyria, while other Scholars equated Akkad and Assyria directly. It all depended on the prognostications the Scholars wished to send. Similarly in x112: 23f the Babylonian Bēl-ušezib writes:

> "If the Moon and Sun are seen together on the 15th; a strong enemy will raise his weapons against the land... Now then, the army of the king, my lord, has raised its weapons against the Mannean and will tear down his royal city..."

He has twisted the ill-boding omen to suit the current military operation. Another example occurs in *Sargon's Eighth Campaign.*[370] Clearly, the Scholars of the late NA period felt free to adapt to their own ends, in certain limited instances, some EAE omens. They did not merely send in the relevant quotations culled from the tablets.

Further to this, within the Reports and Letters are found references to omens that are said to be "from the Scholar's mouth" – *ša pî ummâni.* In 8158: 9f Nabû-mušeṣi writes:

> "If a meteor [flares up] from the rising of the [north wind] to the rising of the [south wind]... This omen is from the mouth of a Scholar, when Nebuchadrezzar (I) broke Elam."

Thus, in the MB period as well, omens were being invented by Scholars that did not find their way into the 70 tablet series of EAE, but which continued to be either written or transmitted orally through to the late NA period. A written extraneous *aḫû* series was also in circulation amongst Scholars,[371] and these experts appear to have held both it and the *ša pî ummâni* omens in equal regard to the *iškaru* series of EAE.[372] Both also applied the same premises that I have discerned lying behind EAE, Mul.Apin and the like.[373] That is, they were as much a part of the EAE Paradigm as was the omen series of that name, but they do indicate that *additions* to the corpus of celestial omens were being made throughout the time from the MB period to the NA.

Equally, *adaptations* to EAE were being made throughout this period as well. The series never became fully fixed as the variant numbering schemes attest to most clearly. The presence in later copies of the names of some MB kings (App.1 §21) demonstrates that the later second millennium redactors felt free not only to gather existing omens and elaborate on them using the rules of textual play, but to *include* omens that could only have been invented at that time.

Adaptations and additions to the EAE Paradigm continued through to the 8th and 7th centuries BC, and incorporated such things as the movement of the ideal vernal equinox from the 15th of month XII to the 15th of month I, the systematic use of the 14th as the ideal day of lunar "opposition" (see §3.1.2), the invention of seasonal hours based on the "ideal year", various versions of the "Pleiaden Schaltregel", the geometric gloss on the Nippur variant of EAE 14 concerning the scheme for lunar visibility during an "equinoctial" month, and the development of many variant omen apodoses. Many more such adaptations and additions could be noted, but their full extent will not become apparent until the series (particularly EAE) are more completely edited. However, while the textual history of the celestial

370 For details see Oppenheim (1960).

371 For an edition of one tablet of this series see Rochberg-Halton (1987a).

372 See n30.

373 Note the use of the regular cardinal directions in the *ša pî* omen quoted. In the published *aḫû* tablet (Rochberg, 1987a), the omens attested concern eclipses. They are arranged by month, just as many of the EAE eclipse tablets are. In particular they concern the appearance of the eclipse, but the watch in which they occur, the prevailing winds, the quadrants obscured, and the direction of shadow are all recorded, albeit with some unusual vocabulary. The same basic categories are attested. The *aḫû* text is "unusual only with regard to its content, not to its form", to quote Rochberg loc. cit. 331.

divinatory material remains to be reconstructed, it is apparent *now* that the core hypotheses of the discipline were not altered.

The variant omen apodoses only rarely altered an ill-boding prognosis to a good boding one or vice versa, and the variations in the ideal schemes never reproduced the behaviours of the celestial phenomena in question to any substantially greater degree of accuracy. The variations did not challenge the fundamental hypotheses of the Paradigm – the basic categories, the decipherment according to the simple code and the rendering of that decipherment in omen form according to the hermeneutic rules of elaboration. The variations employed the same hypotheses as the oldest known celestial omens, and, in a manner akin to what Kuhn calls "normal puzzle-solving science" (see §3.1.4), were simply used to account for more and more heavenly scenarios.

Thus, the traditional view that EAE was "canonised" in the MB period, and that the NA and NB Scholars used it as if it were a sacred text only to be quoted from, must be modified. I suggest instead that the EAE Paradigm developed over the centuries following the OB period, drawing on still older fundamental premises that had first enabled the heavens to be interpreted. Its arena of applicability was ever widened and adapted to the changing political climate, and in the MB period an initial group of the oldest written omens was enlarged to produce a 70-tablet collection including elements of *every* kind of text then derived from the Paradigm. It became very much *the* text of the Paradigm, the basic reference work for anyone entering the discipline, and was called EAE, a title which had been used as early as the OB period. It was transmitted abroad, and copied from generation to generation along with a few related texts, but they and it were never fixed once and for all. Different schools sometimes applied different schemata of the code,[374] and with different rules and local modifications generated variant apodoses. At the level of individuals, different Scholars occasionally derived different interpretations from the same protases depending on their personal agendas *vis à vis* their employers. This too resulted in variant apodoses, and these were either added to the series, or formed part of extraneous series. In the extreme case the impact of single learned scribes can be seen in the preservation of the so-called *ša pî ummâni* omens. In certain cases the meanings of some technical terms in the protases were lost, and in order to make sense of otherwise incomprehensible omens these were reinterpreted.[375] Sometimes, protases that described events that could not happen were made amenable to interpretation by associating certain names (see Ch.2.1.1 MARS). Undoubtedly, the errors caused by miscopying and loss had a further impact on the omen versions used by the Scholars employed by the last Assyrian kings.

Such is the development of a normal, unchallenging, puzzle-solving science, or of a progressive protective belt around a hard-core of basic hypotheses, and this EAE Paradigm of fundamental hypotheses existed in a written form for more than a millennium before its practitioners began to record the data necessary for, and attempt the accurate prediction of, celestial phenomena. Until late NA times the *ideal* period schemes made the regular

---

374 See n210.

375 Very much the view of Reiner & Pingree *BPO*2 and *BPO*3, who assume that the original "uncorrupted" omens must have made good astronomical sense; i.e. if a constellation is said to approach (te) another constellation, it must mean that te has changed its meaning (*BPO*2.2.2.8.2) over the intervening thousand years. I accept the possibility but, as noted above in n295, feel it to be much more likely that the constellations were said to "approach" each other by *analogy* with the movement of the planets. Indeed, some OB texts also describe phenomena that manifestly cannot happen (which I termed the "impossible protases") – see nn273 & 338. Clearly, these were not also "corrupted", but were *invented*.

running of the universe amenable to interpretation, but *thereafter* they were complemented by period schemes whose aim was the accurate prediction of certain celestial phenomena. In many cases these predictive period schemes adhered to the forms attested for the ideal schemes, using the same mathematical techniques, even calling themselves by the same names. They were, however, quite different in purpose, and their production established a further Paradigm, and this forms the subject of the next chapter.

CHAPTER 4

The Prediction of Celestial Phenomena (PCP) Paradigm

The PCP Paradigm is the name I have given to that Mesopotamian cuneiform astrology-astronomy attested in the last centuries of the first millennium BC. It is characterised by the so-called Mathematical Astronomical-Astrological Texts (MAATs) and the Non-Mathematical Astronomical-Astrological Texts (NMAATs), as well as by Horoscopes, zodiacal astrology, and by celestial divination. It is to be distinguished from the EAE Paradigm as it incorporates for the first time, I argue, methods whereby certain celestial phenomena can be predicted to a high level of accuracy. I suggest that this knowledge was useful to those celestial diviners working for the last Assyrian kings – the particular circumstances of their employment developed in them the desire to calculate in advance certain phenomena to an accuracy of at least a day. At the same time the millennial heritage of the EAE Paradigm weighed heavy on their thinking, and this accounts for the particular methods employed to predict celestial phenomena, and for the nature of the material which has survived from the 8th, 7th and later centuries BC.

I will define the PCP Paradigm in §4.1, in particular highlighting the premises that underpin its predictive methods. This will involve a certain amount of technical discussion of the difficult MAATs, but is necessary in order to show that these same premises, and some of the methods derived from them, were being used as early as the period c. 750–612 BC. The textual evidence for this will be provided in §4.2. I will stress in §4.1 that both the intention behind the accurate prediction of celestial phenomena, and the premises and methods used to achieve this, relied heavily on EAE-type divination. This permits me to postulate that texts such as the Horoscopes and those utilising zodiacal astrology (App.1 §49), none of which pre-date the 5th century BC, came about as a consequence of the gradual *development* of the PCP Paradigm, and were not in themselves the *cause* of the subsequent mathematical treatments in cuneiform of celestial phenomena. These texts still relied on the "core hypotheses" of the PCP Paradigm – they came about as a result of the puzzle-solving dynamic of "normal science". That they had a transforming effect on society, only further demonstrates how well suited Kuhn's model of scientific development through Paradigms is to the situation which prevailed in Mesopotamia in the context of celestial writings. I do note, however, that these new forms of astrology may themselves have influenced some aspects of the subsequent development of the Paradigm, and secondly raise the possibility that some of the MAATs, while fulfilling the requirements of the Paradigm in terms of predicting ominous phenomena, contain within them refinements which attest to an "intellectual interest" on the part of their compilers. Their authors were, after all, learned individuals fully capable of excesses of pyrotechnic brilliance in belles lettres and, if *kalûs* for example, were capable of singing dirges in eme-sal, a long dead "women's" sociolect of Sumerian.[376] So while I propose that

[376] Many MAATs were authored by "lamentation priests" or *kalûs*, experts who sang to appease the eclipsed Moon, amongst other things, such eme-sal balag̃s as am.me.amaš.an.na "the bull in its fold". See Brown & Linssen (1997) BM 134701: 7' and comment. Given that (a) the length of an eclipse lasts up to about 4 hours, (b) the singing perhaps only takes place during the period up to maximum obscuration – the so-called "lamentation" phase, see

the initial impetus for, and the premises underpinning the PCP Paradigm can be accounted for on the basis of the needs of celestial divination, I do not suggest that all post-612 BC cuneiform astrology-astronomy can be reduced solely to the requirements of this activity.

4.1 Defining the Paradigm

Copies of EAE and Mul.Apin, for example, which date to the Hellenistic period have been found in Babylon and Uruk. To all intents and purposes, they are identical to their NA counterparts. They were perhaps preserved for antiquarian reasons in some cases, but it is also apparent that they continued to be used. Too little is known of the last NB, Persian and Greek courts in Babylonia to be sure whether or not celestial diviners were employed in the same way as they were under the last NA kings. Some of the Biblical references in the book of Daniel quoted at the start of Ch.1, above, suggest as much for the NB kings, but once Babylon had ceased to be an empire's capital, the group of cuneiform-trained Scholars surrounding and protecting the king may have also come to an end.

Thereafter, it was probably those experts employed by the great temples in Babylon,[377] Uruk and perhaps elsewhere (see n5) who continued to produce and copy works which both predicted and interpreted celestial phenomena – works of the PCP Paradigm. Whether their products were systematically sent to and read to Achaemenid, Seleucid[378] or Parthian overlords is thus far not known. It is, equally, not possible to be sure that the inscription of thousands of NMAATs, MAATs, and the repeated copying of divinatory texts was undertaken solely for the purposes of the "cult". To date, the few horoscopes preserved that attest the names of important private individuals,[379] the evidence that omens from EAE were imported into India during the 3rd century BC (App.1 §21), the extensive nature of the eclipse ritual,[380] and the internal evidence of the texts themselves (which indicates that even the latest MAATs adhered to forms established as early as the 2nd millennium BC – see Ch.5.1.1), all lead me to argue that cuneiform celestial divination was alive and well in Babylonian at least until the Christian era. Even if conducted only by temple employees, it still played a vital rôle in the life of the city. Those texts that considered the sky in ways unattested before c. 750 BC, namely those that accurately recorded and predicted celestial phenomena, must be understood in this light.

---

loc. cit. n14 – and (c) the balağs are immensely long, this may give us some clue as to the manner in which these songs were performed (note courtesy of J. Black).

377 The archive 80–6-17 in the British Museum, which contains some 40 MAATs, was found near the Amran mound in Babylon seemingly in a house near to the Marduk temple. See Reade (1986b) xviif. As Britton (forthcoming) has shown, the earliest of these tablets date to the 7th century BC, and the latest to the late 2nd. The remaining MAATs and NMAATs from Babylon were excavated illegally, and have been assigned to the fictitious "astronomical archive" of Esagila – see I.4 (Babylon) above.

378 Plutarch *Lives Alexander* LXXIII-IV describes "Chaldaean Diviners" warning Serapis, which is perhaps suggestive of this activity. The same is also noted in Diodorus Siculus *Library of Universal History* XVII 116. References cited in full in Parpola *LAS*2 xxix-xxx.

379 Including some Greek names – see Rochberg (1998) 4.

380 On the day upon which the eclipse is predicted to occur (no doubt by methods used in the NMAATs or MAATs) braziers and drums are prepared, the temple enterer, *šangû*, and the *kalûs* are involved, as well as "the people of the land" and "seven soldiers" (*BRM* IV 6: 21'-28') and ultimately the king (BM 134701: 17'f). Many of (Uruk's) temples are included in the ritual in various ways. See Brown & Linssen (1997) 150–4.

Neugebauer's view (1989) was that there was a firm discontinuity between the MAAT material (exemplified by the ephemerides and procedure texts edited in his own 1955 *ACT*) and what came before.[381] In contrast Rochberg-Halton (1993) 31f maintains that the scribes of the *ACT* material were very like the NA scribes, only affiliated to the temple rather than the king, commenting:

> "It may only be incidental that elements with affinities to modern science are to be found in the boundaries of Babylonian mathematical astronomy (quantitative, predictive, aesthetic elegance) – it should not be separated from the sacred, divination and magic."

The connection between non-mathematical and mathematical astronomical-astrological cuneiform texts has been considered before. Aaboe (1980) 27 writes:

> "Speaking very generally, one can say the principle aim of the Babylonian astronomical theories. . . is to reproduce and forecast the astronomical content of the Diaries,"

and suggests a means whereby Diary-like records could have provided the data necessary to generate the parameters underlying a planetary ephemeris (more on this below). Sachs (1948) described the interrelationships of the Diaries, the Goal-Year Texts (GYTs), and the Almanacs and in §54 commented on their relationships to the *ACT* texts. That the Diaries (App.1 §45), or something similar, were an empirical *source* of the parameters found in the MAATs *is* argued here. Rochberg-Halton (1991b) identifies some of the problems with holding this view, without explicitly arguing against it. More recent scholarship, particularly that of Brack-Bernsen and Swerdlow, has, however, dramatically increased the likelihood of its validity.

In Ch.2.2.3 I established that a close relationship existed between the celestial omens used in the 7th century BC and the data recorded in the Diaries, the Eclipse, Mercury and Saturn records. In the following, and in Ch.5, I will argue that the intention, structure and presentation of both the MAATs and the NMAATs relied significantly on EAE-style celestial divination. I argue, then, that accurate prediction of celestial phenomena began in the 8th and 7th centuries BC under the auspices of the last NA kings, and that the means then used depended heavily on celestial divination, resulting in the establishment of certain core hypotheses of the predictive PCP Paradigm that remained in place until the very end of cuneiform writing.

4.1.1 What Phenomena were Predicted by the MAATs and the NMAATs?

As "MAATs" I include the fully-developed ephemerides and procedure texts published in *ACT* as well as the so-called "auxiliary texts".[382] All texts that are concerned with the ac-

[381] Swerdlow (personal communication) assures me that Neugebauer did recognise that the authors of the ephemerides also performed divination etc. He felt, apparently, that the *ACT* texts were the consequence of *more* motivation than that offered by celestial divination – they were the result of "intellectual interest". I discuss this possibility below, but stress that divination did create *strong* needs on the parts of the Scholars to find ways of predicting celestial phenomena.

[382] Ephemerides tabulate in columns those functions necessary for the prediction of successive phenomena. Procedure texts outline the means for calculating the various functions, but are poorly understood. They are the closest examples of texts we have that describe the theories lying behind the MAATs. See App.1 §44. Auxiliary texts again tabulate functions that deal with celestial predictions, but which do not appear in the ephemerides. Some of these

curate prediction of celestial phenomena treated in a mathematical way are MAATs. This differentiates them from the non-predictive, albeit mathematical, period schemes of EAE 14, Mul.Apin, i.NAM.giš.ḫur.an.ki.a, and so forth, treated in Ch.3.1.2, and from the astronomical, but non-mathematical texts such as the Eclipse and Planet Records, Diaries, GYTs and Almanacs – the NMAATs, some of which were very effective at predicting celestial phenomena. Some examples of all these texts are provided in the Introduction above, and the references to their publications can be found in App.1.

The following celestial phenomena were predicted by the lunar ephemerides:

1) The length of the month (29 or 30 days).
2) The date of new-Moon or full-Moon syzygy.
3) The dates and lengths of lunar visibilities near the syzygies – the "lunar six".[383]
4) The mid-time, zodiacal longitude and magnitude of lunar and solar eclipses.

The planetary ephemerides predicted in terms of time and location in the zodiac:

1) The heliacal risings.
2) The disappearances (east and west for the inner planets).
3) The stations (except for Mercury).
4) "Opposition" (acronychal rising – only occurs for the outer planets).

Some planetary ephemerides calculated the *daily* motion of the planets (see below).

The Eclipse Records, Planet Records and the Diaries mostly *recorded observed* data, but where the phenomenon in question had not been seen (often remarked with nu pap/šeš), a calculated value was included in order that a *continuous* record be kept. These *retro-calculated* or possibly *predicted* data included:

1) The luni-solar visibility intervals (the lunar six).
2) Eclipses – the date usually, time and magnitude sometimes.[384]
3) The solstices and equinoxes (in the later Diaries, according to a scheme).
4) The "ideal" date of planetary heliacal rising or disappearance (required because on the date observed the planet was sometimes "too high", meaning that it must have been missed when it first truly appeared or disappeared).

The means by which these predictions/calculations were made will be discussed in the next section. See also Ch.2.2.3, where these texts were discussed in detail and the relationship between what was recorded and what was ominous in the EAE Paradigm made apparent.

The Goal Year texts and the Almanacs were predictive in character, and not merely records of observations. No entries are missing, solar eclipses are mentioned "to be

functions appear simpler than those in the ephemerides and are considered older or even antecedent to the latter. See App.1 §43.

[383] The intervals between the disappearances of the Moon and Sun at the beginning, end, and middle of the month. See Ch.2.2.3 and Hunger & Sachs *Diaries* 1 p20.

[384] Some eclipses in the Eclipse Records are said to occur a certain "rough" time after the Sun has risen (and thus the Moon has set, rendering the eclipse invisible – e.g. LBAT 1414 Iif, dating to –730. They are thus predictions or possibly retro-calculations. See §4.2.4.2. See also Strassmaier *Cambyses* 400 r.19f, for an eclipse predicted *to* occur.

watched for",[385] and non-visible lunar eclipses are noted. They also use the zodiac, unlike the earliest Diaries, Planetary and Eclipse Records, and attested examples date only to the Hellenistic period. The GYTs predict for a given year:

1) The dates and zodiacal signs of the planets' heliacal phenomena, excluding Venus's stations and the longitudes of "opposition" (only the date is given).
2) The dates and distances by which the planets pass by the Normal stars (n250).
3) Date, time and magnitude of eclipses.
4) The lunar six for 12 months (also noting the month lengths, being 29 or 30 days).
5) Šú + NA, me + gi_6 (these are the four mid-month lunar six intervals) for 12 months.

GYTs also contain some remarks about the weather if these interfered with the visibility of the observations, made some years earlier, on which the predictions were based. In those cases the values predicted for the celestial phenomena were based on *calculated* records – on the *unobserved* values in the Diaries, Eclipse and Planetary Records. Sometimes the "ideal" dates of planetary heliacal rising or disappearance are given. The predictions in the GYTs were derived from the Diaries, or from very similar, continuous records of observed celestial phenomena.

The Almanacs included for a year, month by month:

1) The month's length at 29 or 30 days.
2) The date and length of the mid-month NA (mid-month šú, me, and gi_6 were predicted only in some "Normal Star Almanacs").
3) The date and length of the end-month kur.
4) The dates and signs of the planets' heliacal phenomena, excluding Venus's stations and the longitude of "opposition" (only the date is given).
5) Planetary positions by zodiacal sign (if visible) at the beginning of the month.
6) The dates of planetary entry into the next zodiacal sign.
7) Solstices, equinoxes and Sirius phenomena (according to a scheme).
8) Dates, time, and sometimes magnitude of lunar and solar eclipses.
9) The dates and distances by which the planets pass by the Normal stars (only in "Normal star Almanacs").

The GYTs probably provided the bulk of the data in the Almanacs, excepting the solstice, equinox and Sirius data, and the invisible lunar eclipses (i.e. those predicted to occur when the Moon would not be visible). It is also possible that some of the data in the Almanacs were provided by the MAATs. I noted that the GYTs, Diaries, Planetary and Eclipse Records themselves contained a *few* predictions, and we shall see in the next section that some of these were based on mathematical methods. Finally, as we shall also see in §4.1.2, the MAATs themselves relied heavily on Diary-like data for the determination of many of their underlying parameters. The case for the interdependence of the MAATs and the NMAATs is very strong.

[385] The methods employed in both MAATs and NMAATs predicted solar eclipses to occur in the same way as lunar eclipses, just at the other syzygy. Due to parallax, solar eclipses, when occurring, are only visible along a narrow band of the earth's surface, unlike lunar eclipses which can be seen anywhere the Moon is visible.

The most evident characteristic of the Almanacs, the GYTs, the Diaries, Eclipse and Planetary Records, and the MAATs, is that the *bulk* of the data predicted in these texts was that considered ominous in the great celestial divination series EAE. I have already discussed in Ch.2.2.3 how many of the *observations* recorded in the Diaries, Eclipse and Planetary Records were determined by what was portentous. Of those celestial happenings that are amenable to prediction, these included the heliacal phenomena of the planets, their locations and interactions to varying degrees of precision, and those phenomena for which ideal period schemes exist. The predictions included in the Diaries, Eclipse and Planetary Records were designed to fill in gaps caused by observational lapses. We therefore do not need to consider further whether the predictions in these texts were of those phenomena considered ominous – they were.

As to the GYTs and the Almanacs, most of the phenomena they predicted were identical to those recorded in the Diaries, and Eclipse and Planetary Records. These included the length of the month, the lunar six, the dates of the heliacal phenomena of the superior and inferior planets, and the signs of the zodiac in which these occurred, and the date, time and magnitude of eclipses.

Taking each of these in turn, it is apparent from Ch.3.2.2 (xix) that a 29-day month boded ill, and a 30-day month boded well. The dates of lunar "opposition" were similarly encoded – see Ch.2.2.2 (6) – as were the luni-solar intervals. The Moon's presence "not according to its count" boded ill, as did its being "high" when first appearing – for details see Ch.3.2.2 (xix). Put another way, "ideal" values for NA[386] and gi_6[387] – the length of time for which the Moon is visible on the 1st of the month, and invisible after Sunset in the middle of the month, respectively – were derived mathematically in EAE 14, Mul.Apin and i.NAM.giš.ḫur.an.ki.a (see Ch.3.1.2) according to what I termed the "ideal scheme of lunar visibility/invisibility". This ideal scheme was encoded, as were all the others, in such a way that reality cohering with the ideal boded well, and its non-coherence boded ill. Even the unexpected date of lunar disappearance was malefic (8346: 1), which when set against the length of the month, meant that kur – the final lunar six value, and the time between Moonrise and Sunrise at the end of the lunation – was also a time interval whose length had ominous significance. In Ch.2.2.3 I noted that in Diary –567: 4 the mid-month time interval NA was equated with the phrase: "on the 14th, one god was seen with the other". This phrase referred to the Moon's *date* of morning setting, which boded well *only* if it occurred on the 14th. Thus, four of the lunar six (first-day NA, mid-month gi_6 and NA, and end-month kur) were made ominous through application of the "ideal month", and its numerical extrapolation in the "ideal lunar visibility/invisibility scheme". It is for this reason that they, and the two other mid-month time intervals which parallel them, were recorded in the Diaries and predicted in the GYTs, the Almanacs, and indeed in the lunar MAATs.

The dates of the heliacal phenomena of the superior and inferior planets were not ominous *directly*, by which is meant there were no omens for Jupiter's heliacal rising, say, on every day of the year. In general only the month in which the planet appeared for the first

386 Reading unknown – probably na = *manzāzu* – see Hunger & Sachs *Diaries* 1 p21. If this is the case then the use of ki.gub = *manzāzu* ought perhaps to be re-evaluated in certain celestial circumstances. For example Pingree argues in *BPO*3 p18 that ki.gub refers to the location on the horizon above which a planet rises or sets and reads VAT 10281: 106 "Venus changes her *manzāzu* from 9 months in the east and 9 months in the west" as referring to the planet's changing horizon position. It could equally be referring to its lengthening visibility interval.

387 Called gi_6.zal "night passes" in EAE 14.

time was ominous. However, the dates of rising or setting were made significant when compared with those dates of rising or setting predicted by the ideal schemes, such as the "ideal Venus" scheme described in Ch.3.1.2. This was noted in Ch.2.2.3, and evidence was provided in Ch.3.2.2 (xix). It was for this reason that the dates of the planetary heliacal phenomena were *first* recorded in the Diaries, and was at least one of the reasons why they were predicted in the GYTs and Almanacs. In due course, the recording of the dates of these phenomena over many years provided the data base necessary for their prediction, as we shall see in the next section. The aim of prediction characterises the PCP Paradigm, and accounts for why a continuous record of the phenomena was kept for hundreds of years. However, only their ominous significance explains why the phenomena were recorded *in the first place*. The predicted dates of the phenomena may also have come to be used in other types of astrology, but the *original* intention was, I argue, to predict those events deemed ominous by the EAE Paradigm.

The GYTs and Almanacs predicted the signs of the zodiac in which the heliacal phenomena of the planets occurred. In the earliest Diaries the constellations in which these events happened were recorded. These data were ominous according to the EAE Paradigm – see Ch.2.2.2 (17). However, after the mid-5th century BC, the Diaries often recorded the signs in which the phenomena occurred. The relationship between the ecliptic constellation and the zodiacal sign was direct,[388] and the significance attached to a given planetary phenomenon occurring in a constellation was broadly the same as that attached to its occurrence in a zodiacal sign which shared the same name.[389] This accounts, in part, for why these data were predicted in the GYTs and the Almanacs. It was also the case, however, that once the zodiac had been invented and its usefulness to astronomical prediction made apparent, the recording of the sign in which heliacal rising, say, took place was no doubt intended to *improve* the data base upon which the accurate prediction of the same phenomenon relied for its parameters. Equally, the rise in importance of zodiacal astrology may also explain why the signs and not the constellations in which these events occurred were often recorded in the later Diaries, and consequently predicted in the GYTs and Almanacs. The fact remains, however, that the oldest examples of those texts on which the GYTs and Almanacs depended most heavily for their data, recorded the constellations in which the heliacal events of the planets occurred *because these data were ominous*.

It is noteworthy in this regard that the GYTs, Almanacs and the MAATs did not predict the stations of the inferior planets. These events were also not ominous according to the EAE Paradigm, nor were they recorded in the Diaries, as noted in Ch.2.2.3 (5). Also, only the dates of opposition[390] of the superior planets were recorded in the Diaries and predicted in the GYTs and Almanacs – never their locations. Again, no omens specifically concerned with planetary opposition are attested. It must be assumed, then, that the date of this phenomenon was *not* recorded because it was ominous. Perhaps it was done to parallel the record of the dates of lunar opposition. It is also possible that the dates of opposition were recognised to provide a particularly good source of data for the determination of the param-

[388] See, for examples, the constellations listed in Ch.2.2.3 (17). See Rochberg-Halton (1984a) p119 and Van der Waerden (1952/3).

[389] This is made clear from BM 36746 published by Rochberg-Halton (1984a). This text contains omens based on eclipses occurring in zodiacal *signs*, but draws directly on the omens in EAE 19–20, in particular.

[390] Probably "acronychal rising" rather than opposition as we know it. See Ch.2.2.1.

eters in the MAATs,[391] in which case their inclusion in the Diaries, and prediction in other NMAATs was an *artefact* of the *predictive* aspect of the PCP Paradigm.

Finally, the month, day and watch in which eclipses occurred were ominous, as was their magnitude, in so far as this affected the quadrants obscured. Their location in a constellation could also be ominous (e.g. 8300: r.11), as could the simultaneous presence of planets, and whether or not the Moon rose or set eclipsed – see Ch.2.2.2 (11) and Ch.3.2.2. The GYTs and Almanacs predicted all these details, save for the eclipse location,[392] though this was roughly given by the date, since a well-regulated calendar established the position of one of the heavenly bodies involved in an eclipse, namely the Sun. In the MAATs the location of the Sun on any date was given directly in terms of degrees of longitude of the zodiac. These GYTs and Almanacs also provided some additional, non-ominous information concerning eclipses. In particular, they pinpointed the time and location of an eclipse more precisely than was required by the omen protases. The predicting of the precise time and location of an eclipse (and indeed of other ominous celestial phenomena) ensured that it would be observed. This explains why predictions at a high level of accuracy were attempted, when only the watch and constellation in which an eclipse occurred was ominous. A predicted eclipse still had to be watched for, for its full significance to be determined. This is because some of its ominous aspects – the colour, the wind, and so forth – were not amenable to prediction. This applied generally to Mesopotamian astronomy-astrology of the period following the 8th century BC; predicted celestial phenomena were still observed by the diviners in order that their full meaning be gleaned.[393] The predictions simply provided them with data as to when and where to look, prevented them from missing phenomena obscured by poor weather, and provided them with time to arrange the appropriate apotropaic rituals. Astronomical predictions did not alter celestial divination in this regard, it merely provided an *additional service* to celestial diviners.

The GYTs and Almanacs predicted other phenomena whose ominous significance is not apparent. Most notably this included the dates and distances by which the planets "passed by" the Normal stars. Of course, these ecliptic stars do identify constellations, and in this regard they do provide directly ominous data. I also noted in Ch.2.2.2 and n251 that some omens are known which describe in their protases the location of planets next to a *few* of the known Normal stars, and that there are omens in which Venus's presence next to *any* Normal star was considered ominous. This perhaps explains why the presence of the planets close to these stars were recorded, but the reasons for recording and predicting the *magnitude* of the distances and the dates on which this occurred must be sought elsewhere. Just as the earliest Diaries and Eclipse Records recorded eclipses in more detail than was required either by explicit omen protases, or by the indirect way in which reality was compared against ideals, so the earliest Diaries recorded the locations of the planets and their phenomena by means of Normal stars (on certain dates), the distances to which had no significance. And just as

391 In an unpublished appendix to his 1998 book entitled "Acronychal risings in Babylonian Planetary Theory", Swerdlow shows that a record of retro-calculated planetary "oppositions" could lead directly to parameters that correspond closely to those used in the planetary MAATs. It remains unclear, however, that this was indeed the way in which these parameters were derived.

392 This datum *is* given in *MLC* 2195, an Almanac from Uruk.

393 This contrasts with the situation which prevailed in China in the first century AD, according to Sivin (1969) 5: "Celestial phenomena which could not be predicted were ominous in the fullest sense of the word; they were omens. Every solution to a problem of astronomical prediction meant removal of one or more source of political anxiety."

those data lead in the GYTs and the Almanacs to the prediction of eclipses in more detail than was required by celestial divination, so the Normal star data led to the prediction of the location of the planets on certain days of the year. However, while the detailed eclipse data ensured that these ominous events would be observed, there was no particular significance in Jupiter, say, being near to a given star on a certain day. These predictions were a *by-product* of the accurate recording of location in texts such as the Diaries. The reason for the accurate recording of observed planetary locations, and of the locations of their heliacal phenomena, was in order that values for the periods after which they repeated those phenomena or locations could be discovered. How this was done is discussed in the next section. The predictions of the dates and distances by which the planets passed by the Normal stars were perhaps merely *artefacts* of the methods employed to predict those phenomena considered ominous by the EAE Paradigm – that is, they were artefacts of the PCP Paradigm.[394] As I show in the next section, these predictions may *also* have been of particular importance in the art of horoscopy.

The Almanacs also predicted the sign in which each of the planets stood at the beginning of each month, and the dates when they entered the next zodiacal sign. These data were not ominous, but neither could a record of their observations have provided the data necessary for the prediction of ominous phenomena. Their predictions were thus not artefacts of the PCP Paradigm. They did provide, however, some of the information required by the so-called "horoscopes". These texts listed a variety of heavenly data close in time to the date of birth of the client in question, and these included the signs in which the planets were located. It was the growth in the popularity of this form of natal astrology, once the zodiac had been invented, that accounts for why this information was recorded in the later Diaries, and predicted in the Almanacs.[395] The PCP Paradigm, by showing that some ominous celestial phenomena were predictable to a high level of accuracy, permitted the development of an astrology that before the late NA period would have been thought impossible to undertake,[396] since it required the listing of the most important celestial events *due to occur* soon after any given date. Once the practitioners of the PCP Paradigm had mastered this requirement, natal astrology became available to all who could afford the diviners' services. The repercussions of this development in celestial astrology-astronomy continue to be felt to this day, of course.

The dates of the equinoxes, solstices and Sirius phenomena were all predicted in the Almanacs, and indeed in the Diaries, according to a well-known luni-solar calendar scheme.[397] Sirius's rising was *ideally* located on the 15th of month IV, the date of the *ideal* summer solstice, in Mul.Apin Iii42. Its rising in that month boded well. Sachs (1952b) p113 noticed that the intercalary months of the calendar scheme were arranged in such a way that the star rose in month IV in 18 years out of 19. It rose on the 29th of month III in one year – a near-miss caused by technical reasons associated with when the scheme first began. In general, then, it would appear that the significance of the Sirius phenomena (occurring according to

394 It is possible that these predictions may have been used to check the accuracy of the predictive methods, but in the absence of any explicit evidence that this activity was undertaken, it is unwise to speculate on the basis that their interests in the *verifiability* of their astronomical models were similar to ours.

395 See Rochberg-Halton (1989b).

396 Much as Einsteinian-Quantum physics accounts for the same phenomena as Newtonian physics, but with additional or different core hypotheses, and ultimately has created new interests unthought of before this century.

397 Outlined by Sachs (1952b).

a scheme in Mul.Apin that used the ideal year) determined, at least in part, the structure of the luni-solar calendar scheme used in the NMAATs and the MAATs alike.

However, while this ideal scheme of the EAE Paradigm influenced the shape of the later calendar scheme, there is no doubt that that later scheme was very *effective* at regulating the luni-solar year. The regulating of the calendar ensured that the dates predicted for heliacal events (those occurring as a result of a particular relationship to the Sun) could be expressed in days of months (which depended on the Moon). It is undoubtedly the case that the calendar was regulated in order that this might occur. It was regulated to serve the interests of astronomical prediction, and it was not the case that the purpose of the astronomy was to regulate the calendar, as is sometimes suggested.[398] Although the predicting of the lengths of months in the Almanacs and in the lunar MAATs may have assisted in this, it was incidental to the major divinatory purpose of these texts.[399] I note in the next section that the predicting of the dates of seasonal/sidereal phenomena may have been useful in assisting the prediction of the dates of planetary phenomena, when these relied on periods in sidereal and not calendar years.

The phenomena predicted in the MAATs do not need much further comment in this section since they, too, were in the main those considered ominous. The precision with which the times and locations of eclipses and other planetary phenomena were predicted was partly a result of the way in which these texts achieved their ends, but it was also in order that these events would not be missed by the diviners for whom and by whom the MAATs were written.

However, a number of additional comments must be made about the MAATs themselves. Although continuing to subscribe to the core hypotheses of the PCP Paradigm, as we shall see in the next section, it is not sufficient to account for all of their idiosyncrasies to argue that their main purpose was to predict those phenomena deemed ominous by the previous Paradigm. For example, some planetary MAATs predicted when and where the stationary points of Venus would occur. These events were not ominous, so far as we are aware, nor were they predicted in the Almanacs or the GYTs. Equally, a number of MAATs calculate the day-to-day positions of the planets.[400] This information was generally not of significance to celestial divination. However, such data would have been of value to those compiling horoscopes, which required the calculation of the locations of the planets on a given date.[401] Alternatively, these exceptional MAATs were perhaps created for reason of "interest for its own sake",[402] or more likely, for reasons as yet not fully understood.

The MAATs, particularly the lunar and the day-to-day planetary ephemerides, are wonderfully sophisticated in their construction, and attest high levels of mathematical elegance, as well as some remarkably exact parameters. They were often extremely good at predict-

398 E.g. Chadwick (1992) 15: "...most Assyriologists and historians of science...maintain that the real impetus for the development of astronomy came from the need to develop a reliable and workable calendar". See also §4.2.3, below.

399 No calendrical purpose can be attached to any phenomena predicted by the NMAATs and the MAATs, aside from the length of the month. The length of the month was also largely incidental to forthcoming economic transactions, which employed the 30-day month of the "administrative year".

400 *ACT* 654 and 655, Jupiter ephemerides elucidated by Huber (1957), and *ACT* 310, which calculates the daily positions of Mercury. Day-to-day ephemerides are not yet attested for the other planets.

401 E.g. according to the text *MLC* 2190, Aristocrates was born on 4th of month III in 235 BC when Jupiter was in 18°of Sagittarius. This value corresponds well with modern retro-calculations. See Rochberg (1998) 83f.

402 Suggested by Neugebauer in *HAMA* 412.

ing celestial phenomena accurately, and those who have worked on their decipherment have often discussed them as if their intention were to do no more than this. In fact, as we shall see, even the most effective of MAATs continued to adhere to notions derived from celestial divination. So when Britton (1996) 60 writes:

> "The first goal of Babylonian lunar theory was to calculate the time intervals between syzygies of the same type,"

he portrays the lunar MAATs as modern in aim, since calculating syzygies are how *we* proceed in this kind of astronomy. The intention of the MAATs was to predict eclipses, month length, and the lunar 6, and their calculating of the times of syzygy was a means to those ends. Similarly, Aaboe comments in 1958, 244:

> "There is no doubt that System A' (for Jupiter) is an improvement over System A. This makes it difficult to understand why the two schemes were in simultaneous use."

However, the use of two different systems for calculating the times and positions of Jupiter's heliacal phenomena is only difficult to understand if we believe that the aim of these MAATs was to find the *most accurate* way of calculating them. On the contrary, while these texts were by and large effective at predicting celestial events, they were restricted in many ways by the divinatory tradition upon which they drew.

For example, the attested planetary MAATs did not model the variation in latitude of the planets, although we are aware that their authors were capable of doing this.[403] This severely limited their ability to predict the phenomena of Venus, in particular. In the column of the lunar MAATs, in which the latitude of the Moon was modelled, a value for the diameter of the Moon virtually *twice* its actual value was used. This was in order that 12 fingers marked a full obscuration of the body at eclipse, since this had always been the case when eclipses were observed.[404] After the late NA period, however, a finger measured 1/12°, leading to a value for the diameter of the Moon twice reality. Neugebauer in *HAMA* 551 writes on this matter:

> "not only is it difficult to understand how direct observation could result in so gross an error; neither should its consequences for the theory of eclipses have escaped astronomers who developed the most sophisticated methods for the computation of lunar ephemerides."

Describing what the astronomers should and should not have noticed seems an odd way to proceed towards an elucidation of their intentions. These scholars were perhaps not motivated by the same desire for the most exact of theories as are modern astronomers. Aaboe is probably correct when he writes (1977, p1) of those few texts that treat Venus's phenomena mathematically:

> "It may well be that the Venus texts are closer in character to the GYTs than to the standard ephemerides of the *ACT* type (MAATs), and the excellence of the eight-year period and the regular behaviour of Venus prevented the development of more sophisticated theories for this planet by making it unnecessary."

403 See text F in Neugebauer & Sachs (1967). The colophon of this text is the same as those used in the MAATs.
404 See n247.

In other words the predictions made by GYTs were probably *good enough* for the purposes of divination. I argue, therefore, that the intention of the MAATs was not to make the best possible predictions of the ominous phenomena, but to create *legitimate* solutions to the question as to how these events might be predicted. To this end they had to accord with certain underlying rules – rules which incorporated the core hypotheses of the PCP Paradigm, but which also included notions borrowed from celestial divination as to the make-up of the heavens. Some of these I have already alluded to, others I will discuss in Ch.5.

The rules imposed no restrictions on the accuracy of the underlying parameters, or how many different phenomena needed to be modelled to arrive at satisfactory solutions. They did mean, however, that several slightly different values for the time and location of the next heliacal phenomenon could be provided by different, but equally legitimate astronomical texts. These included values given by a variety of different MAATs, but also those provided by the GYTs and Almanacs. These different schemes may once have derived from different individuals in some instances,[405] but there is no doubt that they were frequently used *in the same place and at the same time*.

Continuing to use models that were plainly inferior at predicting astronomical events seems odd to our minds, but it was perhaps not odd to the diviner who was familiar with the fact that any one celestial event elicited more than one *interpretation* (as discussed in Ch.3.1.1). The core hypotheses of the EAE Paradigm could sometimes throw up different, but equally valid, interpretations of any one event described in a protasis depending on the application of what I termed in Ch.3 the "code" and "rules". Similarly, the core hypotheses of the PCP Paradigm meant that there were several ways in which the diviners could predict celestial phenomena, all of which were legitimate, but not all of which were, to our minds, equally effective. However, it must be recalled that even the poorest NMAATs and MAATs still enabled the diviner to locate the next ominous event to the nearest few days, or to the nearest section of the zodiac. Indeed, many of the most precise predictions of the MAATs, once translated into calendar days from the unit of time measure employed in calculations (*tithis*), lost most of their accuracy. Similarly, the inability to observe the abstract zodiac in the sky disguised shortcomings in the spatial predictions. Once, the predictions were good enough to ensure that the diviner would know roughly when and where to look, little further accuracy was needed. Further exactness was merely a by-product of the mathematical methods employed. Of course, the requirements of accuracy for those texts that predicted eclipses, month lengths and luni-solar intervals were very demanding, since the non-occurrence of a predicted eclipse, say, would show up a short-fall in that theory pretty rapidly. So while superficial appearances might lead one to consider the MAATs in the same light as modern astronomical texts, they were instead the most elaborate manifestations of texts belonging to the PCP Paradigm – a Paradigm which came about in the 8th and 7th centuries BC. They were not merely texts whose sole purpose was the most accurate possible prediction of celestial phenomena, they instead predicted events to a level of accuracy that was useful to diviners, and did so using methods that preserved a tradition of divination by then well over a thousand years old.

[405] Britton (1996) 63 remarks that the two basic lunar theories (systems A and B) were so different in their use of functions and parameters that "it is as if two competitors were assigned the same problem, but precluded from using any element of the other's solution". It is remarkable that the vast majority of the system A lunar MAATs were found in Babylon, while it was in Uruk that the majority of the system B lunar MAATs were located, the earliest attested examples of each dating to around 260 BC.

To summarise this section, the PCP Paradigm was characterised by NMAATs and MAATs that recorded and predicted those celestial phenomena considered either directly ominous, or indirectly so, through reality cohering or not cohering with an ideal. Occasionally, other phenomena that were recorded in order to determine the periods after which the *ominous* phenomena recurred, were also predicted. They were *by-products* of the *predictive* aspect of the PCP Paradigm. How the periods were elicited will be the subject of next section. In due course some phenomena were predicted for the purposes of zodiacal astrology, a new departure for celestial divination brought about by the invention of a means of locating calculated positions – the zodiac. Finally, some MAATs attest schemes whose sophistication makes them appear as if they intended to predict phenomena to the greatest possible accuracy. This sophistication is misleading, however, and these texts should not be interpreted in the way that modern astronomy is. This is made clearest by the diviner's simultaneous use of different schemes of varying exactness, but is also apparent from the adherence on the part of the MAATs to age-old assumptions as to the *nature* of the universe, a subject to which we shall return in Ch.5.

4.1.2 How were the Predictions Made?

The most basic hypothesis of the PCP Paradigm is that an accurate record of the times and/or positions of ominous celestial phenomena will lead to values for the periods between their recurrences, and thereby enable them to be predicted. For example, it was discovered that after 59 years Saturn once again rose heliacally, say, in the same part of the sky (which is the same time of the year, since the planet is in a fixed relationship with the Sun). In those 59 years the planet had risen heliacally 57 times and had travelled right around the ecliptic (the path of the Moon, the Sun and the planets) twice. If one wanted to know when and where Saturn would rise heliacally this year, all one would need to know is where and when this took place 59 years ago. This was the major way in which astronomical predictions were made in the NMAATs. In the case of the planets this form of prediction was effective, and it is by no means clear that the methods used in the MAATs achieved substantially better results. Predictions of this sort for the inferior and superior planets were already taking place in the 7th century BC – see §4.2.4.

Eclipses can be treated in the same way. A long enough record of their occurrences will reveal, for example, that eclipses of the same type recur after a period of 223 months, and accurate records will show that after this interval they recur some 81/2 hours later in the day. Eclipses were predicted using characteristic periods from at least the 8th century BC – see §4.2.4.3.

Even the lengths of months and the dates of lunar "opposition" can be predicted using the record of luni-solar intervals, the values of the sums of which repeat after characteristic periods. Brack-Bernsen discovered that the mid-month lunar six values of šú + NA and me + gi_6 repeat after an interval of 223 months.[406] If due to some mishap or meteorological obscuration it had been impossible to determine the length of the lunar six interval "me", say, its value could have been determined using the record of me + gi_6 223 months earlier, and the recent value for gi_6. If both me and gi_6 had been impossible to observe, then it so happens that:

[406] For details see idem 1997, which includes references to her earlier literature.

$me_i = me_{i-223} + 1/3(me + gi_6)_{i-223}$

where me_i refers to the value of me in the current month and me_{i-223} to its value 223 months earlier. A record of me and of me + gi_6 223 months earlier would make the prediction of this lunar interval possible. Analogous equations relate the other mid-month lunar six values to each other. It also happens that:

$NA^*_i = NA^*_{i-223} - 1/3(NA + šú)_{i-229}$ and
$kur_i = kur_{i-223} + 1/3(me + gi_6)_{i-229}$

where NA* refers to the interval between Sunset and Moonset at the beginning of the month, and kur to the interval between Moonrise and Sunrise at the end of the month.[407] These equations and those immediately above are attested explicitly in the Hellenistic period text *TU* 11: 29–38 (albeit doubled), and they explain why in the GYTs the values for the lunar six (and eclipses) 223 months earlier were recorded for a period of a year, and the values for the sums šú + NA and me + gi_6 229 months earlier were recorded for a period of 6 months. All the information necessary (and little else) to enable the calculation of any of the lunar six values during the whole of the goal year was recorded. Not only could the ominous lunar six be calculated, but also the dates of lunar "opposition" and the lengths of the months be predicted using these equations. A small predicted value for NA* for the coming month, say, and it may well be that the Moon will not be seen until a day later than expected, and so forth.[408] The extent to which this was undertaken in the period before 612 BC is discussed below in §4.2.4.2. Two comments can be made at this stage, however. Firstly, the lunar six were ominous *indirectly*, and but for the part of the EAE Paradigm that enabled diviners to elaborate period schemes and compare and interpret reality against ideals, *no* such record of the lunar six would have been made. Secondly, unlike with planetary phenomena or eclipses, the observance of which over a period of time will undoubtedly *give the impression* of their periodicity, no such effect was visible in the case of the lunar six. The fact that it was the values of their *sums* that recurred demonstrates that it could only have been the *close scrutiny* of their record that enabled such periodicities to be recognised.

The luni-solar calendar was also regulated using a period between recurrences. In this case it was discovered that 235 months occurred in 19 years. A record of the lengths of months in combination with a value for the length of the year based on the rising of a bright star such as Sirius might have led to the discovery of this relationship. It might more easily have been determined, however, from the records of eclipses, which are frequently separated by 235 months. Clearly lunar eclipses occur precisely at mid-month, and those separated by 235 months take place in exactly the same part of the sky[409] (beside the same star, say). This

---

407 Brack-Bernsen, loc. cit. 120 and 125. I also note that NA* and kur are calculated in "text K" (BM 36722 in Neugebauer & Sachs, 1969, 96f) according to rules which depend on the longitude of the Moon. The details of these rules are only poorly understood, but they were probably less effective than those cited above, and certainly less sophisticated than those used in the MAATs. Nevertheless, they indicate that many predictive solutions to the problem of determining the lengths of months were attempted, and it may have been just such "solutions" that were employed in the late NA period.

408 This is also alluded to in "text E", BM 41004, published by Neugebauer & Sachs (1967) 200–208, line r.19 "determine the full and hollow months". See the commentary op.cit. p205. Text E probably dates to the 4th century BC.

409 To within $1/5^{th}$ of a degree.

means that the Sun is also in the same place as it was 235 months earlier. In other words a whole number of years have passed.[410] I noted, above, the way in which the 235 months were distributed over the 19 years. We refer to this distribution as the "Metonic calendar", after Meton of Athens who at a later date adopted it. This calendar is still used today to determine the dates of certain religious festivals. See App.1 §40.

Periods noted from the repeat *positions* of heliacal phenomena (and thus marked in years) had to be determined from the record of *dates* expressed in months of the lunar calendar. For example, in the GYT *LBAT* 1285: 11 we find the line:

> "Year 185 of the Seleucid Era... month IV, night of the 1st, Venus was 8 fingers ($3/4^o$) below the rear star of the twin's feet (μ-Geminorum)."

8 years (= solar/sidereal years) later, in the 193rd year of the Seleucid Era (SE) – the "goal year" of this text – Venus will once again be in the same place (at a longitude of approximately 63° of the zodiac in our terminology), and at the same time of the year. Will this be on the same *date* in the lunar calendar, however? It will, of course, only be on the same date if month I of 193 SE begins at the same time of the seasonal year as month I, 185 SE. In fact 8 years last a period of time very close to the length of 99 months, so provided the calendar is so designed that this number of months exist between month I of 193 SE and month I of 185 SE, then Venus will indeed be near the rear star of the twin's feet on the 1st of month IV. 99 months can be distributed amongst 8 years in such a way that 5 years have 12 months each, and 3 years have 13 months each.[411] If the calendar were regulated in this way, then it would have been straightforward to determine the 8 year period for Venus directly from a record of the dates of its heliacal phenomena. No doubt this is one of the reasons why it is one of the earliest attested periods.[412] See §4.2.2.

The heliacal phenomena of Jupiter recur in the same place in the sky after approximately 12 years. 12 years, however, are not a whole number of months, so Jupiter's phenomena will *not* recur on the same date after this length of time. This, and many of the other planetary periods attested in the NMAATs are also not whole numbers of months. The 59-year period for Saturn, for example, lasts about 729 months and 23 odd days.[413] Clearly, the period in seasonal/sidereal years between recurrences of its phenomena at the same place in the

---

410 The small difference between the seasonal year and the sidereal need not concern us here. It did not concern the scribes and Scholars, as Neugebauer (1950) demonstrated.

411 This is usually the case in the Metonic calendar, wherein intercalation occurs at the end of years 3, 6, 9, 11, 14, and 17 and in the middle of year 1. Attested intercalations also show that 3 intercalations in 8 years occurred between 624 and 600 BC, and fairly consistently thereafter. See Fig.2, p67, in Britton (1993).

412 In the text BM 45728 = SH 81–7-6,135: 5f (Kugler *SSB*1 45, van der Waerden *BA* 107) it is said that Venus will repeat a heliacal phenomenon 8 years later, less 4 days. This is also quoted in "text E" (see n408, above) line r.5. Venus repeats a heliacal phenomenon after 8 (sidereal) years minus 2½ days, but 99 months last about a day and a half longer than 8 years. This shows, then, that the 4-day *error* in the 8-year period was determined from the record of the *calendar* dates of the phenomena, and not from seasonally adjusted ones. BM 45728 uses the older names of the planets and is perhaps to be dated to the centuries before the Hellenistic period. It is also noteworthy that the errors are recorded in *days* and not in *distances*, though both are used in "text E" for both Venus and Jupiter. In most cases the errors are noted in days.

413 In 59 years, Saturn is only 1° from its original position. The "59-year" period is thus virtually identical to the length of 59 sidereal years, as one would expect from an accurate period. In "text E" (see n408, above) line r.13 Saturn is said to "lack 6 days to your year". This presumably refers to the number of days by which Saturn's "59-year" period is less than 730 months (= 59 calendar years with 22 intercalary months), rather than the number of days by which 59 sidereal years are shorter than 59 calendar years. Compare n412, above.

sky could not have been determined directly from a record of its behaviour. Given this, it is all the more remarkable that this period was known even before the luni-solar calendar was systematically regulated with the Metonic cycle. It is possible, then, that the calendar dates of its heliacal events were translated into a number of days before and after the dates of seasonal events such as the equinoxes or solstices, or quasi-seasonal events such as the heliacal rising of bright stars. We know that the dates of these events were determined in the late NA period, and were recorded by the Scholars in their Reports (8140–2) – see §4.2.3. Their dates, those of the heliacal phenomena, and a roughly regulated luni-solar calendar were apparently sufficient to determine the values in years for the periods after which the planets repeated their ominous celestial events in the same place in the sky.

In a later period the equinoxes and solstices were predicted according to a scheme which included Sirius phenomena, and these dates were recorded in the Diaries and Almanacs. Probably, the purpose of these *predicted* dates for seasonal events was to facilitate the assigning of dates to those planetary predictions which relied on seasonal/sidereal years. In other words a GYT provided information on Saturn's behaviour on certain dates 59 years earlier. In the goal year these events will not occur on the same dates, but on the same number of days from the equinox. In order to assign dates to planetary events in the Almanacs, the dates of seasonal events in both the coming year and 59 years, say, before that, along with the pertinent planetary observations, would be needed. This, I believe, may explain the purpose of the dates of the equinoxes, solstices and Sirius phenomena recorded in the Almanacs. Of course, the errors in the periods listed in "text E" (see n408), BM 45728 (see n412) and other texts quoted in Neugebauer & Sachs (1967) 206–7 might have been used instead (or as well) to this same end.

In the Hellenistic period GYTs the following periods were used:

Jupiter	71 years (for heliacal phenomena)
	83y (for conjunctions with Normal stars)
Venus	8y
Mercury	46y
Saturn	59y
Mars	79y (for heliacal phenomena)
	47y (for conjunctions with Normal stars)
Moon	223m (for eclipses and lunar six)
	229m (for šú + NA, me + gi_6)

Two periods each were used for Mars and Jupiter – one for heliacal events, the other for distances to Normal stars. The "71-year" period for Jupiter is not as accurate as the "83-year" one. Its inaccuracy means that Jupiter is about 6° behind where it was in the sky one period earlier. This means, in turn, that the heliacal event in question will have occurred some 6 days earlier in the year, since the Sun moves about 1° per day. However, 71 years last some 878 months and about 6 days. Thus, if the luni-solar year were at least roughly regulated (with some 26 intercalations in 71 years), it would have been noticed that the *date* of any given heliacal event of Jupiter repeats after 71 years, even if it has taken place in a part of the sky some 6° distant from that of its earlier occurrence. "In 71 years the same day as before," it says of Jupiter in "text E" (see n408) line r.3. This 71-year interval was thus *not* in seasonal/sidereal years, but in well-regulated *calendar years*. Given that it was usually only the constellations in which Jupiter's heliacal phenomena took place, and not their precise locations, that were ominously significant, the 71-year period would have usually sufficed to predict when the planet would rise in Scorpius, say. This fulfilled a major divinatory need, and it is little wonder, then, that this period is attested long before the 83-year one – see §4.2.2. For predicting sidereal phenomena, such as distances from Normal stars, a period in terms of sidereal/seasonal years was needed. This was the purpose of the period of 83 sidereal years. In the case of Mars the 47-year period was in well-regulated calendar years, and the 79-year period in sidereal ones.[414]

I noted above that I believe the Almanacs to have provided data of use to those compiling horoscopes. The latter required foreknowledge of planetary positions for the purposes of assigning zodiacal longitudes at the time of birth (see n401). It is extremely unlikely that these longitudes were observed, since the abstract zodiac was hard to locate visually, but they may have been interpolated from the locations of the planets with respect to Normal stars, assuming the scribes were able to transform distances from Normal stars into longitudes of the zodiac. In the text BM 46083 published by Sachs (1952c), who considered it to pre-date the Hellenistic period, this is precisely what is done: One line reads, for example:

> ^{Múl}rín *šá* tu$_{15}$.1 20 (UŠ) Rín
> "The southern part of the Scales (=B-Librae) (is at) 20 (degrees) of Libra."

The Normal star ^{Múl}rín *šá* tu$_{15}$.1 was assigned a longitude, thereby permitting locations in terms of distances from Normal stars to be transformed into locations within zodiacal signs, and making the interpolation of a planet's longitude at the date of birth quite possible. If for some reason the observation of planetary distances from Normal stars was not possible, however, using periods in *sidereal* years and the record in the Diaries, say, of planet locations next to Normal stars, the *predicting* of the planets' locations on particular dates close to the date of birth would have been straightforward. We have no explicit evidence that this was the manner in which some longitudes were assigned to the planets in the horoscopes, but I do believe that my suggestion explains, at least in part, the purpose of the predictions in the Almanacs of the dates and distances by which the planets pass by the Normal stars.

If it is not merely chance that the 71-year period for Jupiter is attested earlier than the 83-year one, then this suggests that (a) in the records from which the periods were determined, the *dates* upon which the heliacal phenomena occurred were of greater significance than their precise locations, and (b) the 71-year period proved sufficiently accurate for the purposes of celestial divination, but the further demands of zodiacal or other astrology made the more accurate 83-year period necessary. It is unlikely that the 83-year period was not discovered early merely because for its determination lunar calendar dates had to be transformed into seasonal/sidereal ones, since the 59-year period for Saturn and other periods in sidereal/seasonal years were determined by late NA times (see §4.2.2). We will return shortly to

[414] After one so-called "47-year" period incorporating (in this case) 22 heliacal events of the same type, Mars is some $8^1/_2{}^\circ$ short of its original position. The 22nd event will thus take place 8–9 days earlier *in the year* (w.r.t. the vernal equinox, say) than the first. 47 sidereal years last 17167 days, but 581 months (47 calendar years with 17 intercalations) last 17157 days – 10 days fewer. Again, 47 "calendar years" are an accurate period of time after which Mars repeats the dates, if not the locations, of its heliacal events. In BM 45728 (see n412) Mars is said to re-perform the same heliacal phenomenon after "47 years", except that 12 days have to be added to the original *date*. Was this an attempt to determine the error in the period with regard to the sidereal year? The "errors" in the periods for Mars in "text E" (see n408, above) seem themselves to be in error.

the question of the possible precedence of dates and times over distances and locations in texts of the PCP Paradigm, but in the first instance it would seem *a priori* reasonable that the predicted calendar date for a planetary phenomenon would have been of greatest use to a diviner in his dealings with those for whom the divination was intended. On a simplistic level, knowing *when* an eclipse was due would provide the diviner with the knowledge of when to prepare the apotropaic rituals and warn the king. Its predicted location was relevant only if it showed that the eclipse would be invisible, and for the fine details of its interpretation. There is little doubt that in Mesopotamia, in the first instance, the prediction of eclipses relied on a record of the *times* and *dates* of these phenomena. The same may also have been largely true in the case of the prediction of the phenomena of the other planets. As to the increasing demands for greater accuracy in the latest periods of cuneiform writing, this characterised the achievements of that other group of predictive texts – the MAATs – to which we will now turn.

The MAATs also relied on extremely accurate values for the periods between ominous phenomena of the same type. In general they used longer periods than those employed in the NMAATs, but, significantly, these longer periods were built up from the shorter ones used in the NMAATs. In order to do this the errors in the short periods were determined. For example, in the ephemerides of Mars a period of 284 years was used. This derived from the following combination of the two periods used in the GYTs:

Mars, 284 years = 3(79) + 1(47).

It was observed that the 79-year period was three times more accurate than the 47-year period. Also, after this interval Mars was slightly further along the ecliptic than it had been at the beginning, whereas after the 47-year one the reverse was true. The errors therefore offset each other when the periods were combined in this manner, and so led to the highly accurate, long period value. Other long periods are:

Jupiter, 427 years = 5(83) + 12
Saturn, 265 years = 4(59) + 29
Mercury, 480 years = 9(46) + 2(33)
Venus, 1151 years = 144(8) -1[415]

The errors in the short NMAAT periods could have been determined from a record of the locations of the phenomena by Normal stars, say. This might have been expressed as a distance of, say, 6° in the case of Jupiter, as noted above. It is, however, more likely that they were registered in terms of a number of days by which the phenomenon in question recurred earlier or later than the full interval in *sidereal* years. This appears to have been the case in BM 45728 where the error in Mars's 47-year period is given as 12 days (see n414). In that same text, and in "text E" (see n408), the error in the calendar-year periods for the planets

[415] In this case not another period, but the error of $2^1/_2$° in the 8 year period (see n412) was used to determine the longer period. $2^1/_2$° = 360°/144. It is possible that in this exceptional case, the error was noted as a distance, rather than as an interval of time. However, the use of such a round number ($144 = 12^2$) for the multiplier does suggest that considerations such as 1/12th of 1/12th (a month) of an ideal year may have also come into play. The number 12 plays a large rôle in cuneiform astronomy-astrology – the ideal year, the zodiac, the relationship of the cubit to the UŠ, the number of fingers in an eclipse – and this may be one further example of that play. The 1151-year period for Venus may have been derived from mathematical play with an *appropriate* number. See further in Ch.5.1.1.

are also mostly expressed as days (see nn412 and 415). In the Diaries the dates of the heliacal events are recorded consistently, whereas their locations by Normal stars are not. Of course, if acronychal rising were the phenomenon used to determine the long period (see n391), then only its dates and *never* its locations are recorded.

It is noteworthy that great care was taken by the scribes to record the precise dates of heliacal rising and setting. If a planet was observed to be too high at first or last appearance, dates a few days earlier or later respectively for the two heliacal events in question were noted. The means by which these *true* dates were determined remain to be understood,[416] but, significantly, the planets' excessive altitudes were noted in terms of the *time* that they took to set after Sunset at first appearance, or to rise before Sunrise at last appearance. They were *not* recorded as distances in space. We will see in §4.2.1 that the lengths of just these intervals were recorded in texts uncovered in Nineveh, and which therefore predate 612 BC. We can assume that they were representative of early efforts to determine accurate periods for the planets.

Not only were long periods determined, but the number of phenomena, and the number of times around the ecliptic the planet had travelled in each long period recorded. Thus, in 427 years, Jupiter travels around the ecliptic 36 times, and performs 391 heliacal phenomena of the same type.

Lunar MAATs also relied on long periods between the recurrence of phenomena of the same type, and a count of the number of these phenomena. These included, of course, the number of months in a certain number of years, the best attested being the one which underpins the Metonic cycle: 235 months = 19 years. This was superseded by a more accurate value in one group of lunar MAATs to wit: 2783 months = 225 years. This more accurate relationship was perhaps derived from the interval between eclipses, rather than an observation of the distance by which the period of 235 months exceeded 19 sidereal years, since this was probably too small to measure. Intervals between eclipses of the same type, and a count of the number of eclipse possibilities were also fundamental parameters which underpinned the lunar MAATs. It was discovered, I presume from the written record of their occurrences, that in 223 months 38 eclipse possibilities take place, since during that period the Moon's latitude is small at the month end and beginning 38 times. (A small lunar latitude is necessary for an eclipse to occur – see Ch.2.2.1 under draconitic month). After 223 months, the Moon returns to the same velocity, but during that period it has varied above and below an average value 239 times. Thus 239 cycles of lunar velocity = 223 months (see Ch.2.2.1 under anomalistic month).

In other lunar MAATs the following relationships were used:

251 months = 269 cycles of lunar velocity
1655 months = 282 eclipse possibilities
2729 months = 465 eclipse possibilities.

These are slightly more accurate than the relationships based on 223 months.[417] The periodic undulation of the Moon and the planets above and below the ecliptic was noticed by the

416 See n255 above.
417 See Walker & Britton (1996) 53f.

diviners. Ominous significance was attached to the phenomenon at least by the 4th century BC, as the text *SpTU* = Hunger (1976a) 94: 1f indicates:

> "If you want to make a prediction for the market price of barley [ ] notice the movements of the planets. If you observe the first visibilities, the last visibilities, the stationary points, the conjunctions... the faintness and brightness of the planets and zodiacal signs and their positive and negative latitude... your prediction for the coming year will be correct."[418]

The text is a copy dated to 320 BC, and the reference to the zodiac places the date of the composition in the period after the mid-5th century BC. However, the majority of the phenomena the diviner is recommended to observe are those considered ominous in EAE. I assume that an increased awareness of the movements of the planets (including the Moon) in latitude was brought about by the attempt to predict their behaviour. This phenomenon was then ascribed ominous significance in precisely the same way as other planetary phenomena had been more than a thousand years earlier. It was encoded in the following way:

> A propitious planet's positive latitude boded well, its negative latitude boded ill.
> A malefic planet's positive latitude boded ill, its negative latitude boded well.[419]

This scheme can be added to those in Ch.3.2.2, and attests to the ongoing development of the divinatory side of the PCP Paradigm under the influence of predictive astronomy, just as does zodiacal astrology.

We can assume that if diviners were able to judge the latitude of the Moon at any given moment for the purposes of divination, they were also able to count the number of times that the Moon passed through the line of the ecliptic during any given period, and so discover the relationships noted above between eclipse possibilities and numbers of months. While judging latitude required observations of position, it did not necessarily require any measurements of *distance*. As to the discovery of the number of lunar cycles of velocity in a given number of months, I assume that the Moon's varying velocity was directly observable,[420] but that it took a combination of eclipse records and the summed record of some of the lunar six to discover such facts as the Moon's velocity returning to its initial value after 223 months.

The length of a month, the length of the lunar six values, and the detailed circumstances of an eclipse depend on five things; (1) the length of daylight, (2) the angle of the ecliptic to the horizon, (3) the latitude of the Moon, (4) the velocity of the Sun,[421] and (5) the velocity of the Moon. Because 223 months last 18 years (to within 11°), (1), (2) and (4) are pretty much

418 See Koch-Westenholz (1995) 170–1 and Rochberg (1998) 42–43. Statements about lunar latitude occur in three horoscopes, showing that this development of the EAE Paradigm found its way into this part of zodiacal astrology.

419 Hunger (1976a) 94: 9–14 and 20–21.

420 I suggest that one of the reasons why the Moon's location next to Normal stars was recorded in the Diaries was in order to reveal the extent of its daily movement in longitude and latitude – its maximum, minimum and average daily movement, and the number of days after which it returned to the same velocity or latitude. In "text E" (n408, above) obv.1f, for example, the passage of the Moon beside a series of normal stars and constellations is described. Its positions of maximum and minimum latitude are noted, as are the locations of the *nodes* (latitude = 0) and their movement.

421 Also called the zodiacal anomaly. This is in reality the result of the movement of the earth in an ellipse around the Sun.

returned to their initial values after this period, and because 223 months are accurately equal to a whole number of draconitic months (cycles of lunar latitude), then eclipses separated by this period will be of equal circumstance if and only if the lunar velocity is also returned to its initial value. Since the eclipses are, 223 months must be a whole number of lunar velocity cycles. The same discovery emerges from the record of the sum of the mid-month lunar six (šú + NA + me + gi_6), as is apparent from the fact that šú + NA and me + gi_6 repeat their values after 223 months, as noted above.

While these long periods employed in the MAATs could have been used to make predictions directly in the manner of the NMAATs (though in some cases it would have meant using records going back hundreds, or even thousands of years), there was no particular advantage in doing this. While the error per year or per month in these periods were very low, the total errors over the whole periods were comparable to those in the shorter NMAAT periods. For example, after 256 years Saturn is actually slightly further from its starting point than it is after 59 years, although the error per year is significantly smaller in the 256-year period. How then could the long, accurate period be used in the prediction of planetary phenomena?

This was done by determining a *mean interval*, using the count of the number of phenomena in each long period. For example, in Jupiter's case, the mean interval between successive heliacal events of the same type was calculated to be 427/391 years, or 1;5,32 years. The interval was not expressed in this manner, however, but in terms of a year and 30^{ths} of a month, units we refer to now as *tithis* after the Sanskrit term,[422] but which were simply designated **ud** "days" by the scribes of the ephemerides. The dividing of a month into the ideal, propitious, 30 "days" was one of many legacies of EAE-style divination to be found in the NMAATs and MAATs. 427/391 years was thus expressed as 1 year and 45;14 *tithis*. This value emerges in the following way:

According to the Metonic cycle 1 year lasts 235/19 months.
235/19 months multiplied by 427/391 years = 13;30,28 months.[423]
Since 1 month is 30 *tithis* by definition 13;30,28 months = 12 months + 45;14*t*.

45;14 *tithis* were themselves rendered as 1 month of 30 *tithis* and a further 15;14 *tithis*. Predictions in *tithis* were undoubtedly equated to calendar days, so had predictions for Jupiter's heliacal rising, say, been made using only this average interval, they would have been separated by 1 calendar year, 1 month and 15 days. Predictions on this basis are substantially less accurate than those using a 71-year period in the NMAATs.

The average spatial distance between successive phenomena can also be calculated using the same parameters. In 36 revolutions of the ecliptic Jupiter performs 391 heliacal events of the same type. Their average spacing is thus 36/391 of the ecliptic revolution. This revolution was made equal to 360 UŠ,[424] comprising 12 signs of 30 UŠ each. The signs were given the names of nearby constellations, and thus the zodiac was created. Its

---

422 See *HAMA* 349 for their discovery. They are first attested in BM 36731, which describes the period 616–588 BC – see §4.2.1.

423 There is a tiny error of 2/3600 in this calculation which is not of importance. It is eliminated, in any case, if one performs the calculations using attested parameters for the length of the year derived from the relationship between 235 months and 19 years, but rounded down.

424 For details see Brown *CAJ* forthcoming.

debt to the ideal year of celestial divination is immediately apparent, and its purpose was to enable the expression of *calculated* spatial locations to take place. The mean spatial interval between successive Jupiter oppositions, say, was thus calculated to be 36/391 times 360 UŠ = 33;8,45 UŠ. Predictions using only this mean value would have been less accurate than those derived from the 83-year period, however. Although the mean value is accurate, the actual intervals between successive heliacal phenomena of the same type vary quite substantially about the mean.

The same applies to the long periods used in the lunar MAATs. 38 eclipse possibilities in 223 months led the scribes to a value for the mean interval between successive eclipse possibilities of 223/38 = 5;52,6,18 months. 282 eclipse possibilities in 1655 months correspondingly gives 5;52,7,39 months for this interval. Eclipse possibilities, however, are in reality separated by 6 *whole* months and occasionally by 5 *whole* months.[425] A value for the mean interval is not useful in this form alone.

It is, therefore, extremely probable that the manner in which the *actual* intervals varied about the mean were incorporated into MAATs at the same time as the long periods were used to generate mean intervals. The variations in the actual intervals between successive heliacal phenomena depend on where in the zodiac they occur. This is implicit in the existence of sidereal periods for these events.[426] After 427 years Jupiter once again performs the same heliacal event in the same place in the ecliptic. The cycle repeats. Wherever in the ecliptic one begins the cycle, it repeats after this time period. Thus, the length of the first interval in the 427-year cycle will be the same length as the first one in the next cycle and so on, meaning that each interval between successive phenomena is a function of the longitude at which the beginning of the interval takes place, *or* of the number in the cycle at which it occurs (1–390). In the case of the phenomena for which no sidereal periods exist, the variation about the mean can only be a function of its number in the cycle – its "event number". Both these *arguments* were used by the authors of the MAATs. The variation in the actual intervals about the mean intervals (whether time of distance) were either modelled against longitude or against event number. In the case of the argument of longitude the variation around the mean was described in terms of a step function, but in the case of the argument of event number the variation was modelled with a linear zigzag function. Both these piecewise linear techniques for expressing variation were employed in celestial divination in the mathematical speculation that characterised the creation of ideal period schemes. See Ch.3.2.1. This *appropriate* form of modelling heavenly behaviour came to be used in the MAATs, and, I suggest, helped legitimate the results thereby wrought. Since both forms of modelling were employed in the EAE Paradigm, both could be in the PCP Paradigm, and so predictions derived from zigzag or step functions were equally legitimate.

For example, in one column of the (system B) lunar MAATs the amount by which the length of the month exceeds 29 days due *only* to the variation in lunar velocity was modelled. The amount by which it varies due to other effects were modelled in other columns and summed at the end. The mean value for the length of the month used was 29;31,50,8,20 (a

[425] The 6-month and 5-month intervals between successive eclipse possibilities fall into a scheme which was known to the diviners, and exemplified by what is now known as the "*Saros Canon*" – a table of eclipse possibilities arranged into 24 columns, each lasting 223 months and covering 38 eclipse possibilities. The whole canon lasts 432 years, with the best known (*LBAT* 1414, 1415+, 1419 – see App.1 §39 for references) arranging the eclipses in the period 747–315 BC.

[426] Assuming what are called the "apsidal lines" are fixed – i.e. that the planet's ellipse around the Sun remains in the same place.

very accurate value, probably derived from an eclipse interval), which means that the mean of the numbers in the column is 0;31,50,8,20 days, expressed as 191;0,5 UŠ (1 day = 360 UŠ). The maximum value for the amount by which a month length exceeds 29 days (due only to lunar velocity) is given as 269;27,5 UŠ, and the minimum as 112;34,35 UŠ. The amounts by which the lengths of successive months exceed 29 days rise towards the maximum and then fall towards the minimum by 22;30 UŠ (1/16th of a day). They vary as does a zigzag about the mean value. After 251 months and 18 maxima and minima, the values recur. This is because this column of numbers is based on the empirical discovery that in 251 months a full number of cycles of lunar velocity (in this case 269) take place. This relationship between months and cycles of lunar velocity, this value for the mean length of the month, and the month by month variation due only to lunar velocity of 1/16th of a day, suffice to fix all the parameters in this column, once an initial value had been determined.[427] We will return below to the question of how the extent of such variations was discovered.

In one scheme for Jupiter (system A), the amount by which the intervals between its successive heliacal phenomena of the same type varied about the mean values of 1 year, 1 month and 15;14 *tithis* in time, or 33;8,45 UŠ in distance, were modelled using a step function and an argument of longitude. For example in the text *ACT* 600, quoted in the introduction above, Jupiter's first station is predicted to occur in month I of year 113 of the Seleucid Era at 28;41,40 *tithis* and at 8;6 UŠ of Capricorn. Its next occurrence was calculated to take place 1 year and 48;5,10*t* later, at 16;46,50*t* of month II, year 114 SE, and 36 UŠ further along the ecliptic at 14;6 UŠ of Aquarius, and so forth. However, once Jupiter's predicted location had moved beyond 25 UŠ of Gemini, the time and spatial intervals decreased to 1 year 42;5,10*t* and 30 UŠ respectively. The intervals jump down in value at Gemini 25, and jump back up again at Sagittarius 0. They are fixed in between – only where Jupiter crosses Gemini 25 or Sagittarius 0 will the intervals take values intermediate to 36 and 30 UŠ and 48;5,10*t* and 42;5,10*t*. After 391 values, lasting precisely 427 years and going 36 times around the zodiac, the values all repeat. This relationship, a decision as to the point in the zodiac at which the intervals step up or down, and a figure for the smaller or larger of the two constant intervals means the entire scheme is fixed by one initial value. Not only empirical findings, but the ease of calculation dictated which numbers were chosen. The distance intervals of 30 and 36 UŠ are particularly "round", for example. At all times the difference between the intervals in *tithis* and those in UŠ is 12: 5,10, showing that once this parameter had been determined a record of either times or of locations was sufficient for the construction of the *entire* ephemeris.

Thus, with these two linear techniques, the basic long period relationships, and the discovery of a few more parameters, some of which were chosen for the ease with which they multiplied or divided, the MAATs were constructed. Some planetary MAATs employed not just two zones of the zodiac, in which the intervals between successive phenomena varied, but many zones. Some of these reproduced the actual behaviour of the planets with more success than the 2-zone or zigzag systems. Other planetary MAATs embellished the basic

427 It would appear that some initial values were chosen in such a way that the resulting calculated numbers of the function could be identified at a glance as belonging either to an ascending or descending branch (noted by Britton in an unpublished manuscript entitled "Babylonian Theories of Lunar Anomaly" pp12–13). This facilitated the work of the diviners who used the ephemerides, but does imply that the initial value was not *entirely* determined from observation. This would have impacted on the effectiveness of the MAAT at predicting phenomena, and underlines *again* the difference between these texts and modern astronomical ephemerides.

modelling of one heliacal phenomenon per planet by assigning intervals between successive, but different, heliacal events – say between first station and second station. In this way the times and locations of the *full* series of heliacal phenomena for any one planet could be calculated from one initial time and longitude. I noted above that this embellishment was extended in some cases to the creation of day-to-day ephemerides.

The two systems used in the lunar MAATs employed parameters of varying accuracy. The system B parameters were generally more accurate than those used in system A, but the zigzag techniques and associated parameters used to model the various behaviours were less effective than the step-functions of system A. All the systems, lunar and planetary, were used throughout the Hellenistic period, and it is not at all clear which are the oldest. They all owed a huge debt to the astronomy-astrology of EAE, not only in the purpose to which they were put, but also in their construction. I noted already their use of piece-wise linear techniques of mathematical modelling, the 360 UŠ in the zodiac, and the 30 *tithis* in the month, all of which were borrowed from the "ideal year" period scheme. In the column in the lunar MAATs that modelled the variation in the length of the night through the year, the dependence on the ideal year of celestial divination is most explicit:

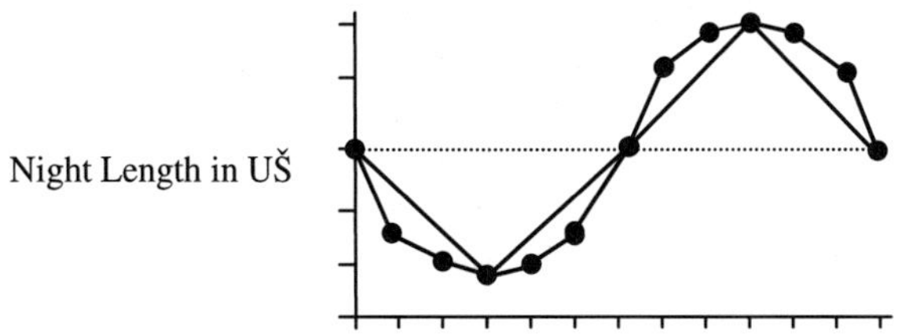

Location of the Sun in degrees of the zodiac from the vernal line (outside line a)
and days of the ideal year from equinox (inside line b)

As the diagram indicates, the length of the night was modelled in the MAATs (line a) in a way that was very close to the way in which it was done in texts of the EAE Paradigm (line b).[428] The equinoxes and solstices were still evenly spaced. The length of the night was measured in UŠ, and varied in a linear manner between points separated by intervals of 30 units (signs of the zodiac rather than months), and repeated after 360 units. Almost no observational data was incorporated into this column of the lunar MAATs. It was borrowed almost entirely from the ideal scheme used in celestial divination, save for a change in the argument from days to UŠ, and a slightly more complicated zigzag. These changes reproduced reality to an extent that was *good enough* for the diviners, but which was also *appropriate*, in so far as their discipline made use of the predictions resulting from that model. More on this in Ch.5.

I also noted the extent of the influence on the MAAT parameters of the NMAAT parameters – the long periods were made up of the shorter NMAAT periods, the Metonic cycle continued to be used in many of the MAATs, and the description of the extent of an eclipse

428 The 3:2 ratio between the longest and the shortest night is not attested in EAE, but in the text BM 29371 published by Brown, Fermor & Walker *AfO* forthcoming. This LB text utilises the ideal year as the argument, and served a divinatory purpose. It was part of the EAE Paradigm, and could *not* have assisted in the accurate timing of celestial events, any more than could the scheme in EAE 14, for example.

in terms of 1–12 fingers found its way into the MAAT scheme for lunar latitude.[429] The majority of the phenomena predicted in the MAATs were those recorded and predicted in the NMAATs, and I described above how some mathematical formulae were used in conjunction with periods to determine the lunar six values in the Diaries and GYTs. The errors in the short periods, upon which the derivation of the long periods depended, were undoubtedly derived from records such as those in the Diaries, and I commented that these errors were usually expressed as days. What of the other parameters underlying the MAATs – the extent of the variability about the mean, and in the case of system A for the planets, the locating of the zones?

Their origins continue to be the subject of controversy. For example, the column in the lunar MAATs of system A known to modern students as "Ø" has been the subject of much recent debate.[430] It models the length of the 223-month period in excess of 6585 days. This is an interval of about 81/2 hours on average, and according to column Ø varies as a linear zigzag[431] function from a maximum of about 9 hours, to a minimum of about 7 hours and 50 minutes. Brack-Bernsen (1980) shows that the real variation in the 223-month period is mainly dependent on the solar anomaly – the variation in the Sun's velocity as perceived from earth. However, since Ø's variation depends on the lunar anomaly[432] (the cycle of lunar velocity), this suggests that the function was not derived from the direct observation of eclipses separated by 223 months. In 1990 Brack-Bernsen showed how column Ø could have been derived from observations of phenomena whose length depended directly on the lunar anomaly. These were the mid-month lunar six values šú, NA, me, and gi_6 – called the 'lunar four'. Their sum[433] generates an oscillating function which varies in a way corresponding closely to the variation in Ø. She writes (op.cit. p45):

> "The linear zigzag function Ø has been derived from the sum of the 'lunar four'."

Britton (forthcoming – see n427), however, argues that Ø was built up from those relationships noted above between 235 months and 19 years, 223 months and 239 cycles of lunar velocity, and 251 months and 269 cycles of lunar velocity, in conjunction with critical observations of the *maximum* variations in the lengths of 235 and 223 months, and the maximum length of 235 months. Britton shows how the "rounding" of numbers at critical points in the construction of the function led to its particular parameters, and that the model could have been constructed without any measurement of position. Importantly, he also argues that the values for the maximum variations in the 223-month and 235-month periods were probably

---

429 See §4.1.1 and nn404 and 247. In Brown *CAJ* forthcoming, I argue that the UŠ and finger units of celestial distance measure are in a fixed numerical relationship with each other. It is this fixed relationship that means that the traditional division of a lunar diameter at eclipse into 12 fingers results in a lunar diameter being defined as twice reality in the MAATs.

430 Brack-Bernsen (1980, 1990, 1993, 1994 and 1997) and Britton (1987, 1990, and forthcoming – see n427).

431 In some MAATs that are not full-blown ephemerides the function was truncated – around its maximum and minimum points it took constant (and slightly different) values – a better approximation to reality. This was postulated by van der Waerden (reference idem, *BA* p228) and then discovered in a text describing the period -474 to –456 known as "Text S" – see App.1§43.

432 We know this because column F is derived directly from column Ø, but gives the varying velocity of the Moon. The formula relating the two columns is preserved in the procedure text *ACT* 200 §5 – an example of the explicit writing out of theory.

433 In 1994, 196f Brack-Bernsen and Schmidt describe the astronomical significance of the sums, and why their summation varies as it does, but they write (206) "we must assume the Babylonians did not know it...they succeeded in finding a purely empirical function".

derived from records of the 'lunar four'.

Thus, although we cannot be sure of the validity of either derivation of function Ø, it would appear that data such as those recorded in the Diaries lay behind the construction of even its most subtle parts. In particular, enough has been done to show that the record in the Diaries of the *indirectly* ominous phenomena of the *lengths in time* of the lunar six undoubtedly played an important part in the derivation of this critical function. The lengths of these intervals were made ominous in EAE through the use of ideal period schemes and mathematical play, and an encoding which ensured that coherence of observation with the ideal boded well, and non-coherence boded ill. The influence of EAE on the MAATs can be seen at every level, from their general purpose to the specifics of the origin of their parameters.

Although complete lunar ephemerides date only to the Seleucid period, a truncated Ø function, in combination with functions for eclipse magnitude and the longitudes of conjunction, is attested for data covering the second quarter of the 5th century BC. This so-called "text S" also includes observational (non-calculated) material. Britton (1989) 1, argues that it must date to a period before the full development of lunar System A had taken place. The existence of function Ø by this date suggests that its formulation may well go back into the 6th century BC – that is prior to the (attested) invention of the Zodiac. It was perhaps the attempt to model the solar/zodiacal anomaly, so that this could be used alongside Ø in the prediction of month length, the lunar six and eclipses, that prompted the invention of the zodiac. In both so-called "text A" and "text L" (App.1 §43), which date to the late 5th – early 4th century BC, two simpler treatments of the zodiacal anomaly are attested, which suggests that the evolution of the system A lunar MAATs took place during the period from at least c.450 – 350 BC. This is significant, for it indicates that even this one part of the full MAAT repertoire was *not* invented by one person, but was the result of a collective effort on the part of the scholars. This collective effort can be traced back to the courts of the NA kings, as we shall see in the next section, and forward towards the very end of cuneiform writing in the 1st century AD.

A number of theories also exist to account for the origin of the parameters which underpin the planetary ephemerides. These MAATs all use the zodiac, and all known examples date to the Hellenistic period. Aaboe (1980) showed how the six-zone system A scheme for Mars could have been derived from Diary-like records of its heliacal phenomena. These include, say, the date on which Mars reaches its first station and the zodiacal sign in which this happens. Using retro-calculated data (since the preserved Diary record is not complete) Aaboe shows that the number of times Mars's first station is located in a particular sign of the zodiac, over a long enough period, will be in close proportion to the values used to designate the zones in the six-zone system A ephemerides. Thus, not only is the mean interval (between heliacal phenomena of the same type) used in these MAATs derived from the attested NMAAT periods for Mars (79 and 47 years), but the manner in which this mean interval varies, and the location of the zones of the zodiac in which it varies are derivable from Diary-like records of heliacal phenomena – records which mainly give the dates and the zodiacal signs of these events, for it was these aspects that were deemed ominous in EAE. Aaboe's model of the origin of these system A parameters thereby takes us from celestial divination, via the NMAATs, to the MAATs.

Swerdlow (1998, and in a forthcoming publication entitled "Acronychal Risings in Babylonian Planetary Theory") has attempted to show how it is possible to derive the parameters of the planetary ephemerides from records of the *times* of the heliacal phenomena by assuming that the "synodic arc" (the spatial interval between successive heliacal phenomena of the same type) can be derived directly from the "synodic time", with the addition of a constant.[434] He writes (1998) 30–31:

> "the Diaries give precise reports of the time of phenomena, as first visibility, to the day of the calendar month, but imprecise reports of locations, by zodiacal sign, which is at best equivalent to giving the date only by month. However, if the limits of (the synodic time) ΔT may be found from such reports, the limits of (the synodic arc) $\Delta\lambda$ then follow from $\Delta\lambda = \Delta T - c$ without any measurement of position at all. And the limits of $\Delta\lambda$, or in some cases of ΔT by itself, along with the 'period' of the phenomenon... suffice to determine the parameters used... in the ephemerides... Date takes precedence over location... (though) it is also necessary to align the variable synodic arcs and times to the zodiac, for which the observation of location by zodiacal sign are adequate."

Swerdlow derives the MAAT parameters in this way for all the planets except Venus. Like Aaboe (1980), he has attempted to show how they could have been gleaned from records of the dates and zodiacal signs of celestial phenomena lasting many years. Also, just as in the proposed derivations of the lunar function Ø, wherein the fine details were determined, at least in part, from the record of the lengths of lunar visibility, Swerdlow argues that the accurate dates of the first and last visibilities were determined using the times for which the planets were above the horizon when first or last seen. These times were recorded for Mercury in the period before 612 BC – see §4.2.1. First and last visibilities were directly ominous, as were the constellations in which these events occurred. Their dates were made indirectly ominous according to the code of Ch.3.2.2 (xix) and Ch.2.2.2 (3 & 4), and significance was attached to their heights at first appearance as suggested by 8093: r.3 and x100: 7f, noted in Ch.2.2.3 and discussed in Ch.3.2.2 (xix). The concerns of EAE influenced which records were made and which periods were derived in the NMAATs, which in turn influenced the parameters – the mean intervals, variations, and locations of zones – in the MAATs. In his forthcoming paper Swerdlow prefers to derive the MAAT parameters from the record of the dates of non-ominous acronychal rising, though this is not to say that the record of first and last visibilities did not provide the data-base from which *some* parameters were determined for *some* of the planets. Whether or not he and/or Aaboe are correct about the origins of the parameters of the planetary MAATs is hard to say, but there is little doubt in my mind that NMAATs such as the Diaries and the Planet Records were the source of the underlying data.

To conclude this section, I have defined the PCP Paradigm as that which is characterised by the prediction of many of those phenomena considered ominous in EAE, but which incorporates celestial divination, and at a later date zodiacal astrology. It is underpinned by the core hypothesis that the accurate record of the locations and especially the dates and times of ominous celestial phenomena will lead to the discovery of *periods* and *parameters* in those data that will permit the prediction to take place of the times and dates (and the locations, though this was less important) of those very same phenomena. I argued that this activity was *useful* to the diviners in their craft, and in Ch.5 I explain why this was the case, particularly under the auspices of the last Assyrian kings.

[434] The "Sun-distance principle", Sonnenabstandprinzip, formulated by van der Waerden (1957), states that heliacal phenomena take place at a fixed elongation from the mean Sun. See Swerdlow' comments in (1998) 65.

I described how the Diaries, Eclipse and Planet Records contained data on the times (and often the locations) of directly ominous phenomena, indirectly ominous phenomena (e.g. the lunar six), and some interactions which elucidated behaviour that needed to be modelled if such things as month lengths were to be predicted (e.g. the Moon's location near to Normal stars). Both the subject matter and the choice of which of any particular details should be recorded accurately were determined by the EAE Paradigm. I noted that a continuous record over many decades, even centuries, was kept, and that the dates and times were recorded with especial accuracy. In order to assign near-seasonal dates to these events, the luni-solar calendar was regulated. Observational lapses were filled with values calculated using schemes which were themselves derived from the observational record. The intention, or at least the result, was to produce a large data base of material that would enable some scholars, perhaps only those in later generations, to elicit periods and parameters relevant to each planet that would result in that planet's ominous behaviour being predictable.

In the NMAATs the predictive power of those periods elicited from the data base was harnessed. NMAAT-type predictions were made as early as the 8th century BC (see §§ 4.2.2 and 4.2.4), but continued to be used until the Christian period. Over the centuries a number of different periods of varying accuracy were used for the planets. It is probable that gradually more and more accurate periods came into use over time, but provided any one period was *good enough*, it continued to be used. In due course the periods were used to predict phenomena that were of use not only in celestial divination, but in the arts of zodiacal astrology.

At least as early as the 6th century BC, mathematical formulae, such as those described by Brack-Bernsen for determining the lengths of the lunar six, came to be used in predictive astronomy. They were used in conjunction with the periods after which certain phenomena recurred, but they also employed parameters and mathematical techniques. These parameters were also determined from the long data-base of observations in the Diaries and related texts, and the mathematical techniques used were mostly those employed in the elaboration of ideal period schemes in EAE. The *mathematisation* of astronomical prediction had begun with these first MAATs. Function Ø, based on lunar velocity, was invented by the beginning of the 5th century BC at least, and in due course the zodiac was invented in order that calculated locations could be plotted, perhaps in particular so that the solar anomaly could be modelled. The importance of lunar latitude for the determination of month length, eclipse magnitude, and the lengths of the lunar six, was recognised. Latitude even became ominous, attesting the continued vitality of EAE-style divination. *Tithis* were invented so that the calculating of future times was facilitated, particularly when those times were far in the future, as was the case with the prediction of the phenomena of the inferior and superior planets. The zodiac, in particular, spawned a whole new discipline that itself modulated the subsequent development of both MAATs and NMAATs.

The MAATs were characterised by long periods that provided accurate mean intervals, and functions which modelled the variations around the mean with linear zigzags or step functions. These were borrowed from EAE 14 and the like. The parameters of the functions incorporated certain *special* numbers, including particularly *round* ones, and initial values that led to instantly recognisable products (see n427). These special numbers indicate that *ease of use* was also of importance to the creators of the MAATs. Some other special numbers preserved the traditional divisions of space and time used in texts of the EAE Paradigm, for example the 12 fingers in an eclipse, the 30 *tithis* in a month, 360 UŠ in the zodiac, and the longest: shortest day ratio of 3: 2. These numbers, the piece-wise linear techniques, the names of the zodiac constellations, and some aspects of their formal presentation to be discussed in Ch.5.1.1, show that the MAATs were also *appropriate* to the art of celestial divination. The MAATs, just as did the NMAATs, evolved over time. Various aspects of celestial behaviour were modelled at different times. They were not the work of one individual, but the results of the collective efforts of many erudite scholars from many periods. I have also stressed that at any given moment during the Hellenistic period, and no doubt before this, *several* different values for the time and location of any one phenomenon were predicted by the various different MAAT systems and by the NMAATs. Provided the predictions were *good enough*, and the methods *appropriate*, those less successful at predicting would continue to be used, I proposed.

So, I argue, the practice of the PCP Paradigm was consistent from its inception in the 8th century BC until the very end of the cuneiform tradition. It incorporated a wide variety of texts, and even some auxiliary hypotheses, such as truncated step functions, and fixed intervals between successive heliacal phenomena of different types. However, the core hypothesis was unchanging throughout the centuries – the accurate record of ominous phenomena provided the data from which could be elicited characteristic periods and parameters which rendered them predictable. If the intervals between successive ominous phenomena were modelled mathematically, the methods, units and even some of the parameters used had to be appropriate to celestial divination, and adequate for the purposes of the diviners in their work. The PCP Paradigm developed through the implementation of this hypothesis very much as does "normal science".

The core hypothesis of the PCP Paradigm cannot be framed as a law in the same way as can the hypothesis of circular motion which underpins much ancient Greek astronomy, or as can Newton's laws. It is not tied to a concept of matter, as is much post-Aristotelian physics, but as I will reveal in Ch.5.1.3, it can be shown to adhere to a notion of the nature (*physis*) of the universe which feeds into it from celestial divination, and from an even more ancient idea of an *ideal* primeaval beginning.

With regard to the thesis asserted in the first line of the introduction that the ability to predict certain planetary phenomena accurately first takes place in Mesopotamia during the late NA period, it is necessary only that there be evidence in material dating to the period before 612 BC of the implementation of the core hypothesis of the PCP Paradigm for this to be the case. The attestation of some of the parameters used in the later texts would be a bonus. The existence of both is revealed in the next section.

4.2 Evidence for the Use of the PCP Paradigm between c. 750 and 612 BC

The PCP Paradigm incorporated the accurate prediction of celestial phenomena into celestial divination. It involved the production of accurate records of celestial events, and the discovery from those records of periods of time after which the phenomena recurred. Evidence for both these exist in texts older than 612 BC, and are discussed in §§4.2.1 and 4.2.2, respectively. To assist in the discovery of these periods the calendar was regulated, and the evidence for this occurring in the late NA period will be presented in §4.2.3. In due course various further innovations connected to the mathematisation of prediction were made, and what little evidence there is for these in the period before 612 BC will be mentioned in

§4.2.1. Finally, in §4.2.4, evidence for the attempts on the behalf of the Scholars to predict the time and occurrence of celestial phenomena, and the limitations of the methods they used during the period from c.750 to 612 BC will be presented.

4.2.1 Accurate Records of Phenomena

Accurate records of eclipses, such as the one quoted in the introduction, dating to as early as the beginning of Nabonassar's reign are known. Walker (1997) 18 proposes that the four lunar eclipses in two years which immediately followed Nabonassar's accession in 747 BC may have triggered the beginning of such astronomical record-keeping. Berossus did claim (see n23, above) that earlier records were destroyed by Nabonassar, but it is most significant that the earliest of the so-called Babylonian Eclipse Records are much *simpler* than the later ones, strongly suggesting that in the 8th century BC the activity of accurate record taking was *in its infancy.*

For example, *LBAT* 1413 which records several successive eclipses, the earliest of which (probably) dates to 747 BC, provides only the dates, watches, and in one case the time before Sunrise of occurrence. Although only a few records can be dated to the 8th and 7th centuries BC, those in *LBAT* 1414, 1415+1416+1417, 1419 (part of the *Saros Canon* – see App.1 §39) that date to the 7th century also indicate the extent of the eclipse, whether or not the Moon rose or set eclipsed, the direction of the movement of the shadow, and from 632 BC on, the duration of the entire eclipse and its location beside stars or constellations (e.g. *LBAT* 1417 obv. II, *LBAT* 1419 obv. III). In *LBAT* 1420 – a collection of a few successive eclipses which date from 603 BC on – the eclipses are also described in a much more detailed manner than the earliest of those recorded in *LBAT* 1413. The time of occurrence is given to the nearest 5 UŠ, the shadow movement is described, the magnitude, total duration and so forth. Eclipse records which date to the 6th century BC consistently include the date, the time to the nearest 5 UŠ, the lengths of the various phases of the eclipse to the nearest UŠ, the direction of the shadow's movement, and the location of the eclipse in the heavens, and records of this sort persisted at least until the 1st century BC. A few further details were occasionally included, such as the varying velocity of the shadow in *LBAT* 1413 rev. V (317 BC), and the distance in cubits of the eclipse to a star in *LBAT* 1366 (251 BC). The final innovation was to give the time of the eclipse in terms of the culminating of a *ziqpu* star. This is first employed in *LBAT* 1436, 324 (194 BC), and used thereafter.

I argue that the eclipses were recorded in increasing detail over the centuries as the means by which they could be predicted became increasingly apparent. Even the very oldest records include some data that were not directly ominous, and we must suspect that these extra details were recorded because it was felt, or recognised, that they would provide some of the means by which subsequent eclipses might be predicted. Some of these details filtered into the Scholars' correspondence, demonstrating that they were familiar with, or even participants in, the accurate recording of celestial phenomena. The Assyrian Letter x149, for example, records the culminating *ziqpu* star at the time of an eclipse, and the eclipse's magnitude, showing that even Assyrians were familiar with these Babylonian methods. As we shall see in §4.2.4.3, the Eclipse Records included some predictions. The manner in which such predictions could have been made in the 8th and 7th centuries will be discussed there in the light of the evidence in the Letters and Reports.

The earliest Diaries also recorded fewer details of celestial behaviour, and with less accuracy than the later ones, suggesting that they too were part of a *new* activity in astronomy-astrology in the 7th and 8th centuries BC. The earliest known Diary dates to 652 BC. It remarks only of the first day in month I that the Moon was "bright and high", and, as noted in Ch.2.2.3, comments that on the 15th "one god was seen with another," when referring to lunar "opposition". By the date of the next attested Diary in 568 BC, however, values for the lunar six intervals were being systematically noted and even *predicted* (obv.11, and see §4.2.4.2). This is indicative of substantial development in the treatment of lunar behaviour in this period.

However, not only did the -651 Diary anticipate developments in lunar prediction, it accurately recorded the dates of the superior and inferior planets' heliacal events, the separation of the planets in fingers and cubits, and the location of Mars to the nearest finger (iv.15'). Such accuracy is also attested in the 7th century BC records of Saturn's heliacal risings and settings,[435] discussed in Ch.2.2.3. This planet was precisely located beside Normal stars, and once using the units UŠ. On one occasion (line 23') the "last appearance, because of cloud, was computed (*muš-šúḫ*)". The means by which this prediction was undertaken will be discussed below.

Evidence of the accurate recording of planetary phenomena can be found in the 7th century Letters and Reports as well. For example:

> "[Last year], it (Jupiter) became visible on the 22nd of month II in [Perseus], it disappeared in month I of the [present] year on the 29th" (x362: 3'f, see also x100: 5).
> "Mars which stands inside Scorpius is about to move out; (not) until the 25th of (month IV?) will it move out of Scorpius" (8387: 3).
> "Mars was sighted in month V; now it has approached within 21/2 spans (*ūṭu*/*rūṭu*) of Libra" (x172: 4').

The first shows that the precise dates of earlier heliacal events were recorded by the Scholars, no doubt so that they could interpret the visibility and invisibility periods according to the code of Ch.3.2.2 (xix). The last two examples show that the constellation boundaries were well defined, and that the rough rate at which Mars (and no doubt the other planets) moves after a stationary point was known. The distances between the planets and the constellations were carefully measured in fingers and spans,[436] but never in UŠ (as a unit of celestial distance).

In one group of texts, however, the UŠ *is* used to describe what appear to be distances, as indeed it is in the Saturn Records. These are the Mercury Records, published by Reiner & Pingree (1975), and discussed briefly in Ch.2.2.3, above. They cannot be dated precisely, but three of the four published can be assigned to the Kuyunjik collection, and so pre-date 612 BC.[437] UŠ are used for the interval between Mercury's first appearance and Sunrise, and dates are given. This is stated in two forms, depending on whether heliacal rising in the east, or in the west is being described:

---

435 Walker (1999).

436 E.g. 8082: 8, 8489: r.6, 8500: 4, x047: r.1 & x084: r.7.

437 Of the texts from Nineveh, K6153 is in Babylonian script, Rm2303 and 2361 are in Assyrian script. BM 37467 is not from the Kuyunjik collection.

Diš *ina* iti MN [d]gu$_4$.ud ud.x.kam *ina* [d]utu.è 10 UŠ gùb [d]utu gar-*ma* igi-*ir*.
Diš *ina* iti MN [d]gu$_4$.ud ud.x.kam *ina* [d]utu.šú.a ki.ta [d]utu 10 UŠ gar-*ma* igi-*ir*.

"- Mercury became visible in the east in month MN on day x, (and) the Sun is 10 UŠ to the left."
"- Mercury became visible in the west in month MN on day x, (and) the Sun is 10 UŠ below."

The difference in the forms no doubt expresses the fact that at heliacal rising in the east Mercury is obliterated by the Sun as it rises, where in the west the Sun has set before Mercury sets. The references to "the left" and "below", suggest that the intervals were perceived of as distances. In the later Diaries the following form is used to describe the intervals between heliacal rising and Sunrise:

gu$_4$.ud *ina* šú *ina* lu igi 14 NA-*su* (Diary -346: r.26)
"Mercury became visible in the west in Aries. 14(UŠ) Sunset to the setting of Mercury."

11 UŠ NA *šá* gu$_4$.ud (Diary -346: r.5)
"Rising of Mercury to Sunrise, 11 UŠ"

The term NA has replaced the "Sun is so and so UŠ below/to the left", but the Mercury Records certainly anticipate this form of record in the Diaries. Because of the parallel with the NA of the lunar six, it seems probable that the UŠ referred to in the Diaries in this context described *times*, though these may still have been perceived in terms of distances.[438] The NA interval first appears in the -418 Diary (r.3'), as well as a corresponding interval related to heliacal setting. As noted in Ch.2.2.3, were the NA interval too long the author, using (we presume) a mathematical formula as yet not determined, wrote down an earlier date on which the planet was thought to have truly risen.[439] The reverse was done for the dates of last appearance. This is perhaps what was being referred to in the 7th century BC Saturn Records in line 23', when the date of last appearance was said to have been "computed", *muššuḫ*. The Diaries also record if a planet is "high" nim, *elû* (e.g. -382 Diary: 17'). A planet being "high" (nim-a) on first appearance is also noted in the Saturn Records, and in the Reports and Letters references were made to the planets being "high", using *šaqû* (e.g. x100: 7 for Mars). The Scholars were aware of a planet's usual altitude at first appearance, and ominous significance was attached to its being "high" at heliacal rising, as well as to its visibility and invisibility periods.

The Mercury Records indicate that the means by which the 5th century BC compilers of the Diaries determined the *actual* dates of heliacal rising and setting were probably being employed in Nineveh in the 7th century BC. Also, the distances by which, or the times for which, Mercury was visible at first appearance were only recorded to the nearest 10 UŠ. This is less precise than was the case in the later Diaries (see n439), but suggests only that this method of determining the actual dates of heliacal events was again *in its infancy*. On the one hand these imprecise data would have provided the Scholars with the means to calculate the visibility and invisibility periods of Mercury regardless of the weather. They would have permitted comparisons to be made with the ideal periods of the EAE Paradigm. On the

438 Distances in "right ascension", perhaps. This issue is discussed in Brown, *CAJ* forthcoming.

439 E.g. in Diary -366: ii 38, Mercury's first appearance in the west after Sunset, and the planet's subsequent setting some 16 UŠ (64 minutes) later, provided the compiler with the information necessary to calculate the day upon which Mercury *actually* rose – in that case two days earlier. See also n255, above.

other hand these data enabled the Scholars, or others, to create an accurate and continuous data-base of the actual dates of heliacal risings and settings from which such things as the characteristic periods of the planets could be determined, and in turn the long periods and parameters of the MAATs. What fulfilled a divinatory need impacted directly on the form of predictive astronomy produced. I also note in passing that those Mercury texts in Assyrian script demonstrate that under the last Assyrian kings there is every reason to believe that the Assyrian Scholars were very much the equals of their Babylonian counterparts in the practice of both the PCP and EAE Paradigms.

One further text deserves mention in this section. This is BM 36731 (see App.1 §38). It describes the period 616–588 BC, but is a late copy. It concerns the last and first visibilities of Sirius and equinoxes and solstices, and can be dated securely through the attested intercalations. It uses parameters for the year that are different from those found in the ephemerides or in the Metonic cycle, and are less accurate. They are, however, substantially more accurate than any known from earlier periods. This suggests that BM 36731, too, might reflect an *early* attempt at determining an accurate value for the length of the year. The later Diaries record the dates of the phenomena of Sirius and the equinoxes and solstices, often not observed but calculated according to a scheme.[440] BM 36731 shows that they were accurately recorded as early as the 7th century. Importantly, this text uses *tithis*, a unit of time measure used in the planetary MAATs. Their use suggests that long intervals of time were calculated, since the expression of these in *tithis* is much easier than in calendar months. It is not beyond the realms of possibility that *tithis* were used to facilitate the calculations of the dates of planetary phenomena at this early period. This small piece of evidence, and the still unknown rules by which the actual dates of helical rising and setting were determined from the visibility periods, suggests that the earliest MAATs may also have come about in the late NA period. Aside from the evidence it presents of accurate measurement in the 7th century BC, BM 36731 attests an early attempt to regulate the luni-solar year. This itself may well have been driven by a desire to predict phenomena, which requires calendar regulation. More on this in §4.2.3.

4.2.2 Knowledge of Characteristic Planetary Periods

One set of texts that has received little attention by those writing histories of Mesopotamian "astronomy",[441] is the cryptic astrological-astronomical series DT72+78+81–6-25,136.[442] I date the composition of this short series to the period before 612 BC, though DT 72+ may itself be a late copy. Its colophon reads: "of Assurbanipal king of the lan[d of Assyria]". Hunger (1967 & 1975) dates the text to the LB period on the basis of its script and internal criteria. One of these is the so-called Seleucid period order of the planets (J,V,Me,S,Ma). However, this same order is found in the -651 Diary: 7–10, though possibly this was governed by the order of observation.[443] Similarly, the planet names are the same as those used in the -651 Diary, except GENNA (Saturn) – a name first attested in the -567 Diary. So, even if Late Babylonian, DT 72+ may well predate 612 BC, and there is some possibility that the texts themselves were found in Nineveh. In any case the colophon is all important – the first

440 Hunger & Sachs (1988) p27 and Sachs (1952b).
441 E.g. Neugebauer *HAMA*, van der Waerden *BA*, and Britton (1993).
442 See n52, above.
443 See Hunger &Sachs (1988) p26.

version of the text was written during or before Assurbanipal's reign. The reason DT 72+ is of interest in the history of predictive astronomy is because of lines 56'-66' (according to Gadd's 1967 numbering) or 2'-12' of the reverse of DT 78 (according to Hunger's 1967 numbering). They read:

^{d}sag.me.[gar	Jupi[ter
^{d}*dele-bat ana* š[à	Venus [...] into [
^{d}gu$_4$.ud *ana* šà[	Mercury [...] into [
^{d}genna *ana* šà ^{d}[	Saturn [...] into [
^{d}*ṣal ana* šà ^{d}30[	Mars [...] into the Moon [
ú-te-ṭè 1,11 mu.meš	he darkened 71 years
ú-šaḫ-rib 1 mu.meš	he laid waste 60 years
ú-kin 59 mu.meš	he made firm 59 years
uš-te-šir 8 mu.meš	he set in order 8 years
ú-lab-bir-ma 15 mu.meš	he prolonged 15 years
ú-li-iṣ-ma 12 mu.meš	he rejoiced 12 years

The five planets are listed, but it is unclear what it is they are described as doing. Below this appears a list of years with a variety of epithets. The number of years and the epithets undoubtedly pertain to the planets. 71 years and 12 years are periods of Jupiter, 8 years is a period of Venus, 59 years a period of Saturn.[444] These periods are familiar from the Goal-Year texts (see §4.1.2) and from "text E".[445] 15 years is a pretty good minimum sidereal period for Mars. 60 years is a reasonable sidereal period for Mercury (1/8th of the long period used in the MAATs). The order of the planets in this section is thus J,Me,S,V,Ma,J, which further shows that no one order was of particular significance in this text.

These periods are (mostly) in calendar years, and not in sidereal years.[446] They seem to show that the periods were elicited from the record of the *dates* of heliacal events, and not from the record of the locations in the sky where they occurred. Provided that the date of

444 That the 59 years pertained to Saturn is also suggested by the verb *kânu* "to be firm" which alludes to Saturn's name, Kaiamānu. See Ch.2.1.1. This is noted by Gadd op.cit. 62.

445 See n408. In "text E" the following characteristic periods in years are attested: Jupiter (12, 71, 83); Venus (8, 16, 48, 64); Mars (32, 47, 64, 126); Saturn (59, 30, 147); Mercury (13, 46, 125). Some of these numbers are merely multiples of others – a learned play reminiscent of the invisibility and visibility periods of Mul.Apin IIi44–67 – see Ch.3.1.2.

446 The "71-year" cycle of Jupiter is made up of 65 successive heliacal phenomena of the same type. In that time Jupiter travels c. 6° *less* than 6 revolutions of the zodiac. This is the "error" in the period. 71 calendar years (= 878 months, if properly regulated) last c. 6 days less than 71 sidereal years. Since the Sun moves c. 1° per day, 65 phenomena last 71 calendar years.

Planet	"Period"	No. of Events	Error in "Period"	A "Period" of Sidereal Years – Calendar Years
J	71	65	–6°	71 sidereal years – 71 calendar years = 6 days
J	12	11	+5°	(149 months) = – 17 days
V	8	5	$-2^{1}/_{2}$°	(99 months) = $1^{1}/_{2}$ days
S	59	57	+1°	(730 months) = –7 days
Me	60	189	–14°	(742 months) = 4 days
Ma	15	7	–19°	(185 months) = 16 days

As can be seen by comparing the last 2 columns, except for Saturn and perhaps the 12-year "period" of Jupiter, the characteristic "periods" were meant to describe calendar years. In other words, equating days and degrees, the sum of the values in the last two columns is less than the value in the penultimate column.

recurrence was nearly identical to the date of the first recorded event in question, then a period in calendar years was elicited regardless of how close this was to a whole number of sidereal years. Nevertheless, the discovery of calendar periods of the magnitude of 71 years is suggestive both of a long data record (longer than any one individual could make), and of a well-regulated luni-solar calendar, the evidence for which we will turn to next.

DT72+ provides, I argue, good evidence that planetary periods akin to those used in the Goal-Year texts were known before 612 BC. The fact that some of the periods were different from those found in the Seleucid period texts, and in general inferior (so far as their ability to predict phenomena goes – see the last two columns of the table in n446), suggests strongly that this text dates from a time when the characteristic periods had only recently been discovered.

We lack explicit references in texts dating to before 612 BC to characteristic eclipse periods, and to whole number ratios between months and years, between months and cycles of lunar velocity, or between months and cycles of lunar latitude, as we find in texts from later periods. Nevertheless, predictions of eclipses and the lengths of month *were* made by the NA and NB Scholars, and the calendar *was* regulated. It would appear, then, that some periods characteristic of these phenomena were known to the experts in the Ninevite court.

4.2.3 Calendar Regulation

I proposed in Ch.3.2.1 and Ch.3.2.2 (xix) that the "ideal intercalation schemes" of Mul.Apin and related texts do not attest to a particular interest on the part of the Mesopotamian astrologer-astronomers in regulating the luni-solar calendar, despite many such opinions to the contrary.[447] Texts such as *The Babylonian Diviner's Manual* (Ch.3.1.2) indicate that the "ideal intercalation schemes" served a largely divinatory, rather than calendrical purpose, and merely evoked an existing rule of thumb whereby one additional month every three years of so sufficed to keep the vernal equinox in months XII and I. The record of intercalary months, from the earliest times right through to the period of concern here, shows that the calendar was also regularly affected by other forces, such as royal whim. Indeed, some Mesopotamian lunar calendars of the MA period were not regulated against the Sun at all, further demonstrating how *insignificant* to the general populace regulation probably was.

I argued in Ch.4.1.2 that the regulation of the calendar occurred in order that the astronomical prediction of ominous celestial phenomena could occur. To argue that cuneiform mathematical astronomy was (even in part) developed in order that the calendar be regulated seems absurd,[448] considering how complex mathematical astronomy became by comparison with the relatively modest requirements of the Metonic cycle, say. What possible purpose could the non-lunar planetary predictions have played in this, for example? The calendar was regulated by the Scholars in order to facilitate astronomical prediction, the purpose of which was to assist them in celestial divination, at least at first. Indeed, there is some evidence that the astronomers utilised a regulating luni-solar scheme *without* it being used by the general populace. For example, Walker (1997) 24 notes those economic documents from the 5th and 4th centuries BC which attest intercalary months different from those implied by the 19-year cycle, which has long been thought to have been in use by then. The 19-year

447 E.g. Chadwick (1992) 18, and Horowitz (1998) 164, *inter alia*.

448 Chadwick (1992) 15, quoted above in n398.

cycle is attested in some MAATs and NMAATs from that period, but these are, in general, later copies in which the relevant material may have *retrospectively* been ordered into this luni-solar cycle. This, perhaps, has given us a false impression of the *actual* intercalations made by the rulers in these centuries.

Irrespective of the importance, or otherwise, of intercalation and calendar regulation in the wider Mesopotamian community, calendar regulation was necessary for astronomical prediction, which relied on periods between the recurrences of planetary phenomena expressed in years. Using an interval of, say, 59 broadly sidereal years in order to make a prediction to the nearest day or so, required there to be in that period of time 59 *calendar* years comprising 730 lunations and 22 intercalary months. While this occurs in the Metonic cycle, so precise a luni-solar cycle was not necessary. Provided 22 intercalary months were added in over the course of the 59 years, predicting the date of Saturn's heliacal phenomena would be quite possible.

This is the situation which prevailed in the late NA period. In *LAS* II App.A Parpola lists the attested and inferred intercalations for the 33 years from 684 to 651 BC. There are 13. They would have regulated the luni-solar calendar extremely well. This can be demonstrated easily from the calculated dates for Venus's heliacal phenomena in *LAS* II App.C2. Every 8 years the same phenomenon recurs at the same place in the sky to within $2^1/_2{}^{\circ}$. Checking the dates, it can be seen that these also fall within the same Mesopotamian months, showing how well regulated the calendar was. Whole-year periods of the planets, such as those found in DT 72+ (above), *could* have been used successfully in the NA and NB period to predict planetary phenomena to an accuracy governed only by the accuracy of the periods. Except for Mercury, the characteristic calendar year periods in DT 72+ could have been used to predict the dates of the planets' heliacal phenomena to within 4 or 5 days.

Balasî's writes to the king in x042: r10:

> "Concerning the adding of the intercalary month... this is indeed a leap year. After Jupiter has become visible, I shall write again to the king... it will take this whole month. Then we shall see how it is and when we have to add the intercalary month."

It appears as if the date of Jupiter's rising in some way determined whether an intercalary month VI or XII should be added. Parpola *LAS* II p45 r17 states that the rôle played by the planet is obscure. However, it is not improbable that the 71 or 12 year period of Jupiter was being alluded to by Balasî. A glance at *LAS* II App.C4a & b will show that during the period in which the surviving Letters and Reports were written, after every 12 years Jupiter's heliacal phenomena returned to their original longitudes and, *according to the Scholars*, recurred in the same Mesopotamian month. I suggest that Balasî was waiting to see in which month Jupiter was going to rise in order to decide whether an intercalary XII should be added immediately (ensuring that Jupiter rises in the same month as 12 years earlier), or whether the king should wait until month VI. Of course, Balasî's motivation may have been divinatory. I proposed in Ch.3.1.2 that intercalation may have been used in order to change the month in which the phenomenon occurred, thus changing the prognostication.[449]

[449] E.g. one of the only good-boding Marduk planet (=J or Me) omens occurs when the planet rises in month I (Mul.Apin II: GapB1).

Not only was the calendar broadly regulated in the late NA period, the dates of the vernal equinoxes were noted, for example by Nabû'a in 8140–2.[450] In 8140 the days were said to be in balance (*šitqulu*) on the 6th of month I. This date was probably determined by observation,[451] since it does not correspond to that generated by any known luni-solar scheme, and the "ideal" equinox was located on the 15th of month I. Since there appears to be no ominous significance attached to these dates, it is probable that the date of the equinox was communicated to the king in order to provide him with further information on whether or not he should intercalate a month.[452] The appearance of stars "not according to the ideal", provided him with further justification for this (8098: r8f, quoted in §3.2.2 xix above). The tiny difference between the seasonal and sidereal year meant that both pieces of information, ominous or not ominous, corresponded. Although the evidence is lacking, it seems possible that the noting of the dates of equinoxes was part of a new effort in the late NA period, more connected with the accurate prediction of celestial phenomena than with celestial divination. I argued in §4.1.2 that knowing the dates of seasonal phenomena would have assisted the scribes in determining and using characteristic planetary periods expressed in *sidereal* years.

4.2.4 Pre-612 BC Predictions of Celestial Phenomena

4.2.4.1 Prediction of the Movements of the Inferior and Superior Planets

Many examples of predictions of celestial phenomena are to be found in the NA and NB Letters and Reports. Some of these have been discussed by others, particularly by Parpola in *LAS* II. The following will consider all the material together for the first time:

i) "Afterwards, in month III, (Mars) will turn and move forwa[rd...]" (8052: r.2).
ii) "During this month (Venus) will become visible in the east in Leo" (8246: r.2).
iii) "Venus...does not complete her days (of visibility) but sets" (8145: 2).
iv) "The man who wrote to the king [my lord] is ignorant. He does not k[now] the [...], the period (*adannu*) [...] (or) the revolutio[ns[453] (*da-a-a-la-*[*te*) of Venus]" (x051: 10).
v) "[He] who wrote to the king, my lord: 'Venus is visible, it is visible in [month XII]' is a vile man, an ignoramus, a cheat! Venus is not yet visible" (x072: 6).
vi) "Venus will presently make a good omen" (x074: r.16).
vii) "Venus has risen [*at*] the time of its (computed) [appearance]" (x031: r.2).
viii) "One of your colleagues wrote to me: 'The planet Mercury will be visible in month I'- we take the present month to be XII, and this day to be the 25th... [The person who] wrote this...to the king my lord...] An incompetent one can frustrate [a j]udge, an uneducated one can make the mighty worry..." (x023: 7f).
ix) "[On]e day (Mercury) might be too early, [anoth]er day it might lie flat (to the horizon). [To see it], our [e]yes sho[uld have f]allen on it" (x050: r.9).
x) "Saturn will 'push itself' (retrograde) this very month" (x008: r.25).

---

450 Also, in 8207 the solstice appears to be noted in r.8. This text contains no omens.

451 Parpola *LAS* II p360 suggests the equinox was determined using a gnomon, and notes that its determination had no known ritual significance.

452 In x253: 15f it is made clear that the king decrees when an intercalation should be made. That this was done at the recommendation of the Scholars is made clear by 8098: r.8.

453 See Parpola *LAS* II p72 n15.

Some of these predictions were based on the correspondence of observed reality with an ideal, e.g. (iii), and see also x362: 5. Some of the predictions required little more than a broad familiarity with the planets' behaviour – e.g. (i),[454] (ii),[455] (vi), (viii) & (x). Example (ix) indicates an awareness of Mercury's erratic behaviour.[456] However, (iv), (v), and perhaps (vii) demonstrate that periods for Venus were being used by the Scholars to predict heliacal phenomena. In (iv) this methodology is referred to as "periods" and "revolutions". Parpola *LAS* II p72 n15 suggests that "revolutions" between heliacal phenomena of the same type are meant. He argues (p73) that not only was the 8-year period for Venus known and used by the Scholars, but that the variation about the mean interval between heliacal phenomena may have been approximated by something like a zigzag function.[457] He notes (p70) that in (v) the Scholar could not have ruled out the possibility of Venus having risen in month XII unless something equivalent to the 8-year period had been used. He writes:

> "A person ignorant of this cycle could not have ruled out the possibility of the planet's being Venus, for the interval of invisibility of Venus at superior conjunction can vary from 54 to 74 days and the planet had already been 50 days absent at the time the observation was made."

There is no doubt that the 8-year period was known and used by the NA and NB Scholars, as DT 72+ makes clear (see §4.2.2). However, it is less likely that a mean interval and a model of the variations around that mean, as in the later MAATs, were being used by this time. No evidence exists for such a development in predictive astronomy at this early time, save the use of *tithis* in BM 36731 (see §4.2.1).

4.2.4.2 Predicting Month Lengths and the Dates of Lunar "Opposition"

(i) "The Moon will be seen with the Sun (lunar "opposition" = morning setting) in month X on the 14th. The Moon will complete the day in month XI (month X will be 30 days long), on the 14th it will be seen with the Sun. The Moon will reject the day in month XII (month XI will be 29 days long); [on the xth] it will be seen with the Sun. The Moon will complete the day in month I" (8060: r.1f).
(ii) "In month I and month IV (the Moon) will complete the 30th day" (8257: r.8).
(iii) "In month VII the Moon will become visible on the 30th. From this day until month VI of next year the Moon will not be seen with the Sun on the 13th day" (8266: r.1).
(iv) "On the 1st, I wrote to the king: "On the 14th the Moon will be seen together with the Sun. (Now) on the 14th the Moon and Sun were seen together" (8271: 1, cf. 8410: 7).
(v) "The king must not say: "Clouds; how did you see (anything)?" This night, when I saw (the Moon's) coming out, it came out when a little of the day (was left), it reached its region (*qaqqaršu*) where it will be seen (with the Sun on the 14th). This is the sign (*idat* < *ittu*) that it is seen" (8293: r.1).

454 See also x048: 6.
455 Venus is only invisible for a few days at inferior conjunction. The constellation with which it sets will probably be the same one with which it next rises.
456 See Parpola *LAS* II p60 r.3f.
457 He cites the existence of just such a function in the *Sargonid Ivory Prism* (op.cit. n146), a photograph of which is now to be found in SAA8 p78. Zigzag functions are also in EAE 14, Mul.Apin and the like, and some concern the variations in lunar visibility around a mean. They are *all*, however, *ideal* period schemes and were not intended, I argue, to model actual celestial behaviour accurately. The *Sargonid Ivory Prism* includes an "ideal seasonal hour" scheme – see §3.1.2.

(vi) "[the Moon] will be close to rejecting (*ana turra qerub*) (the day) in month III" (8516: 5').

The Scholars, both Assyrian and Babylonian, felt able in some way to predict the lengths of months, sometimes long in advance. Beaulieu (1993) 72f & n20f discusses the prediction of month lengths in the Reports.[458] Example (vi) does indicate that the predictions were not merely the apodoses of omens, but were presumably the result of observation and calculation. How could this have been done? In example (v) the Scholar explains that even though it was cloudy, just before Sunset he saw that the Moon had moved into that part of the ecliptic he *knew* was to rise with the Sun the following morning. Was this more than a good guess?

The length of the month and the date of the Moon's morning rising are no easy matters to predict. As noted in §4.1.2, they depend on the velocity and latitude of the Moon, the angle of the ecliptic to the horizon, the velocity of the Sun, and the length of daylight. The effect of the last three can be eliminated by predicting a year ahead. Perhaps this reasoning lies behind (iii). It is also possible that those equations, attested in *TU* 11: 29–38 and discussed above in §4.1.2, that relate sums of the lunar six separated by 223 or 229 months were used. A prediction of a kur value in the –567 Diary: 11[459] suggests as much for the early 6th century.

Certainly, a connection between month length, the dates of lunar "opposition", and the periodicity of eclipses was recognised by the NA and NB Scholars:

> "The eclipse of the Moon which took place in month VIII... the Moon was seen three times with the Sun on the 16th day. In Months VII, VIII & IX the Moon at its appearance..." (8469: r.6)
> "(The Moon) is seen on the 16th day (with the Sun)... Within one month the Moon and Sun will make an eclipse: on the 14th one god will not be seen with the other" (8320: 1f).[460]

This connection may be no more than an ominous one since lunar "opposition" on a date other than the 14th bodes ill, and so does an eclipse. However, it may also have been derived from the connection between eclipses of the same type being separated by 223 months, and an awareness either that šú + NA and me + gi_6 repeat after the same interval, or less formally that the morning and evening visibility periods of the Moon at mid-month are together the same length as those one Saros cycle earlier, or even that month 224 will be approximately like month 1. In "text E" (see n408, above) it states in lines r.18f that after 18[461] years an eclipse will recur in the same place, and that the Moon will return to the same latitude, and that one can thereby "determine the full and hollow months". Although "text E" dates to the last few centuries BC, it probably preserves methods similar to those employed in earlier periods. It does not employ the zodiac, for one thing.

Alternatively, the connection between eclipses and month length may have derived from some physical notion that a lunar eclipse provides a date and time when the Moon is precisely at mid-month, which could perhaps then serve as the starting point for the attempted prediction of the length of the next few months using a simple scheme for lunar velocity and/or latitude. In "text E" again (lines 1–22), one such scheme for lunar latitude exists. In it the Moon is said to increase in latitude by 1/9 UŠ per month for a maximum of 9 years, and

458 They are also made in the Letters – see x048.
459 "The 26th (Moonrise to Sunrise) 23 (UŠ). I did not observe the Moon".
460 See also 8346 & 8382: "If the Moon is carried off at an inappropriate time; there will be an eclipse."
461 Text has "19", but this is clearly a mistake.

then decrease at the same rate. In the LB "text K" (see n407, above) a scheme for computing the time between Moonrise and Sunrise on the last day of the month (kur), and the time between Sunset and Moonset on the first day of the month (NA*), is outlined that depends on the kur and NA* of the previous month, and on the longitude of the Moon. Although this longitude is expressed in degrees of the zodiac in "text K", it was perhaps a scheme like this, used to predict the lengths of successive months, that was being alluded to in example (i) above.

While it is difficult to believe that the NA and NB Scholars used methods akin to those found in the MAATs to predict the lengths of the months and the days of lunar morning setting, we recall that function Ø was invented as early as the mid-5th century BC, and perhaps some time earlier (see §4.2.2). Function Ø depends heavily on the 223 month "Saros" period, and as I will show in the next section, we have reason to believe that that at least was indeed known by the 8th century BC.

When, however, in (iv) the Scholar boasts about having correctly informed the king two weeks earlier that the Moon would be seen with the Sun on the 14th, we are witness, perhaps, to the limits of the NA and NB lunar prediction methods. In (vi) the Scholar states that the Moon will be *close* to rejecting the day (i.e. the Scholar is unsure whether the current month will be 29 days long), which suggests that whatever scheme existed to predict the lengths of months, it could do so only to a limited degree of precision. In the late NA period the art of accurate prediction was *in its infancy*, I argue.

4.2.4.3 Predicting Eclipses

A number of studies of the eclipses predicted in the NA and NB Letters and Reports have been published. These include Schaumberger's *SSB* Erg.III 251f, van der Waerden's *BA* 115f, Parpola's *LAS* II on numbers 42 and 63 particularly, and Rochberg-Halton's *ABCD* 40f. See also Beaulieu & Britton (1994). The material has never been brought together before, however.

Eclipses Predicted, but which Failed to Occur

> "Concerning the watch about which the king, my lord, wrote to me, the Moon let the eclipse pass by, [it did not occ]ur" (x132: 6). See also x133: 6, x135: 7, x159: 4, x220: 10 (solar), x224: 8 (solar), x363: 7 (solar) & the -567 Diary: 17.

A "watch" for an eclipse implies that it was predicted. Lunar eclipses can occur every 5 or 6 months. That this was known by the late NA period is made clear from the Ninevite commentary text *ACh.* Sîn 3: 26 cited by Van der Waerden *BA* 116, Rochberg-Halton *ABCD* 41, and Schaumberger *SSB* Erg.III 251:

> "If the Moon is eclipsed not at its appointed time (*ina lā minātišu*) – six months have not elapsed, or alternatively, an eclipse occurs on the 12th or 13th day."

It is also suggested by x071, quoted below. At the most basic level lunar eclipses were predicted to occur at mid-month during a 2-month interval every 5–6 months. This form of prediction may even have occurred before the mid-8th century, though there is no reason to date the composition of *ACh.* Sîn 3: 26 to a period before c. 750 BC, and we have no other evidence for such predictions. In any case this type of prediction was not accurate. Indeed, *ACh.* Sîn 3: 26 is reminiscent of the ideal periods of the EAE Paradigm discussed in Ch.3.1.2. Much more accurate predictions of eclipses occurred in the late NA period, and it is these that are indicative of the new achievements of the PCP Paradigm.

Eclipses Predicted, but not Stated if they Occurred

"On the 14th Adad-šumu-uṣur enter[ed] Nineveh (saying) "Let him sit (on the throne) before the eclipse occurs"" (x377: 8). See also 8250: 1, 8251: r.6, 8320: 8, 8346: 4, 8382: 5f, x026: r.1, x114: 3f, x170 (solar), x216 (solar).

Eclipses Predicted not to Occur

"The Moon will be seen [with] the Sun in month VI on the 15th, it will let the eclipse pass by (Š of *etēqu*)" (8042: 1). See also 8046: 1, 8067: 1, 8087: r.1 8321: r.1, 8344: r.2.
"The 13th [day], the night of the 14th day is the [da]y of the watch, and there will be no eclipse. I guarantee it seven times, an eclipse will not take place. I am writing a definite word to the king" (8447: r.1). See also x046: 6.

There is no mention in these examples of why the Scholars were so sure that an eclipse would not take place. Some may have written to the king in months in which it was known eclipses could not occur. In the second quoted instance (and in x046), however, the Scholar was confident that an eclipse would not occur even though a "watch" for an eclipse was underway. Was this merely unfounded bravado? If not, it must have been based on a knowledge either that the eclipse would occur in the daylight and thus be invisible, or that the period between the last eclipse possibility and the one coming was now 6 months and not 5 (or vice versa). Alternatively, both pieces of astronomical knowledge may have been contained within the dictum that each eclipse possibility is preceded by the same 223 months earlier, but some 8 1/2 hours later in the day (see §4.1.2). This, of course, is the Saros interval, which could have been determined directly from the accurate records of eclipses.

Similarly, in the Eclipse Records from Babylonia (App.1 §32), eclipses are predicted to "pass by" (also *etēqu*, dib) in 731, 686, 684, 677, 668, and in 649 BC in the *Saros Canon – LBAT* 1414+. In *LBAT* 1414: 1f it is written:

1,50 mu-1-kam du-numun, bar *šá* dib *ina* 1 me nim
"110, Year 1 of Ukin-zēr (731 BC), (an eclipse) which passes by at 60 (UŠ) after Sunrise (literally, day risen)."

In other words a lunar eclipse was predicted to occur roughly 4 hours after Sunrise, and therefore in the middle of the day, when the Moon was below the horizon. The eclipse therefore "passed by". The same is noted in 684 BC (*LBAT* 1415: obv. II):

"Year 5 (of Sennacherib), month I, the 15th (an eclipse) which passes at [1]0 UŠ after Sunrise."

In these examples not only were the months and dates of these eclipses calculated, but the times of their occurrences – to the nearest 10 UŠ, or 40 minutes. Although the *Saros Canon* was written much later, and arranged eclipses into groups of 38 eclipse possibilities in 223 months, it would appear that *original* records were copied to that end. It is this that explains the developments over the decades seen in the details of what was recorded of the eclipses in the Canon, described above in §4.2.1. So, although it is conceivable that the eclipse predictions dating to 731, 686, 684, 677, 668, and 649 BC were actually *retrocalculations* undertaken by scribes in the 4th and later centuries BC, it is much more likely that they were *predictions* made and recorded shortly before each of those years, and that they were only later incorporated into the *Saros Canon*. How then were these predictions made?

Two possibilities exist, I believe. One is that the Saros interval was known. This would permit a prediction to be made of a forthcoming eclipse, provided the record of one 223 months earlier had been preserved. The forthcoming eclipse would be of the same type as the one 223 months earlier, but occur 81/2 hours, some 130 UŠ, later in the day. This might mean that the forthcoming eclipse would occur in the daytime, and so be invisible. These eclipses would be marked down as "passing by". Those predicted to occur *and* be visible would be watched for, and their *observed* details recorded.

The second possibility is that the predictions were made, but a few days in advance of the anticipated eclipse, using some simple scheme of lunar velocity. If on the 12th of the month, say, it was noticed that the Moon was in advance of its normal position *vis à vis* the Sun for that time of the year, then the Scholar might estimate that the Moon will reach opposition (the position for eclipse) during the day of the 13th and not during the night of the 14th, and so be invisible. Inconclusive evidence for both forms of prediction are present in the NA and NB Letters and Reports and the Eclipse Records.

Eclipses not Predicted, but which Occurred

> "If there is an eclipse in month III not (occurring) at its appointed date/hour/time (*ina lā minātišu*)" (8004: r.14).

This is an omen derived from the commentary text *ACh.* Sîn 3: 26, quoted above, but nevertheless may indicate that this eclipse did occur unexpectedly. It is noteworthy, however, that there are no other references, to my knowledge, of non-predicted eclipses taking place. The omen, no doubt, refers to the eclipse occurring after 5, rather than the *ideal* 6 months (see n354, above). This may have been an ideal period long known to the practitioners of the EAE Paradigm, but it may also have come about in the late NA period in the wake of the efforts at predicting celestial phenomena accurately for the first time. It was perhaps a result of the same feedback of the PCP Paradigm on the EAE Paradigm exemplified by those omens concerned with lunar latitude, discussed in §4.1.2.

Eclipses Predicted and Seen

> "Concerning the watch for the eclipse... on the 14th, during the morning watch, the clouds dispersed, and we were able to see. The eclipse took place" (x147). See also x351: 5f, 8279.

Many of the reported eclipses may also have been predicted.

The Days On Which Eclipses were Watched For

> "The night of the 13th... 14th... 15th... 16th (when) the Moon made an eclipse" (8279).
> "On the 28th, 29th, and 30th we kept watch for the eclipse of the Sun" (8363: 8).

These are indeed the possible days of the month on which eclipses may occur. There is no evidence that they were watched for on the 21st, for example, despite the existence of omen protases which described lunar eclipses occurring on this and other impossible days of the month (see Ch.3.1.1).

Period of Time During Which Eclipses Were Watched For

> "Concerning the watch of the Sun... we will keep watch twice, from the 28th of month VIII (and) from the 28th of month IX. Thus, we will keep the watch of the Sun for 2 months" (x045: r.1). "Concerning the lunar eclipse... it was observed... I shall keep watch for the solar eclipse" (x347: 5f).

These Letters show that solar eclipses were not always predicted to the month. Probably solar eclipses were watched for either side of a predicted lunar eclipse. See Parpola *LAS* II p51 and n385, above.

What Aspects of the Eclipses were Predicted?

> "If it should occ[ur], what is the word about it? – the 14th (signifies) Elam, month III (signifies) Amurru, and its decision is for Ur, and if it occurs, the quadrant[462] it afflicts and the wind blowing will be quoted as well" (x026: r.1).

This example indicates that a lunar eclipse was predicted by this Scholar to the *month* and *day*, but that he was unable to predict which quadrants would be covered. In 8250: 1f Nergal-eṭir gives a very confident prediction of an eclipse, indicating the month, day, the *watch*, and the *presence of the planets*:

> "In Adar (month XII) on the 14th day the Moon will make an eclipse... In the Moon's eclipse, Jupiter and Venus [will not? stand] there... If there is an eclipse in Adar in the evening watch... When the Moon has made the eclipse, let the king write and have dikes cut in Babylonia at night as a substitute for the king".

We cannot be sure that the eclipse occurred, though it seems likely. In x071: 6f we find:

> "Concerning the watch for the lunar eclipse... its watch will be [on the deci]ded [night]; [whether] its [wat]ch should be during Sunset [we have not been able to decide]. [Eclipses] cannot occur [dur]ing certain periods. [...] 4 months there was a watch in month VIII and now in month IX we will (again) keep watch."

This Letter is fragmentary, but enough remains to show that this lunar eclipse was predicted only to within an interval of 2 months, separated by a period of probably 4 months, during which time an eclipse was known never to occur. This 2-month interval was the time of the eclipse "watch". The prediction in this text apparently relied on no more than the fact that eclipses can occur every 5–6 months, or even on the ideal period of *ACh.* Sîn 3: 26. Importantly, the Letter also indicates that *sometimes* the watch during which an eclipse might take place was predictable, though not in this case (assuming the reconstruction is good). This is confirmed in x078: 1':

---

[462] *Kaq-qu-ru*, for which see Parpola LASII p29 r7'.

> "[Concerning the watch for the lunar eclipse]. . . its watch will be tonight, in the morning watch. The eclipse will occur during the morning watch,"

which indicates that once the month, and perhaps the date of an eclipse were established the watch in which it was thought to occur could then sometimes also be predicted. This is shown here because this Letter is a response to the king's enquiry about the watch, *once* the king had heard that an eclipse was anticipated. Predictions at this level of accuracy are comparable to those recorded in the Babylonian Eclipse Records discussed above, and the methods used there were probably the same as those used by the Scholars employed by the Ninevite court. The above example suggests that in this case, at least, methods other than those using the Saros interval were employed.

In *LAS* II p68 Parpola argues that the prediction of the watch was made only a few days in advance, perhaps from calculations derived from the Moon's visibility periods sometime before opposition. As proposed above, if the Moon were visible after Sunset for longer than expected on the 12th, say, it meant that that the Moon had travelled further in its path than usual. Opposition, and any eclipse, would therefore occur earlier than expected, enabling the Scholars to assert that the eclipse would take place in the evening watch, say. In x240: 23f, on the 13th, or thereabouts, an eclipse is predicted for the night of the 15th:

> "I shall now look up, collect and copy numerous – 20 to 30 – canonical and extraneous tablets, (but) perform (the prayers) tomorrow evening and on the night of the 15th. . . I am also worried about the impending observation of the Moon; let this be [my] advice: If it is suitable, let us put somebody on the throne. (When) the night [of the 15th day] comes, he will be afflicted [by it (the eclipse)], but he will *sa*[*ve your life*]!"[463]

Some Scholars did, then, estimate the watch and the day of forthcoming eclipses a few days in advance of lunar opposition, probably relying of a simple notion of lunar velocity, as described. Some predictions, however, were also made long in advance:

How Long in Advance were Eclipses Predicted?

In 8388 Rašil writes that an eclipse will occur on the 14th and boasts:

> "(Already) when Venus became visible, I said to the king my lord, "An eclipse will take place.""

This was, presumably, sometime earlier, though as I explain in App.2 (year 667) we cannot be sure how much earlier this was.

In x224 (dating to 667/VII/30) and x363 (669/IV/1) eclipses were predicted which did not occur. In both instances a solar eclipse was predicted 2 weeks after the lunar eclipses of 667/VII/14 and 669/III/14 respectively, demonstrating that solar eclipses were watched for either side of lunar eclipses, as suggested above. In x351 a predicted eclipse (that it was predicted is shown by the fact that in line 5f a substitute is described as being already prepared before the date of the eclipse) took place on 671/X/15. It was preceded by an eclipse in 671/IV/14, 6 months earlier. Perhaps this was the basis of the prediction. In x358: r.7 it is written:

---

463 See also Parpola, *LAS* II p176f.

> "They (the more expert Scholars?) say "(Concerning) the watch for the Moon, he (the Moon) will make (the eclipse) pass by in the intercalary month VI_2; it will take place in month VII,"

which shows that the predictions were made at least some months in advance. See my discussion of the redating of this letter to 670 BC in App.2. In 8502: r.7 we find:

> "All the signs which have come concern the land of Akkad... An eclipse of the Moon and Sun in month III will take place. These signs are of bad fortune for Akkad... and now, in this month IX, an eclipse will take place... and Jupiter will stand in its eclipse."

8502 is datable to 679 BC, and shows that *some* eclipses were predicted at least 6 months in advance, and with enough detail to be sure that at the time of eclipse Jupiter would be visible. While estimating Jupiter's approximate location 6 months hence is trivial, since it moves so slowly, to be sure that it will then be above the horizon at the time of an eclipse requires knowledge *now* of the time of day or night at which the phenomenon occurs. Only the Saros interval could provide such knowledge.

8502 provides the clearest piece of evidence that the Scholars used the Saros interval. It could, of course, be argued that its Babylonian author merely guessed that Jupiter would be above the horizon in the forthcoming eclipse. One further tiny piece of evidence from the Babylonian Eclipse Records, however, suggests to me that the Saros interval of exactly 223 months, incorporating 6585 days and about 81/2 hours was known as early as the mid-8th century BC. This is the writing of the numbers 1,40 and 1,50 at the very beginning of *LBAT* 1413: Ii and *LBAT* 1414: Ii, respectively – the last quoted above. These numbers are obscure, but appear to pertain to the year in question and to be in UŠ, since all other large numbers in the Eclipse Records are in this unit. I tentatively suggest that these numbers indicate the amount of time by which 223 months were thought to exceed 6585 days. In other words they indicated to the compiler of the records how much later in the day an eclipse similar to one 223 months earlier would occur. 1,40 and 1,50 are a little smaller than the mean value of the function Ø, which models how much longer than 6585 days 223 months are (see §4.1.2), but only by some 15 to 20%. Indeed 1,50 UŠ (=110) is close to the minimum value for function Ø. As noted above, in *LBAT* 1414: Ii a prediction was made as to when in the day, April 9th 731 BC, an invisible eclipse was to occur. The figure of 1,50, and a record of an eclipse 223 months earlier was used to make this prediction, I suggest. The eclipse was predicted to take place at 60 UŠ after Sunrise, in which case the text implies that 223 months earlier an eclipse occurred some 50 UŠ *before* Sunrise. Modern records show that on the 28th March, 749 BC an eclipse did indeed occur between c.4 am and 6 am local time.[464] The eclipse set as it ended, so it began about 2 hours or 30 UŠ before Sunrise. Depending on the precise definition of the length of Sunrise, these figures indicate that it was quite possible that the invisible eclipse of 731 BC was predicted using the Saros interval in which the amount by which 223 months exceeded 6585 days was then considered to be 150 UŠ.

The evidence for the use of the Saros interval before 612 BC is not conclusive. It must be asked why it was not used more generally by the Scholars. Why were watches made for eclipses that never occurred? Careful use of the Saros interval would have shown whether or not an eclipse would be visible. I argue, though, that the accurate prediction of celestial phenomena was an art in its infancy in the late NA period. Undoubtedly, not all the Scholars

464 Steele & Stephenson (1997/8) 200.

were equally competent at it. It is noteworthy that the author of 8502, the text which provides perhaps the best evidence of the use of the Saros, was a very senior Babylonian Scholar. Although, his name is lost, he states to the king in r.12f:

> "Let the king do this, (and) what ever Bēl-ušezib will write to the king his lord, and I will guarantee it the king, my lord."

Bēl-ušezib was perhaps the most senior Babylonian Scholar employed by Esarhaddon and his father Sennacherib. A large number of his Letters and one Report have been preserved. The author of 8502 was either senior to Bēl-ušezib, or felt himself to be his equal. Perhaps he was amongst the few who knew of the predictive power of the Saros. In Ch.5.1.2 I will discuss the fact that many aspects of the PCP Paradigm were kept secret. If I am correct about the significance of the number 1,50 in *LBAT* 1414: Ii and of 1,40 in *LBAT* 1413: Ii, then since the latter possibly dates from as early 747 BC, this would mean that accurate records of eclipses were made at least some years before 750 BC. At least 223 months of records would be needed to determine a rough value for the amount by which the interval between eclipses separated by that time interval exceeds 6585 days. Without such records, however, this is merely speculation.

To sum up, the Scholars were interested in predicting eclipses. Many used the simplest possible period for eclipse prediction, of 5 or 6 months, but some, I argue, knew and used the longer 223-month period we call the Saros.[465] The 5/6-month period provided the Scholars with the knowledge that a lunar eclipse might happen in the middle of a particular month. Sometimes the prediction of the precise day may have been no more than a statement that eclipses *usually* happened on the 14th. Knowing the month in which a lunar eclipse might occur also ensured that the Scholars looked for solar eclipses at the beginning and end of that month. Predictions of eclipses to days other than the 14th and to watches, predictions of invisible eclipses, and the prediction of the visibilities of planets during an eclipse, were probably often done only a few days in advance of each event, but some Scholars, at least, were able to undertake them long in advance.

This concludes our summary of the current understanding of the extent of the Scholars' ability to predict planetary phenomena in the period c. 750–612 BC. Most important is the evidence that they *intended* to predict them, and that they felt confident enough to write to the

465 The Saros period may have been built up from shorter periods after which eclipses recur – see Britton (1989) 5–9. Parpola *LAS* II p51 argues that "the scholars of this period certainly had recognised the 47-month period and probably the 18-year Saros." He points to van der Waerden *BA* 118, but van der Waerden in fact argues that the "eclipse not at its appointed time" in 8004: r.14 (quoted above) can only be accounted for on the basis that the Scholars were aware that *total* eclipses were normally followed by eclipses 6 months later, where *partial* eclipses were usually followed by eclipses 5 months later. This is true, but the *ina lā minātišu* omen in 8004 most likely simply indicated that it was remarkable that the eclipse occurred after 5 months and not 6 because of the ideal nature of 6, irrespective of the type of eclipse seen 5 months earlier. The 47-month eclipse period is thus unattested, and its merely being shorter than the 223-month one hardly constitutes grounds for believing it was discovered or used first. Speculating how the Saros period was determined is pointless in the absence of more evidence. We know that it was in use by the mid-5th century BC as part of function Ø, and probably used to determine the lunar six by the early 6th century BC, as the -567 Diary indicates. The determination in the late NA period of month lengths, eclipse times months in advance, and the presence of parameters comparable in magnitude to those in function Ø in texts whose purpose was also eclipse prediction, suggests that if one long eclipse period *was* known then it was undoubtedly the Saros.

king and tell him of their calculations. Their reputations were at stake when they did this. The Scholars were interested and capable of regulating the luni-solar year, and they had available to them both the accurate records of eclipses and planetary phenomena, and the characteristic periods after which they recurred. Their Letters and Reports show that some of them made predictions of planetary behaviour, using, at the very least, a characteristic period for Venus and some model of lunar velocity, and probably the Saros for eclipses. They also predicted lunar phenomena with an accuracy that remains difficult to evaluate as yet, but which may also have depended on the Saros and/or models of lunar velocity.

At the end of §4.1.2, I argued that it was necessary for the thesis of this book, only that the core hypothesis of the PCP Paradigm was being implemented in the late NA period. The core hypothesis was that the accurate record of ominous phenomena would enable the same phenomena to be predicted through the use of characteristic periods and parameters. There can be no doubt, now, that this premise was embraced by those authors of the Letters, Reports, Diaries, Eclipse and Planetary Records and related material. This hypothesis was their *revolutionary thought*. They were interested in predicting phenomena in advance, and they did so. Some of their methods can be reconstructed, and these were directly ancestral to the NMAATs and MAATs of later centuries. Mesopotamia was famous in the ancient world for its predictive astronomy (see nn5 & 9, above) and for its celestial divination, in other words for the PCP Paradigm. The Paradigm came to have an enormous impact on subsequent Western astronomy and astrology. The reasons why this facility to predict celestial phenomena came about in the late NA period, and the significance of this to a study of the history and philosophy of science will be the subjects of the next and concluding chapter.

CHAPTER 5

A Revolution of Wisdom

5.1 From the EAE Paradigm to the PCP Paradigm

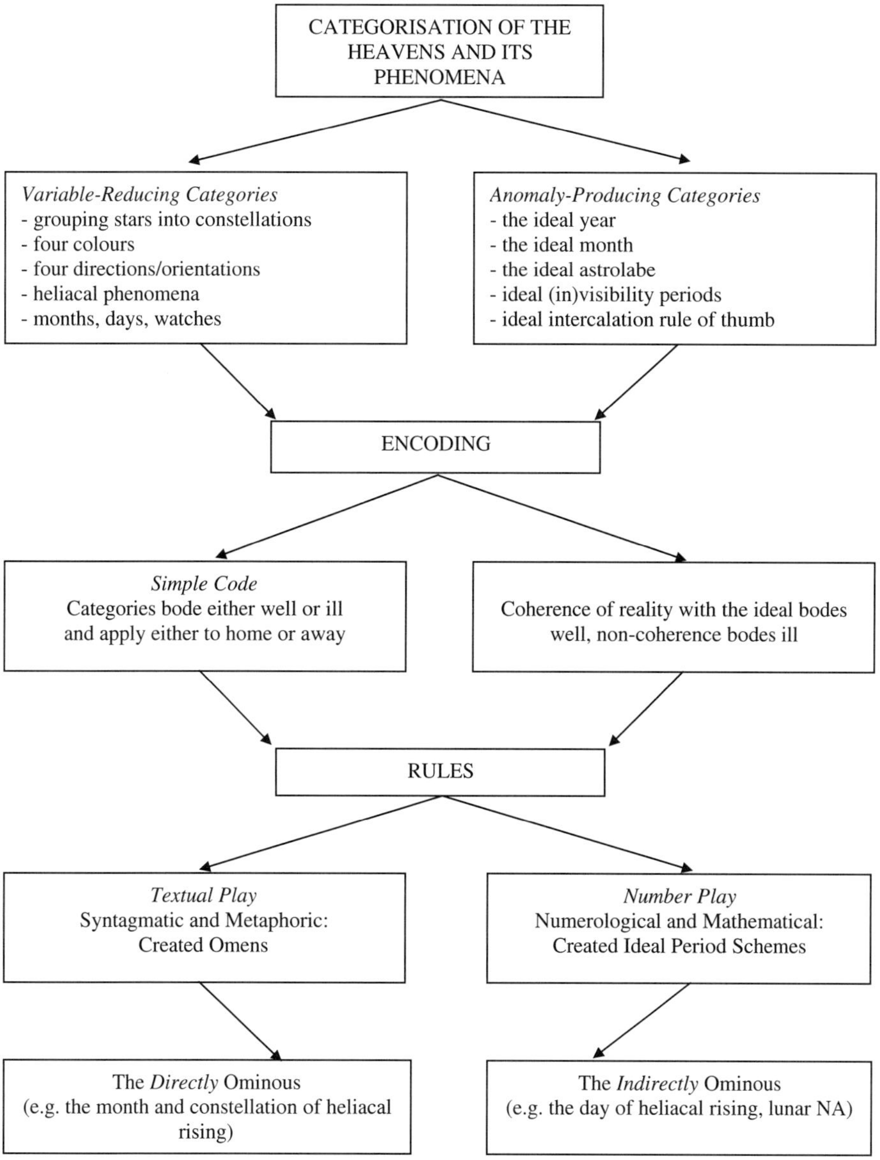

CATEGORISATION OF THE HEAVENS AND ITS PHENOMENA
Variable-Reducing Categories
- grouping stars into constellations
- four colours
- four directions/orientations
- heliacal phenomena
- months, days, watches
Anomaly-Producing Categories
- the ideal year
- the ideal month
- the ideal astrolabe
- ideal (in)visibility periods
- ideal intercalation rule of thumb
ENCODING
Simple Code
Categories bode either well or ill and apply either to home or away
Coherence of reality with the ideal bodes well, non-coherence bodes ill
RULES
Textual Play
Syntagmatic and Metaphoric: Created Omens
Number Play
Numerological and Mathematical: Created Ideal Period Schemes
The Directly Ominous (e.g. the month and constellation of heliacal rising)
The Indirectly Ominous (e.g. the day of heliacal rising, lunar NA)

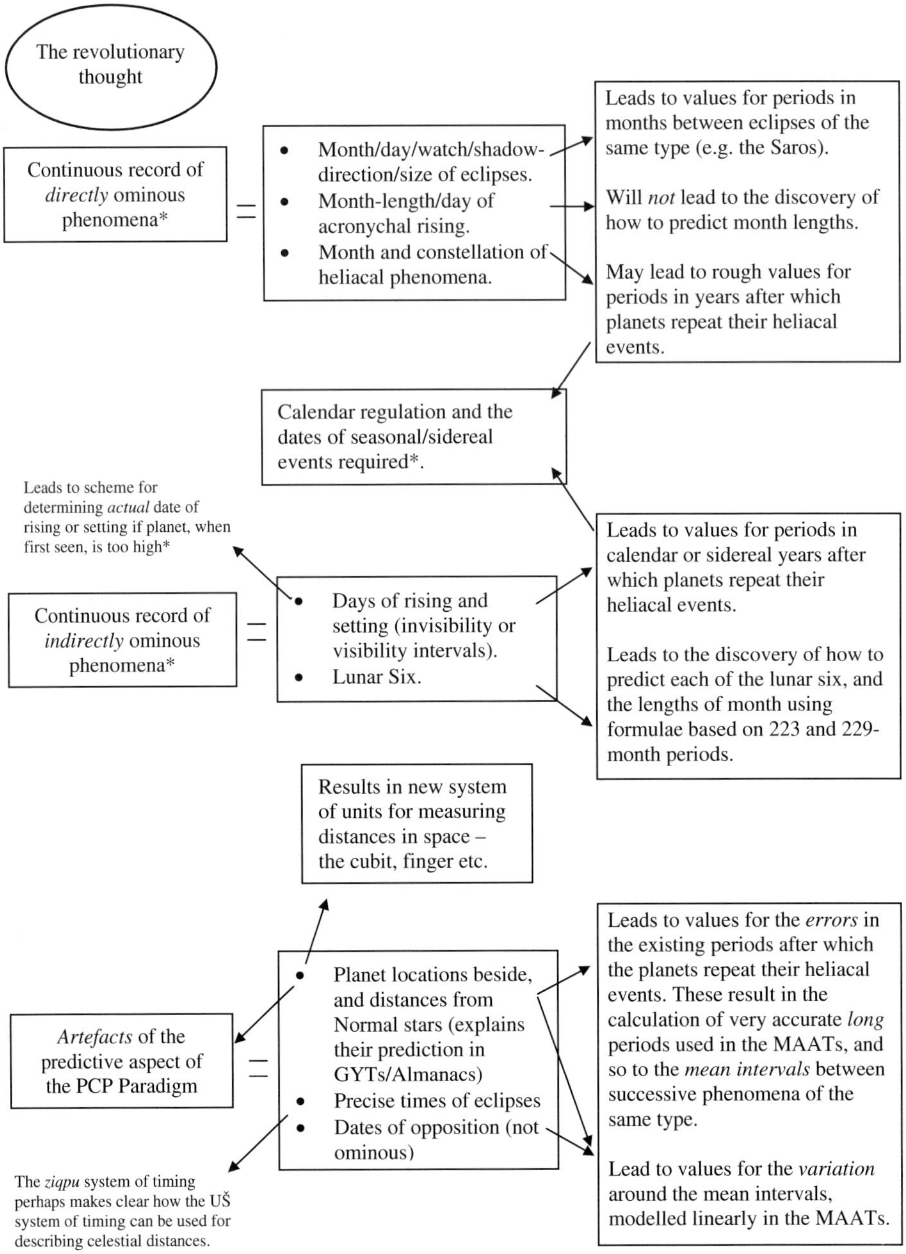
The revolutionary thought
Continuous record of *directly* ominous phenomena*
=
• Month/day/watch/shadow-direction/size of eclipses.
• Month-length/day of acronychal rising.
• Month and constellation of heliacal phenomena.
Leads to values for periods in months between eclipses of the same type (e.g. the Saros).
Will *not* lead to the discovery of how to predict month lengths.
May lead to rough values for periods in years after which planets repeat their heliacal events.
Calendar regulation and the dates of seasonal/sidereal events required*.
Leads to scheme for determining *actual* date of rising or setting if planet, when first seen, is too high*
Continuous record of *indirectly* ominous phenomena*
=
• Days of rising and setting (invisibility or visibility intervals).
• Lunar Six.
Leads to values for periods in calendar or sidereal years after which planets repeat their heliacal events.
Leads to the discovery of how to predict each of the lunar six, and the lengths of month using formulae based on 223 and 229-month periods.
Results in new system of units for measuring distances in space – the cubit, finger etc.
Artefacts of the predictive aspect of the PCP Paradigm
=
• Planet locations beside, and distances from Normal stars (explains their prediction in GYTs/Almanacs)
• Precise times of eclipses
• Dates of opposition (not ominous)
Leads to values for the *errors* in the existing periods after which the planets repeat their heliacal events. These result in the calculation of very accurate *long* periods used in the MAATs, and so to the *mean intervals* between successive phenomena of the same type.
Lead to values for the *variation* around the mean intervals, modelled linearly in the MAATs.
The *ziqpu* system of timing perhaps makes clear how the UŠ system of timing can be used for describing celestial distances.

The long (accurate) periods after which the phenomena recurred in the same place incorporated the older periods*. They were, however, sometimes longer than the data base, and could not be used in this (the NMAAT) way to make predictions. Knowing the number of events occurring in this period, and the number of times around the ecliptic the planet had travelled, *mean* intervals between successsive intervals could be calculated.

In order for these intervals to be expressed the *zodiac** and *tithis** were invented, the former for the distances, the latter for the times.

Zodiac itself leads to the development of astrology such as horoscopy. This in turn results in the recording and predicting of the dates of planet entry into zodiac signs etc., in GYTs and Almanacs. It also probably explains why day-to-day ephemerides were composed.

The *actual* intervals between successive phenomena of the same type were discovered to vary about the *mean* depending (a) on their location in the zodiac, or (b) on the number of the event taking place. The variations were modelled using step-wise linear mathematical techniques*.

Extent and manner of the variations determined from the records of the *dates* and *longitudes* of the phenomena, in the case of the superior and inferior planets, and from records of the lunar 6 *times* and the *locations* and *times* of eclipses in the case of the Moon.

Restrictions on the models were imposed by the use of simple numbers and the demands of minimising calculations. Some numbers used were traditional, rather than deriving from observation*. Calculated dates were expressed in the Metonic calendar and *tithis*, calculated planetary and solar longitudes in UŠ of the zodiac. Month length and the lunar six were expressed in UŠ. Much of the apparent accuracy was in fact perceived precision.

Several slightly different predictions for any one event co-existed. As long as the methods by which they were achieved were *legitimate*, and they were sufficiently accurate for the purposes of divination and astrology, the models that predicted the most accurately were not adopted *at the expense* of the others*.

5.1.1 Internal Considerations

The scheme above summarises the development of Mesopotamian planetary astronomy-astrology outlined in Chapters 3 and 4. It describes an evolution from a celestial divination that did *not* incorporate the accurate prediction of celestial phenomena to a celestial divination/astrology/astronomy that did. A concern in cuneiform with interpreting the phenomena of the heavens is attested at least as early as the first few centuries of the second millennium BC, but the transition to a discipline that included the prediction of those same events takes place, I argue, during the 7th and 8th centuries BC. During that period a core hypothesis connected with astronomical prediction was added to the repertoire of divinatory wisdom, and this axiom of the predictive Paradigm remained in place until the very end of cuneiform writing. The axiom connected the record of ominous events with an astronomy that in its most sophisticated state of development was capable of modelling the movements of the moon in latitude and longitude, and combine these in such a way as to enable the accurate predictions of the times and magnitudes of eclipses to be made.

I argued in Chapter 3 that the determination of what was and was not ominous in the heavens depended very little on empirical observation. The skies were decoded by diviners, which meant that they had already been encoded, and the part played by the simultaneous occurrence of celestial and terrestrial happenings in that process of encoding was minimal. A close study of the material that together constitutes the EAE Paradigm – the Paradigm that enabled the phenomena of the heavens to be deciphered – revealed that in order for those phenomena to be interpreted they were analysed in two main ways. Firstly, the infinite number of locations and times at which a celestial event might occur, and the infinite variety of colour and shades it might manifest, were fitted into a few broad categories. Aside from meteors, comets and meteorological effects, only the heliacal events of the planets, and planet-planet and planet-stellar interactions were deemed worthy of inspection, and the many possible planet-stellar approaches were reduced in number through the treating of most stars as members of larger constellations and using but a few levels of separation. If we term the colours, locations, separations and so forth the "variables" of the celestial event, then the first means by which the skies were rendered interpretable by the third millennium BC (or earlier) diviners was by reducing the number of these variables. The resultant broad categories are attested in some of the oldest divinatory and non-divinatory cuneiform texts, and survive until the end of cuneiform writing. Variable-reducing categorisation was a core hypothesis of the EAE Paradigm.

The heliacal events of the heavens are cyclical in pattern, and this phenomenon provided the diviners with another means by which readings from above could be gleaned. Adopting or deriving ideal, round-number values for the lengths of the year, the month, the periods of time for which the planets were visible or invisible, and for the rough association between the months and certain rising stars, permitted the diviners to compare what was observed with what was anticipated by such ideals. These ideal, largely temporal, categories generated anomalies and coherences, and the interpretation of one was antithetical to the interpretation of the other. Some of these ideals, the ideal year and month, and ideal intercalation scheme, for example, were known at least by the mid-third millennium BC, those connected with the planets were perhaps only discovered in the OB period. Indeed, I suggested in Ch.2.1.2, on the basis of evidence connected with their name associations, that Mercury and Saturn were discovered significantly later than the other planets, and long after divination had already associated many star and planet names. One ideal period, the one connected with the interval between successive eclipses, was not perhaps formulated until the late NA period, but as a central axiom of the EAE Paradigm, the application of round-number periods to divinatory ends did not change until cuneiform itself died out.

The variable-reducing and anomaly-producing categories were encoded simply. Each boded either well or ill, and applied either to the land of the diviners or to the lands of foreigners, which by the OB period invariably included Akkad and three others in a fourfold division. The binary division of *pars hostilis* and *pars familiaris* perhaps characterises the earliest form of celestial divination, as reflected in the opposite significance attached to brightness and dimness, left and right, above and below, and so forth. There are even some hints that the planets were originally either benefic (Jupiter and Venus) or malefic (Mars), and only the discovery of Mercury and Saturn led to intermediate positions of significance.

Such broad encodings make ridiculous any idea that observation played a part in the assignations of value to the heavenly bodies and their phenomena. Much of the encoding drew instead on "traditional" notions as to the rôle played by those particular gods linked to the heavenly bodies, or on other such folklore, the analysis of which is all but impossible. The encoding of Venus with the benefic qualities associated with Inana, latterly Ištar, for example, was *basic* to cuneiform divination, but little more than this can be said. Jupiter's basic association with Marduk may have been preceded by a basic association with Šulpae, but without the written record of the decipherments of this planet's behaviour when it was associated with Šulpae, little can be said about the earlier encoding of that heavenly body aside from a suspicion that Jupiter may always have been considered benefic. From what has survived, however, all one can say is that by the time Jupiter was associated with Marduk it was considered to be a good-boding planet, and that this encoding must be understood to be an *a priori* fact, so far as all subsequent celestial diviners were concerned. It was on the bedrock of core "facts" like these that learned scribes came to build the edifice known as EAE, and were able to render the heavens readable. One such core divinatory axiom of the encoding was that if an event occurred according to that predicted by the ideal period which modelled its behaviour this boded well, and if it did not this boded ill. I made much of this discovery, for to my mind all previous studies of cuneiform divination and astronomy have misunderstood the rôle played by these ideal periods, treating them as examples of "primitive astronomy", and not appreciating that they were not intended to enable the diviners to establish the future movement of the heavens, but were used instead to make its cyclical, regular-running nature amenable to interpretation.

Once the basic associations with deities had been made, and the variable-reducing and anomaly-producing categories had been assigned the simple code, the way was paved for the elaboration of omens using the rules of textual play, and the creation of ideal period *schemes* using the rules of number play. It was the application of these rules that led to the rich and complex collection of divinatory material exemplified by EAE, Mul.Apin and others. It is they that account for the variant apodoses, the multiple readings, the learned allusions, the historiettes etc. To a great extent only these omens and ideal schemes of celestial divination have survived. Very occasionally particular schemata of the code were written out, attesting to a degree of abstraction on the core hypotheses of the EAE Paradigm, but in most cases it is only a close analysis of the texts that can reveal the underpinning syntagmatic play, the metaphoric cross-references to other texts, the code, and the core categories.

A small amount of corroborating evidence for the existence of these underlying premises comes from the styluses of the Scholars themselves. Aside from those examples of the abstracted code noted in Ch.3.2.2, in Ch.3.1.1 I quoted and referred to a number of texts that described the constellations as "drawings/designs", and the year as having been "measured" in order that "signs" be established. Some texts explicitly described "writing" on the sky, and with this recommendation from the very Scholars who composed the incipits of EAE itself, I see no problem in interpreting cuneiform celestial divination as a *creation of the early literate* (see n286). While traditional, basic encodings of some heavenly bodies very likely predated the oldest written omens, and the simple code could easily have been used by the non-literate (see n343), the rigorous application of four-fold, three-fold, or even binary divisions in the code, the uses of the technology of listing (see n203), the elaborations based on the multiple readings of signs, the quoting from other texts, the numerological speculation, and the mathematical play manifested in extant divinatory texts from the OB to the

Hellenistic periods all attest the efforts of a literate, learned few. They, the scholars, were the creators of the EAE Paradigm, and as I also argue it was another such group that created the later PCP Paradigm.

The few texts known that indicate that the learned scribes considered celestial divination to be a form of celestial writing, and thereby show that they thought of their discipline as a creation of the literate, date from the NA period, though some may have first been composed in the second millennium, particularly the incipits to EAE. The idea of accounting for the heavens in terms of celestial "design" is far older, however, as we shall see in §5.1.3, and thus I argue that the literate nature of celestial divination was being alluded to as early as the third millennium. The premise that the heavens were a slate upon which were written signs that could be deciphered, providing one knew the code, stood behind all celestial divination in Mesopotamia. It was alluded to by the authors of Sumerian literature, as we shall see below, and persisted until the Christian period.

This description of cuneiform divination is far from one which sees in it a collection of the record of those heavenly phenomena that were followed by remarkable events, and which when recurring were thought to predict the same event once more (see Ch. 3.1.1). Texts of the EAE Paradigm did not evolve in this way, by accreting to the body of omens ever more accounts of simultaneous celestial and terrestrial happenings. Its core hypotheses were established at least by the OB period, and these hypotheses indicated how, at any time, the heavens could be observed and decoded. For reasons poorly understood, but perhaps connected with the demise of the Sumerian language or changes in the political arena, some of these decipherments came to be written down, and the texts produced then helped preserve the discipline for centuries thereafter. They became *the* texts of the EAE Paradigm, and came to be treated with a degree of reverence. The core hypotheses which underpinned the decipherments did not alter over time, and this is probably most clearly demonstrated by the persistence in the latest omens of the same four-fold, three-fold and binary divisions of the code found in the earliest, by the continuity in the values attached to the planets, by the existence of "impossible protases" in OB eclipse omens that attest textual play in omen creation at this early time, and so forth. In App.3 I have outlined the extent to which omens in the standard version of EAE were in large part invented, and this was true of the OB omens, *and* those used in late NA times. It was also true for those few new omens in LB times that dealt with the significance attached to lunar latitude (see Ch.4.1.2). The core hypotheses influenced the creation of celestial omens for at least 2 millennia, and were transmitted along with the texts of the EAE Paradigm, either explicitly or implicitly. The few allusions to celestial writing, and the few examples of abstraction just referred to lead me to suspect that they were transmitted *explicitly*, and formed part of the "wisdom" known at least by the most senior Scholars.

The means by which the core hypotheses – the premises – were elaborated into omens and ideal period schemes *did* vary over time, however, and it is in these variations that the influence of individuals, schools or eras can be felt. The particular schemata applied to the code differed somewhat, and the use of syntagmatic rules in conjunction with the schemata could lead to more than one *legitimate* interpretation of one set of heavenly phenomena (see n348 and Ch.2.1.2, for example). Which of a stock of apodoses should be used with a given protasis was also sometimes a matter of personal choice, as the variants in both EAE and in the Letters and Reports indicate. Certain political expediencies led some Scholars to reinterpret existing omens in rather favourable ways. These subtle variations, caused by the normal work of Scholars over the centuries, formed what one might term the "protective belt" of celestial divination, a belt around the core "wisdom" in which innovation was still possible, in which the pyrotechnic brilliance of these learned scribes could still shine without any challenges to the core premises being necessary.[466]

The EAE Paradigm was an extremely important cultural achievement in Mesopotamia, used by kings, referred to in literature, transmitted abroad, preserved in temples, and leading to the employment by the late NA kings of a large number of professional celestial diviners. I discussed in Chapter 1 the heavy investment made by the last NA kings in maintaining a personal entourage of Scholars. In due course many of the core hypotheses of the discipline were transmitted both to the West and to the East. These included the largely benefic or malefic nature of the planets, the constellations (albeit some transformed into zodiacal signs), the concept of planet-planet interaction, and such things as the three and four-fold divisions of the heavens, and perhaps the significance of brightness and dimness, left and right etc. Other technical aspects of celestial divination (non-core hypotheses) were also borrowed into European astrology, in particular the hypsomata,[467] the 12-fold divisions of the zodiacal signs,[468] and of course the zodiac itself, and the many constants used in the NMAATs and MAATs that came to be used in the writing of horoscopes. Much in cuneiform celestial divination appears not to have been used elsewhere, however, including the significance of reality cohering with ideality and vice versa, or the central place given to heliacal phenomena. Nevertheless, the wide-spread use of astrology today owes a huge debt to celestial divination, and thus to the learned elaborations of a few literate Scholars living in the centuries around the turn of the third millennium BC.

Returning to a consideration of the internal structure of the EAE and PCP Paradigms, one consequence of the simple code and the significance attached to events occurring according to, or not according to ideals was the production of what I have termed "direct" and "indirect" ominous events. For example, the variable- reducing categories and corresponding code ensured that Jupiter's heliacal rising in a given month had a certain significance. This planet, this event and the month in question were directly ominous variables, and omens were constructed that employed these variables in their protases. A record of the observations of Jupiter, made for the purposes of divination, would record only the month in which each heliacal rising occurred. Even a very long record of these variables would probably prove inadequate to provide the data necessary to discovery periods after which Jupiter repeated its heliacal phenomena.[469] For this a record of the *dates* would be necessary, but there are no omens that attach significance to every date upon which Jupiter might rise. The anomaly-producing categories, however, encoded as described, did mean that the date upon which Jupiter rose was ominous, albeit indirectly. The amount of time for which the planet had been invisible could be compared with the ideal time for that occurrence, and interpreted accordingly. It was for this reason, I suggest, that any record of the observations of Jupiter,

466 In a similar vein Oppenheim (1978) 642 writes of omenology that it is: "an example of the process of additive rather than structural changes (for prestige purposes) that is evidenced in nearly all types of Mesopotamian literary production," but I use Lakatos' (1978) terminology in order to emphasise the "scientific" nature of the EAE Paradigm. See §5.1.3.

467 The "exaltation" of a planet – see n193, above.

468 See App.1 §49 for a brief summary of literature on these transmissions.

469 One of Venus's characteristic periods is short and accurately equal to 99 months, however, so may have been discovered from a record solely of the months in which it rose heliacally.

made for the purposes of divination, would have included the dates of these events as well as the months in which they took place. Only on the basis of these records could the characteristic periods of Jupiter have been discovered, and the possibilities of accurate prediction realised by the diviners.

The dependence of the means by which astronomical predictions were made in cuneiform on the EAE Paradigm's ideal period schemes is clearest in the case of the record of the lunar six. I argued that the lunar visibility/invisibility scheme rendered these times ominous, and that their record led to the discover of a number of periods and parameters crucial to both NMAATs and MAATs, as shown in Ch.4.1.2. While it is possible that in the later NA period the interest in predicting celestial phenomena led directly to the production of long accurate records of the dates and times of heliacal phenomena, it is unlikely that the Scholars should have recorded the lunar six if these times did not have some divinatory significance. It is thus highly plausible that the dates of the planets' heliacal events were also recorded *because* they were ominous, and not because they were known to provide the data base necessary to derive characteristic periods, say. This last would make an unjustified assumption that those in ancient Mesopotamia interested in astronomical prediction would have gone about the task as would we today. I thus suggest that it was a record made *without knowledge of its astronomical potential* that first led to astronomy in Mesopotamia. The production of these records for the purposes of divination was a critical step in the invention of predictive astronomy. The revolutionary thought – the establishing of that central core hypothesis of the PCP Paradigm – did not appear out of thin air, but emerged out of the process whereby the divinatory industry functioned. That is, the industry employed scribes to "make astronomical observations" (mul.meš an-*e ṣubbû* – see Ch.1.4), and they endeavoured to produce a continuous record of ominous phenomena. The so-called Eclipse Records, Planet Records, and Diaries are examples of these. In order to *maintain the continuity* of the records in the case of inclement weather or equivalent, certain periodicities in the data record were exploited. This led to the discovery of equations such as those in *TU* 11: 29–38 (see Ch.4.1.2), which related lunar six values separated by 223 and 229 months, and to the discovery of the characteristic periods of the other planets. Maintaining a continuous record of celestial phenomena in the absence of direct observation was, however, astronomy, of course.

Over time the means by which astronomical predictions were made were refined and improved. The continuous record of indirectly ominous data provided values for the errors in the characteristic periods of the planets, and so led to the long periods used in the MAATs. It also provided the parameters used in the MAATs to model the variation of the actual intervals around the mean intervals between successive heliacal phenomena. To assist in determining and using the characteristic periods the luni-solar calendar was regulated, and in order to simplify the calculation of the predicted times and locations of future celestial events *tithis* and the zodiac were invented. Many of these developments depended heavily on specific aspects of the EAE Paradigm, as well as on the continuous record of ominous phenomena. These dependencies have been noted with asterisks in the chart above. The ominous significance of Sirius's rising in month IV played a part in the regulation of the calendar into the so-called Metonic pattern. The numbers of *tithis* in a month and UŠ in the zodiac were determined by the ideal month and year respectively. The importance of a planet being "high" at first visibility led to the record of the times for which planets were visible at first and last appearance, and so to a calculation of the actual dates for these heliacal events.

The step-wise linear techniques used in the MAATs were anticipated by those used in EAE 14, Mul.Apin and similar. Some numbers, such as 12 fingers in an eclipse (see n415) that had played a part in celestial divination for centuries, came to be used in the MAATs even though subsequent developments had by then made them wildly inaccurate.

There were, however, some parts which cannot thus far be accounted for on the basis of a database lying behind celestial divination. They are *artefacts* of the predictive nature of the PCP Paradigm. While much that exists in the NMAATs and MAATs can be explained as resulting from a continuous record of directly and indirectly ominous events, the same cannot be said for the record in the Diaries, for example, of the locations and distances of the planets from Normal stars. Only a few planet-Normal star interactions were ominous, and while it is not hard to imagine a database of *purely* ominous phenomena extending this to include a record of all such events, I believe that the Diaries, Eclipse and Planet records did *also* include data *for the purposes of astronomical* prediction. These include the accurate records of the times of eclipses and the dates of opposition, neither of which was ominous, as well as the Normal star data. Indeed the Normal star data probably led to the development of the kùš or cubit-system for measuring celestial distances. I argue in a forthcoming paper[470] that this system was itself related to the UŠ system, and thus also owed a debt to the EAE Paradigm.

Those parts of the texts that together constitute the PCP Paradigm, and which are artefacts of its predictive and not of its divinatory aspect are important for they show that the prediction of celestial phenomena was based on a new and revolutionary idea. While the original stimulus behind an *unbroken* record of celestial events was one caused by the needs of the divination industry (see §5.1.2), the discovery of the possibilities for accurate prediction came to serve an additional purpose. This was that it allowed the diviners to know in advance when and when not to look, and where in the sky to look. This allowed them to prepare the appropriate counter-measures to a forthcoming ill-boding event, such as an eclipse, and permitted them and their assistants to reduce the amount of time spent in fruitless observation of the heavens. In time it allowed those Scholars who wrote horoscopes more or less to reconstruct the *state* of the heavens at any given time.

The revolutionary idea that accurate astronomical prediction was possible did mean that some at least of the messages sent by the gods (for it was they who created the signs[471]) could be known *in advance*. This perhaps had theo-philosophical implications, which I will touch upon in §5.1.3, though I propose now that the predictions remained palatable to the clergy firstly by the fact that for the full implication of an anticipated event to be determined it had to be observed, since its colours and accompanying meteorological effects had meaning and are unpredictable, and secondly on the basis that the core hypothesis of the resulting Paradigm meant that predictions were achieved through the judicious use of the continuous record of *ominous* phenomena, supplemented by a few additional data points that were recognised to fill in the missing gaps in the database, and mathematical modelling using techniques already familiar to practitioners of celestial divination. Cuneiform astronomy was constructed in such a way that its predictions continued to be legitimate for the purposes of divination.

---

470 Brown *CAJ* (forthcoming).

471 E.g. x056: r.18f quoted in Ch.1.3.

The PCP Paradigm thus evolved under the auspices of the millennia old celestial divination industry, but cast off a central axiom of the EAE Paradigm that characteristic (ideal) periods and accompanying ideal period schemes served *only* a divinatory purpose. More accurate characteristic periods and *their* accompanying schemes (MAATs) could be used to predict forthcoming ominous phenomena to the accuracy of a day or better. This shift had begun with the creation of a continuous database of ominous events, evolved to include the recording of other data necessary for prediction, led some scholars towards a mathematisation of the problem of prediction, and others to a use of NMAATs and tables of errors in the periods. It created the zodiac, which itself spawned a whole new industry concerned with celestial interpretation – astrology, which in its turn led to a demand for the calculation of locations of the heavenly bodies at or near a date of birth. This fed back into MAATs and NMAATs, and resulted in the creation of Normal star Almanacs and day-to-day ephemerides. Such were the consequences of this revolutionary thought, this challenge to a central axiom of cuneiform celestial divination, this gestalt shift in Paradigms.

I dated this shift to the 8th and 7th centuries BC in Ch.4.2, and I account for it in §5.2 on the basis of the consequences of a large network of Scholars all seeking to please the last NA kings, the kings of the four quarters. I dated the revolution partly on the basis that continuous cuneiform records of ominous phenomena, and indeed of more than ominous phenomena, begin around c.750 BC, but mainly on the fact that in those records the earliest examples are significantly more primitive than the later ones. Here is the central point about the surviving data. In the 8th and 7th centuries BC the accuracy of the records, the predictions, and the means by which the calendar was regulated *were in their infancy.*

So, a close consideration of the internal structures of the astronomical-astrological texts of Mesopotamia has shown that the deservedly famous MAATs and NMAATs of the PCP Paradigm could never have occurred but for EAE, and that they also incorporated a revolutionary idea at their heart. The resultant astronomy both successfully predicted celestial phenomena, and adhered to a conception of the universe that also underpinned celestial divination (see further in §5.1.3). The solutions to a problem of prediction had to be legitimate in so far as they employed mathematical techniques and parameters from EAE, and relied on periods and parameters derived from the record of ominous phenomena, but were otherwise only limited by the ingenuity of the scribes in question. Some solutions were more successful than others, but the less accurate models were used alongside their more sophisticated (to our minds) relations. This is perhaps strange to us, wedded as we are to the idea of science evolving to achieve ever more and more accurate results. However, despite the tremendous achievements of the MAATs, NMAATs such as Almanacs and GYTs continued to be written by the very same scribes throughout the Hellenistic period and beyond. It must also be recalled that we have evidence outlined in Ch.4.2 that the lunar MAATs themselves were developed over a period longer than the lifetime of any one individual. They came about as a result of a collective effort. Perhaps, this collaboration made it more difficult to abandon one system of prediction in favour of another, since all the systems were recognised to be in a state of evolution, and were, in a sense, possessed by the schools or guilds to which the scribes in question were attached. Respect for authority perhaps played a part in the preservation of these rival solutions – see further below. The result, in any case, was an astronomy that was simultaneously *legitimate* and *good enough*. It accepted the existence of more than one solution, and probably deliberately limited its predictive potential by preserving elements of EAE, and by what can perhaps be termed "saving the phenomena" – a notion used by some Greek astronomers and discussed further in §5.1.3. Cuneiform astronomy never became wholly like our science, and should not be analysed accordingly.

5.1.2 External Considerations

Textual evidence in texts that do not belong either to the EAE or PCP Paradigms, as well as non-textual evidence, can tell us a great deal about the purpose and evolution of cuneiform astronomy-astrology.

It is self-evident that my model of the transition from the EAE Paradigm to the PCP Paradigm during the late NA period has been governed by what texts have survived. The existence of many hundreds of texts directly relevant to this subject found in Nineveh, Babylon and Uruk in particular, and dating to the period after c. 750 BC can in large part be accounted for on the basis of the 612 BC layer of destruction wrought on the Assyrian capital, the water table under Babylon that has destroyed the oldest material in that city, and the near-abandonment of both Babylon and Uruk around the beginning of the Christian era. Nevertheless, Nabonassar is alleged to have destroyed records predating him, archaeological surface surveys indicate that the period before c. 750 BC in Babylonia was particularly harsh, and it is known that the Assyrians began at that time to involve themselves directly in the affairs of their southern neighbour (see I.2). So while the mid-8th century BC is a convenient point with which to date the beginning of the period of transition from one Paradigm to another, so far as those texts that have survived suggest, it is not without wider geo-political significance.

The growing might of Assyria in the 8th and 7th centuries BC is another such *fact* derived from non-astrological-astronomical sources that can brought to bear in accounting for the revolution in Paradigms. The destruction of Nineveh in the late 7th century was fortuitous, in so far as it resulted in the survival of the Letters, Reports and much else. It is, however, clear from the internal evidence of the astronomical-astrological texts recovered from there, and from Babylonia, that much of the transition to a predictive divinatory Paradigm had taken place long before c. 612 BC. It is also apparent merely from the *numbers* of texts recovered that the end of the reign of Esarhaddon was one when a great deal of astrological-astronomical activity was undertaken. 612 BC is, therefore, another archaeologically *convenient* date with which to end the period of change, but as I associate that change with the particular circumstances brought on by Neo-Assyrian might in the area, it too is not without justification. I suggested that the "revolutionary thought", that accurate prediction was possible through the use of characteristic periods and parameters derived from (in the first place) the record of ominous phenomena, was a mid-8th century BC (or slightly earlier – see Ch.4.2.4.3) event, but the means by which this thought was realised evolved rapidly through the 8th and 7th centuries, particularly during the reigns of Esarhaddon and Assurbanipal. The result was that by the end of the 7th century BC the vast majority of the methods employed in the later Babylonian and Urukean NMAATs and MAATs were already in place, albeit still in their infancy. To this extent the external and internal evidence combine to suggest together the critical importance to the development of predictive techniques of the rule of the last Assyrian kings.

After 612 BC the focus of attention is on the south alone, and any reconstruction relies on the contents of two "archives", one of which was in large part excavated illegally.

Still, evolution *can* be seen over centuries in the records from these archives, and enough internal evidence of adherence to the norms of celestial divination etc. exists to show that the surviving texts do illuminate, even if only in patches, the changing *state* of cuneiform astronomy-astrology in Mesopotamia. The external archaeological and physical evidence that the texts whose contents date to the 8th century BC, and those whose contents belong to the 1st century AD belonged to the same temple (even to the same temple employees – see n377), *in combination* with the internal evidence of a continuity in names, scripts and methods over the centuries, shows that Mesopotamian Hellenistic astronomy and astrology belonged to a tradition that can be traced back unbroken to the late NA period, and in part to the third millennium BC. The MAATs, for example, were not innovations of the 3rd, 4th, or 5th centuries BC, but of the 8th and 7th centuries BC. Their "origins" cannot be accounted for on the basis of foreign – Persian or Greek – influence, say, but on the basis of the circumstances that pertained under the Neo-Assyrian empire.

Further evidence in non-astrological-astronomical texts comes from Sanskrit, Greek and Egyptian material. Pingree has dated the export of omens in EAE to India to the period after the 4th century BC, and omens, predictive techniques and parameters appear unchanged in papyri, and even in some mediaeval sources[472] – in addition to the already mentioned export of astrological values, the zodiac and so forth. External evidence such as this suggests that EAE was still important in elevated circles in Mesopotamia, and that cuneiform astronomy-astrology was utilised, perhaps in a modified form, by Greek overlords of the Seleucid period. On the one hand this shows that astrology did not wholly supplant the discipline of celestial divination, the demise of which should perhaps be connected with the demise of cuneiform, and on the other that the MAATs, NMAATs, astrological and divinatory texts were not merely used for and by temple employees, but in the wider community (see Ch.4.1 and nn378–9). The rôle of the temple in developing and preserving this industry, in developing new predictive techniques, and in preserving a legitimating continuity with the past was perhaps greater than it was under the Assyrian kings, with their large, personal entourages of Scholars, but I believe we should not think of cuneiform astronomy-astrology of the last few centuries BC as being in the preserve only of backward-looking, intellectual clergy. Such a view tends to suggest that the most sophisticated MAATs were created "for intellectual interest", whereas I have tried in Ch.4.1 to ground the invention of even the most difficult texts in the functional requirements of an elaborate, vibrant divination/astrology industry, tied to a cuneiform tradition millennia old, but nevertheless closely linked to important aspects of contemporary life and politics (e.g. see n380). I have linked those texts that predict celestial phenomena with those that predict their meaning, and criticised those who treat the MAATs, in particular, separately from EAE-style divination. Indeed, I would argue that any reconstruction of the LB world, its temples and literate citizens, must incorporate those contributions to history offered by these predictive texts. It should include the fact that they attest to a continuity of tradition, but accept innovation within a limited framework, that they indicate that temple intellectuals kept one foot in the mundane as well as one in the sky, and so forth. This material should not remain the concern only of a few historians of astronomy, but form a more central part of general Assyriological study.[473]

---

472 For references see nn264 and 266, above.

473 We may now move on from Oppenheim's pessimistic statement in *AM* 305 "it is to be regretted that such an essential aspect of Mesopotamian science as mathematical astronomy cannot be utilised more directly in the presentation of Mesopotamian civilisation...we are completely at a loss as to the nature of its development."

The situation prior to the late NA period presents more difficulties. Very little astrological-astronomical material survives from the period between the early second millennium BC and the mid-8th century. There are but a few highlights – 9th and 8th century copies of EAE from Nimrud, the MA astrolabe from Assur, and a few related MB texts, some OB celestial omens from a few sites, the title of EAE in a catalogue from Ur, and allusions in Sumerian sources to the background to celestial divination (see App.1), all without context. A reconstruction of the creation and evolution of the EAE Paradigm, the "normal science", say, of its development, is thus based largely on internal, and very much less on external evidence. It is consequently less secure. We know so little about the state of astronomy-astrology in Assyria at the turn of the first millennium BC, for example, that we cannot be sure if *Astrolabe B*, say, is characteristic of the products of the discipline at that time, or an exception to them. Were the MA kings surrounded by Scholars, as were their NA counterparts, and as were the OB kings of Mari, to a limited extent (App.1 §5)? Can we really talk of a borrowing from Babylonia into Assyria of the "wisdom" of celestial divination around 1200 BC, merely on the basis of the internal evidence of EAE – the references to a few MB kings, and to Subartu as an enemy (see App.1 §21)? Just because a few Babylonian scribal families became Assyrianised in the late NA period (Ch.1.1), does not mean that this had not taken place throughout the 2nd millennium as well. In the absence of external evidence more extensive than the *existence* of OB celestial divination etc., I have tried to be methodologically sound in my approach. I have worked backwards from the situation that pertained in the NA period (see I.2), which was based on extensive records, internal and external evidence, and argued the following points: (a) that celestial divination as practised by the late NA Scholars showed that an (EAE) Paradigm could be described, as was done in Ch.3, and that the surviving material from earlier periods paralleled texts of that Paradigm extremely closely, and (b) that predictive astronomy was in its infancy in the late NA period, perhaps even beginning, and was directly ancestral to all subsequent Mesopotamian astronomy, and to zodiacal astrology. Perhaps there were times before the mid-8th century BC when accurate predictions were attempted that have so far been lost to history. The "revolutionary idea" may have had many precursors, but I doubt this strongly because I link the shift in Paradigms to the circumstances of the late NA period, which were particularly special in Mesopotamian history. My hypothesis is falsifiable, however, for one day an earlier collection of astronomical cuneiform texts may be discovered, which, perhaps, cannot be accounted for on the basis of socio-political circumstances, but only on the brilliance of one scribe. We have to begin somewhere,[474] however, and I have been explicit about asserting that I believe cuneiform astrology-astronomy to have been significantly influenced by the changing world in which its practitioners worked.[475] This, connecting the earliest NMAATs with celestial divination, noting that period schemes had a divinatory purpose and that the omens were mostly invented, and analysing texts from Nineveh and Babylon in which the earliest accurate predictions were attempted, has allowed me to reconstruct cuneiform astronomy-astrology in a way never done before.

---

[474] As Oppenheim (1969) 114 writes: "I am therefore going to use the above statistics…to the utmost, even to the extent of resorting to *argumenta e silentio*. My conclusions may be debatable, but I do not concede that my right to utilise the evidence of this kind in this way can be disputed. After all, Assyriology is where one finds it."

[475] See Lakatos's comments on the explanation of scientific change by Kuhn and Polanyi, for example, in terms of what he calls "social psychology" in (1978) p31 nn2–3.

For the most part temples and palaces have been the focus of archaeological interest in Mesopotamia. The vast majority of cuneiform astronomical-astrological texts come either from royal or temple-based archives and libraries. Despite this slant on the database, it is known that celestial divination of the EAE variety was performed for kings, and that Scholars who were associated with temples undertook both divination and astronomical prediction. What has survived still does then, I suggest, reflect the complete nature of Mesopotamian cuneiform astrology-astronomy. It is extremely unlikely that such celestial divination or astronomy existed outside the royal courts or temples.

Similarly, it has not been possible to incorporate into this model of Mesopotamia planetary astronomy-astrology those texts written on perishable materials. Cuneiform has survived because of the materials on which it was written, but it is not *prima facie* obvious that Mesopotamian astronomy-astrology was a cuneiform-only discipline. Indeed, those texts written in cuneiform may have, in part, been written on clay merely because they survived for centuries, and those *theoretical* texts that backed them up have perhaps been lost to the ravages of time. This may be true of the texts which together formed the continuous record of ominous (and other) phenomena – the Diaries, Eclipse and Planet Records. If this were the case, one could then argue that by continuing to use clay when other materials were available (see I.6), it was recognised that the characteristic periods were, in some cases, so long that a durable medium was required to make progress. One might suggest that the "technology" of writing on clay, when finally consistently applied to celestial records, played a not-insignificant part in the evolution of Mesopotamian astronomy – see n24. Equally, the robustness of clay tablets may have been one reason why those texts that needed to be consulted repeatedly were written in cuneiform, such as the Almanacs and MAATs. Despite these cases, however, the presence in the surviving record of texts that describe only a few functions used in the MAATs, or of procedure texts that describe the workings of the ephemerides, suggests that Mesopotamian royal or temple-based astronomy was a largely, if not entirely, cuneiform affair. I also noted that so far as celestial omens were concerned, the very sign forms used contain part of the message imparted, so I would suggest that just as they borrowed heavily from EAE in order to *legitimate* their MAATs and NMAATs, the scribal astronomers used clay and cuneiform to that same end.

I discussed in Ch.1 the names applied to those Scholars who undertook celestial divination in late NA times – the exorcist, chanter and scribe of EAE in particular, whose work together constituted what I termed there a "wisdom". I also noted there that these same names described the authors of NMAATs and MAATs unearthed in Babylon and Uruk. Many of the authors of the texts from Uruk published in *ACT* claim Sîn-lēqe-unnini to be their ancestor. His name usually identified a group of scribes, according to Lambert (1957),[476] though some chanters claim his ancestry, and he is said to have been an exorcist in the *Catalogue of Texts and Authors*.[477] Ekur-zākir was another such ancestral figure, and his name identified in Uruk a group or "guild" of scribes and exorcists, some of whom wrote MAATs. In Babylon, Egibi, an exorcist, is cited in two surviving ephemerides.[478] In the Hellenistic period text *BOR* 4, 132, Bel-aḫḫe-iddina and Nabû-mušetiq-uddi, the sons of the "scribe of EAE" Itti-Marduk-balaṭu, are said now to be "capable of making observations (Diaries)" *(mala naṣāri ša naṣār maṣû* – line 16), and will consequently replace their father in the guild of

[476] See also Beaulieu & Britton (1994) 84.

[477] Lambert (1962).

[478] *ACT* I 11–25, Lambert (1957) n19.

such scribes. They will be paid (by the assembly of Esangila in Babylon) 2 minas of silver and "will make observations (Diaries)" (*ša naṣār inaṣṣarū* line 24) and "they will deliver the computed tables every year" (*tersêtu ša šattussu inandinū*[479]) with Belšunu, Labaši, Muranu, Iddin-Bel and Bel-naṣiršu. These were presumably those Scholars employed by the Marduk temple in Babylon, Esangila, to perform some at least of those tasks connected with producing Diaries, and probably annual Almanacs. It is highly probable that in Babylon these same "scribes of EAE" produced the other attested NMAATs and the MAATs.

The use of these three names to identify those who wrote NMAATs and MAATs in Babylon and Uruk connects these texts yet again with that material used by the late NA Scholars – texts of the EAE Paradigm – and connects the practice of the PCP Paradigm with that of the EAE Paradigm. It is as if a *continuity* in the tradition of "wisdom" behind cuneiform astronomy-astrology, albeit incorporating a significant change in one core-hypothesis of that wisdom, was unbroken from NA to Hellenistic times – or was believed to be, so far as the later scribes were themselves concerned.

It is noteworthy in *BOR* 4,132 that the new "scribes of EAE" were to be paid for their work, just as were the late NA Scholars. In *CT* 49 144: 24 the scribes of EAE, who were also said to perform calculations/measurements (*mišiḫiti*), had free use of certain land. The parallel with the situation that prevailed in the 7th century BC is again close, except that the temple and not the king was the paymaster. Similar pressures perhaps prevailed, and McEwan (1981) 20 comments that the text *BOR* 4, 132 appears to have been a request for the ratification of a decision made by the temple assembly, and is thus suggestive of *state* involvement in the employment of such scribes.

McEwan (loc. cit.) also notes on p20 that *BOR* 4,132 and *CT* 49 144 show that the "scribes of EAE" were organised in Babylon in the Hellenistic period into a largely self-governing guild, in which the crafts of observing, measuring, and computing tables were passed on father to the son. It seems likely that the same occurred in Uruk in this period. In Ch.1.1 I outlined the familial relationships amongst Scholars in the late NA period, arguing that a few families dominated in the kings' entourages, and that many of these families ascribed to themselves ancient, illustrious ancestors. In other words the "wisdom" of, and behind, celestial divination, appeasing the angered gods, and exorcising portended and present evil was also passed on in families, or guilds, or was certainly perceived to do so by the late NA scribes. This wisdom was believed to be very ancient, as the ascription to divine or near-mythical authorship of many of the relevant series indicates, and we have no evidence to doubt that it did indeed go back, in an unbroken line to the OB period and perhaps beyond – passing from father to son, or at least from scholar to apprentice. It was also extremely involved, and there is some evidence that in the 8th and 7th centuries BC the part of the wisdom connected with the heavens became the sole concern of some specialists, and was of particular concern to the Sargonid dynasty (see Ch.1.3). This dynamic, external to the concerns of the texts of the EAE and PCP Paradigms, probably also played a part in prompting a revolution in this wisdom at this time, and thereby make the accurate prediction of future ominous events a real possibility for future scholars.

Clearly, then, the tradition of keeping celestial divination, and the emerging art of astronomical prediction "in the family" employed under the auspices of the NA kings persisted into later centuries. To some extent this is true of most forms of writing undertaken in

[479] *CAD* Š/III 206.

the Hellenistic period – they were monopolised by a small number of experts grouped into guilds, in which the expertise was passed on from generation to generation, often through the family. It is, nevertheless, extremely important to keep this in mind when trying to understand how and why cuneiform astronomy-astrology developed.

Lloyd (1996) Ch.2, for example, explores the institutional background to ancient Greek and Chinese science, and points up the relative importance of adversarial confrontation between "scientists" in the former, and of appeals to authority in the latter.[480] In particular he compares the respective rôles of the *hairesis* or "sect" (literally meaning "choice", but commonly translated as "school") in Greece with the *Jia* or "family" in China, and approaches the question of to whom the resultant scientific texts were directed. It would appear that a Chinese *Jia* commonly preserved and transmitted the teachings of a master in the form of a canon or *Jing*. This canon had to be memorised by each new generation of scholars, and only when recitation was achieved would explanation and mastering of the *Jing* follow (loc. cit. 33). Sometimes access to the text was only offered after initiation into the *Jia*, and thereafter the scholar owed his "family" filial respect. Much of this is reminiscent of the education of Scholars in the NA (and undoubtedly later and earlier) period, outlined in Ch.1.4.

The Greek *haireseis*, on the other hand, were populated by students who had selected *them*, having attended lectures given by a variety of institutions, and thus also had had access to important texts without first having been initiated into one particular "school" (loc. cit. 35f). Criticism of teachers was possible and frequent, and debate about the subjects covered served to maintain a student body, and to differentiate the "sects". Scholars did not feel life-long commitments to their *haireseis*, or to their founders' views (with important exceptions). Lectures and debates between the views espoused by sects were sometimes public and open. The same was not true of the wisdom possessed by the families of scholars in Mesopotamia. Particularly in the first millennium BC, many cuneiform texts contained colophons expressly forbidding their reading by the uninitiated.[481] This also applied to some astronomical-astrological texts – see below.

As Lloyd notes (loc. cit. 39f), the *prima facie* contrast between Greek and Chinese scientific and philosophical work is evident from the fact that in China the preferred audience for the work was the (perceived) benevolent and wise emperor, whereas the works of Greeks, who could never agree on what constituted the best government, were directed at colleagues and rivals. In the case of celestial studies, this perhaps explains why different city states in Greece used different calendars in the 4th century BC, whereas in China any new astronomical ideas, once accepted by the so-called Astronomical Bureau that surrounded the emperor, were implemented throughout the empire. Such is the situation that also prevailed in Mesopotamia, at least from the beginning of the second millennium BC.

Texts *BOR* 4,132 and *CT* 49 144 show that under Greek auspices perhaps only a half-dozen or so scribes were employed at any given time by what was the largest LB temple. It was these few scribes who wrote the famous MAATs and NMAATs, I suggest, and the size

[480] This is not to imply that these were mental traits characteristic of the respective cultures – that the Greeks were agonistic and the Chinese irenic, say – but that such behaviour emerged out of the particular prevailing socio-political structures – loc. cit. p31. Against ascribing "mentalities" see Lloyd (1990) and (1996) 3f.

[481] Colophons containing this warning are attested as early as the MB period. See Beaulieu (1992) 98f. It is hard to know to what extent such colophons indicated that the texts were *fully* prevented from being seen by outsiders to the guild, or merely "express the tendency of the scribes to keep the knowledge of their arts within their own circles" (Neugebauer *ACT* 12). Beaulieu loc. cit. publishes *NBC* 11488, which suggests that the restrictions referred to in the colophons may have existed in practice.

of the retinue compared with that kept by Esarhaddon, say, shows how relatively rich with experts the late NA period was. It was, I have said, a period of change, and the large number of Scholars employed at that time undoubtedly had an impact upon that change. The preponderance over the centuries of certain families of scribes, whilst not precluding premise-challenging intellectual activity, suggests perhaps that preservation and adherence to old norms might have been a more dominant force at play in their work, as it was amongst Chinese astronomers associated with the Astronomical Bureau. I noted that some of the characteristic periods of the planets exceeded any individual's lifetime, and that the evolution of lunar theory occurred over more than a century. The production of the parameters underlying cuneiform astronomical theory had then to be a collaborative endeavour in which the long-term goal of recording observations continuously for decades for the purposes of accurate prediction in texts such as the Diaries could not have been achieved by those who first began those records. This is a remarkable side to the achievements of some of the Mesopotamian scribes – that they participated in a project that could not have been realised in their own lifetimes. It is also shows how the achievements in "science" of a "guild", or *Jia*, might differ from those of an *hairesis*, whose methods of astronomical prediction, say, will probably depend on a short, or imported, data record, and may perhaps survive or fail on their *immediate* ability to predict accurately. The gradual means by which texts that adhered to the PCP Paradigm became ever more accurate in their predictions depended, in part I argue, on the ability of that Paradigm to survive in particular guilds or families of scribes over generations without fundamental criticism.

I proposed that, in the first instance, a continuous record of ominous events was merely a by-product of divination. No doubt this record was made by junior Scholars under duress from senior Scholars to whom they were apprenticed (e.g. Marduk-šapik-zeri's comments on his training to the king in x160, quoted in Ch.1.4). Once its predictive potential was realised, however, the ultimate purpose of recording some data *more* accurately – an artefact of the predictive nature of the PCP Paradigm – could not have achieved fruition by those who first recorded those data in this way. These pioneers, however, did establish forms of recording data that were adhered to by their sons, nephews and grandsons, and no doubt a respect of one's elder, particularly of one's father or uncle, played a part in this. Only a few changes in what was recorded, and how, took place over the years. These included the invention of the zodiac, *tithis* etc, but even these were construed in such a way that they harked back to antique forms. In other words these were the non-hard-core challenging "progressive, protective belt" developments of the PCP Paradigm. With the lunar theory of the MAATs, for example, the Hellenistic period cuneiform scribes drew on the achievements of their 5th century BC colleagues, incorporating the parameters used in those texts, sometimes even if they were extremely inaccurate. This did not matter, as noted, provided the techniques predicted phenomena to an accuracy that was *good enough* for the purposes of divination or astrology, and were *legitimate*.

Scholars in a guild probably had a duty to certain systems prevalent within that guild, even to poor ones, simply because they were created by the institution and not by an individual. Where an individual's work could be challenged, an institution's could not be so easily by a member of that same institution. Differences *between* guilds that had diverged gradually over the centuries in their approaches to astronomical prediction, without challenging the shared core hypotheses, did exist, however, as the system A and system B MAATs, predominately written in different cities, indicate. As noted by Britton (see n405, above), the Uruk and Babylon lunar MAAT schemes differ significantly, whilst still adhering to the same central axioms.

Similarly, whilst any individual cuneiform expert interested in interpreting the signs of the sky worked together with his colleagues in a guild, school or entourage, and respected the work of his predecessors, differences between guilds and schools are apparent. Some (see n207) employed different schemata, but all used the same core hypotheses of the Paradigm. This is a phenomenon perhaps common to the evolution of much science – a new idea catches hold across a wide area, but is thereafter exploited slightly differently depending on location – and we must not argue that *explicit* competition between schools in Mesopotamia accounts for the differences in Scholarly approaches. "Divergence" born of the internal logic of "normal science" better explains the differences between Mesopotamian guilds when it comes both to celestial divination and to cuneiform astronomy.

The audience of cuneiform celestial divination was the king. Divination was used to protect him, and thereby the state, and served to legitimate his rule (see Ch. 1.3). This accounts, of course, for the royal nature of the archives that have survived. However, as the Letters and Reports found in Nineveh indicate, different interpretations of given celestial events *were* sent by different *individual* Scholars within the entourage. Some can be accounted for on the basis of the different temple-based schools or guilds from which the king (probably deliberately) drew his entourage, but some do appear to be the work of individual genius. See for example the interpretations of some Scholars quoted in Ch.2.1.2. I suggest, then, that the very scale of the entourages surrounding Esarhaddon and Assurbanipal in particular, promoted a measure of personal scientific endeavour. It both accelerated the gradual embellishment of EAE, but more significantly played a leading part in leading some Scholars towards astronomical prediction, regardless of their long education in schools that demanded filial adherence. It is as if the slight "secularisation" of these Scholars caused by the king employing them directly created a unique situation in the history of Mesopotamian scholarship – one in which personal rivalry, in a manner not dissimilar from that described between Greek *haireseis* by Lloyd (loc. cit., and see further in §5.2) was able to coexist briefly with the traditional preservation of a corpus of work. The result was that, although a revolution was created, the innovative texts, practices, and techniques produced were carefully couched in such a way as not to conflict with the existing tradition. The new techniques, therefore, borrowed heavily from EAE, as outlined in §5.1.1, above. And, once the hiatus had subsided, the *new* practice of astronomical prediction was performed in the traditional way. It was undertaken by scribes with the titles of those Scholars employed by the last NA kings to protect themselves, was passed-on through the generations, as in earlier times, father to son, and was paid for by the temples, probably with some direct state involvement for whom the products of the PCP Paradigm still fulfilled an important need.

Even the innovative texts were sometimes described in ways which tied them to the past. For example, the colophon to the ephemeris *ACT* 135[482] reads:

> "*Arû*, the wisdom of Anu-ship. . . a secret of the scholar,"

[482] See also the colophon to *ACT* 180.

which resonates with the LB divinatory water clock text BM 29731,[483] whose colophon reads:

> "*Arû*, the wisdom of Nabû...",

and line IIi15f of i.NAM.giš.h̯ur.an.ki.a that reads:[484]

> "*Arû* of words of wisdom... a secret of the scholar".

In Ch.3.2.1 I discussed the meaning of the term *arû*. Whether or not in *ACT* 135 it was intended to mean "mathematical table" or "number play", or both, its use, particularly followed by "the wisdom of Anu-ship", shows that its author was intent on *presenting* his work in a form also used by texts of a divinatory nature.

Colophons describing the texts as "wisdom" of one god or another are not uncommon, but it is worth noting that the ephemeris, although a text *first* written in the closing centuries of the pre-Christian era, was still attributed to a divinity. It was, I suggest, thereby considered legitimate to play a rôle in the functioning of the temple, and of that state sustained by the temple. It is also noteworthy that the MAAT was considered to be a "secret of the scholar". This is, again, by no means an unusual comment to be found in colophons dating to this period,[485] but it undoubtedly shows that the products of the guild were not considered suitable for the uninitiated, as noted, and also, perhaps, that the means by which ominous or astrologically significant phenomena might be predicted was worth keeping secret from rival guilds, or even perhaps from rival scholars. In section 2 of "Text E",[486] another text containing information useful for making astronomical predictions, the characteristic periods of Venus, Mercury and Mars and the associated errors are described as mí.urì *niṣirtu* "secret". Similarly, text DT 72+, discussed in Ch.4.2.2, presents the characteristic planetary periods in a cryptic manner. I suggest that this may have been both because they would have otherwise appeared unsuitable for Assurbanipal's library, to which I argue the text belonged (see n52), and/or in order that another Scholar may not have been able to use such valuable information to further his own career. See §5.2.

5.1.3 Philosophical Considerations

I have, at various points, discussed the development of cuneiform astronomy-astrology, as I see it, in terms of categories more commonly used to describe ancient Greek or more recent science. I would now like to bring these descriptions together in order to show, if nothing else, that Mesopotamian celestial study is a subject that should be of serious concern to philosophers as well as to historians of science. I will touch, therefore, both on the question of the scientific status of cuneiform divination and astronomy, and on its philosophical background.

The term "science" is used by Assyriologists to express something as broad as a rationalistic tendency,[487] to something as narrow as the ability to predict future phenomena

483 See Brown, Fermor & Walker *AfO* (forthcoming).
484 See Ch.3.2.1.
485 References in Beaulieu (1992) 110f.
486 Neugebauer & Sachs (1967) p206. See n408, above.
487 Bottéro (1992) 29f, Limet (1982) 19 §3, Jeyes (1991/2).

using only one initial observation.[488] This in itself has limited the apparent relevance of Mesopotamian achievements to questions as to the nature and practice of science. To my knowledge only Bottéro (1974 & 1992) has considered the cuneiform legacy so far as "modern scientific thinking" goes. However, the achievements of both the EAE and PCP Paradigms can contribute to a discussion of scientific thinking, particularly since the latter Paradigm developed under the auspices of the former, and both evolved under social and political environments quite distinct from those more usually studied.

I do not wish, nor am I competent, to attempt any serious discussion of what constitutes a science.[489] For Pingree (1992) 559, "systematic explanation" suffices, and indeed both celestial divination and cuneiform mathematical astronomy can be characterised in this way. It is apparent in the very structure of EAE, which systematically discusses lunar phenomena, then solar, and so forth, and from that of the MAATs, which treat longitude, then latitude etc. in separate columns. Nevertheless, such a largely relativistic understanding of science does not take us very far.[490] "Science" in modern parlance is recognised by some to have more specific meanings, and it so happens that these in some cases correspond closely with approaches undertaken by the Mesopotamian scholars.[491] This is an important result, I argue, and derived using sound methods since my reconstruction of cuneiform astronomy-astrology has drawn on all the available material, placed in the relevant context to the best of my ability, and no attempt has been made, thus far, to isolate and emphasise those practices which are reminiscent of those of contemporary scientists. While it is completely laudable to argue against "seeking" elements of modern science in the ancient world and treating them as precursors,[492] if it so happens that elements akin (and possibly ancestral) to modern science happen to be recovered in the ancient sources they cannot be ignored. My approach has been to be "up front" about the means of addressing the ancient material,[493] and to take a middle road between an awareness of the possibility of our minds and their minds being "incommensurable", and the belief that there must be some communication between 'them' and 'us' or else translation itself is doomed.[494] I have not assumed "unidirectional linear progress"[495] from them to us, from their intellectual achievements to our modern scientific

---

488 Aaboe (1974) p21f identifies a *primitive pre-scientific* astronomy (already isolating it from the astrology), which involves the naming of stars and planets and connecting seasons with stars, a *pre-scientific* astronomy which is based on periodic cycles of the planets and of eclipses, and a *scientific* astronomy which is free from the need (repeatedly) to consult observations. This last differentiates the NMAATs from the MAATs, but I have argued that both form part of the same PCP Paradigm. This distinction between pre-scientific and scientific astronomy is also espoused explicitly by Neugebauer (1946 & 1957a), and by Britton (1993), and frequently assumed elsewhere in the secondary literature. Nowhere is it combined with what is meant by "science". The distinction is more representative of the division in Assyriology between those who understand these late texts and those who do not, I suggest.

489 See, for example, the arguments of Lloyd (1996) 1f who treats "science" as a "place-holder for a variety of specific inquiries...(including)...astronomy, mathematics and medicine."

490 Pingree loc. cit. 554 argues that Mesopotamian astrology is a science *in its own context* and should be treated as such.

491 Similarly, French (1994) xiii writes "why should we use a modern term to denote ancient usage, when the categories and terms of the past are better?" I agree, of course, but when they happen to coincide this should also be stated.

492 Lindberg (1992) 3.

493 Lloyd (1992) 565: "Since methodology is inevitable, it is better to be self-conscious about it", 576: "whatever terms we use must be treated as provisional and revisable".

494 See n161 and Rochberg (1992) 549.

495 Rochberg loc.cit. 553.

methodology, but I argue that the possibility that the Mesopotamian approach to celestial phenomena was ancestral to today's must be considered.[496]

By using the Kuhnian term "Paradigm" it may appear that I assumed *a priori* that both celestial divination and the post-8th century BC astronomy possessed many more elements akin to "modern science" than mere "systematic explanation". However, I decided to use this term only when I discovered that the ideal period schemes, long thought to be examples of "primitive astronomy", were in fact divinatory in aim. Again, a description of the underlying structure and practice of celestial divination was only *subsequently* compared, and briefly, with some definitions as to what constitutes a "Paradigm". Celestial divination, I argued in Ch.3.1.4, relied on an underlying set of beliefs concerning the form of the heavens, and the means by which its "signs" could be deciphered. The heavens were thought of as a page upon which were written divine messages, and, as we shall see below, their behaviour related always to the ideal form in which it was believed they were first constructed. These underlying beliefs – or to use Lakatos's description, "core hypotheses" – did not change throughout the time celestial divination was employed.[497] Also, celestial divination was used by a specific group of scribes working in specific circumstances, and involved the use of specific texts and apparatus. It evolved from its (probable) 3rd millennium oral precursors in ways similar to those described for other scientific Paradigms. The core hypotheses were not challenged, but the code and rules were employed to create variant apodoses, ideal period schemes, and so forth. This evolution can be compared with what Kuhn terms "normal science", or what Lakatos (1978) 48f calls the:

> "negative heuristic of the scientific research programme, which forbids us to direct the *modus tollens* at this 'hard core' (but) instead... to use our ingenuity to articulate or even invent 'auxiliary hypotheses' which form a *protective belt* around the core."

Similarly, a careful analysis of the attempted astronomical predictions and background methods articulated in the late NA Letters and Reports, and of the earliest Planet and Eclipse Records, Diaries and DT 72+, in combination with an understanding of the NMAATs and MAATs of later centuries, allowed me to *deduce* the existence of a second, consistent scientific enterprise, which I entitled the PCP Paradigm.[498] I was able to isolate its new core hypothesis, and show that it remained unchallenged until the demise of cuneiform itself. The GYTs, Almanacs, MAATs, including day-to-day ephemerides, all depended in part, and at first largely, on data derived from the continuous record of ominous phenomena

[496] For Lindberg loc. cit.3 "the historian requires a very broad definition of "science" – one that will permit investigation of the vast range of practices...which lie behind...the modern scientific enterprise." If it so happens that a narrow definition, based on modern science's "privileged way of knowing" (loc. cit. 2), also matches this cuneiform evidence, this is not without importance.

[497] Lakatos (1978) p48 n4 argues that the core hypotheses do not emerge "fully armed", but only slowly over time. This reflects significantly on the prevalence of explanatory texts in the NA and LB periods. Perhaps only then were the practitioners fully aware of the hypotheses underlying their discipline. Many of these explanatory works were regarded as "secrets of the Scholars", indicating perhaps that wisdom normally transmitted orally was being committed to writing, with ramifications as to the resulting "discovery" of the core hypotheses of the EAE Paradigm.

[498] Admittedly, an initial suspicion (the hypothesis) that cuneiform astronomy of the Hellenistic period had roots in earlier centuries may have guided my initial enquires, but in the course of this study I was able to consult virtually everything published that pertained *directly* to astronomy-astrology in Mesopotamia, and it was only through doing this that I determined *when* those roots began, and to what extent they provided all the nourishment for the Hellenistic flowering.

– that is they all relied on the premise that the accurate record of locations, dates and times of ominous celestial phenomena will lead to the discovery of periods and parameters in that record that will enable the prediction of the times, dates and locations of *future* ominous phenomena to take place. In particular the characteristic periods were recognised to serve *not only* a divinatory function, but a predictive one. All subsequent developments in predictive astronomy, the mathematisation, use of the zodiac and *tithis* etc., were all generated through the practice of normal science – they were "auxiliary hypotheses" – and were also as closely tied as was the new core hypothesis to celestial divination. The MAATs and NMAATs formed part of the same scientific endeavour whose origins are situated in the late NA period. There was no *transition* from NMAATs to MAATs – from "pre-science" to "science" – as Aaboe and others have argued (see n488). They were both complementary realisations of the same fundamental premises.[499]

In Ch.4.2 I isolated the period of revolution from one to the other of these so-called Paradigms. Again, it is of concern from a methodological point of view that this is the period of time from which most texts discussed in this work have come. However, it is the case that I was originally intending to provide a detailed synchronic slice through Mesopotamian astronomy-astrology, comparing it arealy, but not in order to discover that this was a period of infancy in astronomical prediction, of particularly high employment of Scholars etc. In other words, I argue that a feedback mechanism between socio-political circumstance and numbers of relevant texts characterises the late NA period. It was a time of scientific revolution in part because of the domination and wealth of the Assyrians, which included the employment by the king of large numbers of Scholars, which itself resulted in the production of a lot of texts, which we, by chance, have been able to recover in part.

In Ch.3, and in §5.1.1, I argued that the EAE Paradigm contained many subordinating, abstract elements. These are exemplified by texts that spelled out both the "code" and the "rules", which together related categories in the protases to words in the apodoses, and which permitted the decipherment of the sky to take place. I showed in Ch.3.2.1, in particular, how hypothetical protases were invented, and their resultant apodoses deduced through "textual play", and how the ideal period schemes were deduced through "number play" from the hypothetical ideal periods which underpinned the behaviour of the universe (see further below). In other words, between phenomenon and prognosis a *hypothetico-deductive* method was basic to the EAE Paradigm. One could argue that the approach of the Paradigm was analytic, based on a supposed arrangement, and provided what, in part, could be called knowledge for its own sake. Such an approach is characteristic of other compositions of the OB period, such as the "Grammatical Texts",[500] in which Sumerian verbs are parsed in elaborate, sometimes impossible ways. It is, perhaps, the case that the hypothetico-deductive technique was employed in order to "cover as many options as possible." By ensuring that many possible scenarios (celestial events, verbal forms) were anticipated, many impossible scenarios were considered resulting in what I have termed the "impossible protases", for example. It is also apparent that the technology of listing (see n203) played an important part in these endeavours. In the case of the OB and later celestial omens, however, not only was "play" in the vertical axis performed, but so was a syntagmatic, horizontal play on

499 We can, I trust, now move away from such statements as that of O'Neil (1986) in his introduction: "Early Babylonian astronomy…was pre-scientific or at least proto-scientific. By about 500 BC it was approaching a genuinely scientific status".

500 E.g. Black (1984) 129f.

words and numbers. The parallel with the OB Grammatical Texts is, therefore, only good up to a point. Not only were the categories and simple code of an oral discipline being preserved in written form, and extended through "Listenwissenschaft"-like activity to cover many scenarios, the implicit *literate* nature of celestial divination was being made apparent in the first texts in which it was written down. That is, celestial divination was *already* the hermeneutics of the "writing on the sky" long before it was first inscribed in clay. Its *bi-axial* hypothetico-deductive method was thus quite different from others in the OB period.

The PCP Paradigm also applied such methods to the phenomena it treated.[501] In particular, the variation about a mean of the actual intervals between successive phenomena was hypothesised to vary as do piece-wise linear functions, and the resultant times and locations deduced accordingly. This hypothesis was, I argue, as much grounded in an idea that the planets' behaviour varied linearly (since their ideal behaviour did so according to EAE and other related compositions), as it was determined by prevailing mathematical orthodoxy, ease of use, availability of multiplication tables, and so forth. One crucial difference between the deductions of the PCP and EAE Paradigms is that those of the former could be *falsified*.[502] The predictions could be compared with observations, and undoubtedly in the case of mis-predicted eclipses, and so forth, some parameters, even some techniques, were abandoned for new, more successful ones. More often, though, less accurate methods were not abandoned in favour of more accurate ones, which suggests that accuracy of prediction was not the sole, or major, criterion of a successful MAAT or NMAAT. Some very inaccurate parameters were also preserved, and as with the less accurate methods, this was done, I suggest, for reasons of *adherence* to the authority of the relevant guilds, and to the core hypotheses of the Paradigm. The prognostications of the EAE Paradigm could not be falsified, since their non-occurrence could always be counted the result of "mis-reading", or the success of the appropriate apotropaic ritual.

Both Paradigms, therefore, incorporated models that came about through a detailed confrontation with experience, were driven by a hypothetico-deductive style, and included attractive theories of underlying simplicity and coherence. The ephemerides, in particular had a high informative content – they were highly testable and are highly falsifiable (if mainly by modern scholars), and are scientific according to some modern narrow, epistemological definitions, as well as in their own contexts. It is apparent that both celestial divination and cuneiform astronomy were different from other systematic, subordinating, analytic, hypothetico-deductive cuneiform genres, for example the lexical material, or Grammatical Texts – texts which are also called "scientific" by some. Probably because of its unfalsifiable aspect some do not consider the EAE Paradigm to have been a science,[503] and on the basis of a tight, epistemological definition this is undeniable. However, the importance of

[501] That it was subordinating is exemplified by the GYTs, for example, which contained *only* those lunar six records necessary to predict lunar opposition and month length in the goal year. Nothing redundant was included in these compositions. See Ch.4.1.2.

[502] Popper in his books of 1959 (orig. 1936) and 1963 offered a celebrated definition of the process whereby science "discovered" more about the universe through the positing of hypotheses and deducing of phenomena, which only then were compared with observations. Broadly this describes the manner in which omens were generated, except in so far as no observation was allowed to undermine the core premises of the system, only to *enrich* them. See also Popper's article (1970) concerning Kuhn. Popperianism and Kuhnianism can broadly be reconciled by assuming that the periods of revolution between Paradigms are those times when theories are falsified, but that "normal science" is characterised by the ongoing verification of theories.

[503] E.g. Rochberg-Halton, *ABCD* 9: "the systematic presentation of natural phenomena in omens *while not yet qualifying as science...*" (my italics).

this "non-science" to the "science" (defined similarly) of the PCP Paradigm was enormous. I argued that the collection of data for celestial divinatory purposes stimulated the creation of a hypothesis that just such a record, if long enough and precise enough, would make the accurate prediction of ominous phenomena possible. This was how the falsifiable hypothesis underlying the NMAATs and MAATs emerged. Only then were more accurate observations and other "artefacts of the predictive nature of the Paradigm" recorded. This explanation is superior, I believe, to one which suggests that the passive observations of the heavens made it apparent that the planets behaved in periodic, and accurately predictive ways.[504] Firstly, I argue, a variable-reducing categorising of the heavens occurred in order that the hypothesis that it was decipherable could be realised. Probably only as a result of this were the cyclical natures of some the planets' movements noticed, and this too was seen to be part of the message imparted by the gods in the sky. Later, only the aim of keeping a continuous record of ominous phenomena made it apparent that certain precise characteristic periods and parameters could be elicited from such a database, and accurate prediction of future events realised. This led to the *accurate* record of ominous phenomena and some other events, upon which all subsequent cuneiform astronomy depended. The scientific revolution was not, therefore, some "mystical experience of purely socio-psychological dimensions",[505] but was tied functionally to the requirements of the industry that preceded it.

Another approach to the question of the scientific character of the EAE and PCP Paradigms is offered by Lévi-Strauss's idea of the *science du concret*.[506] The "science of the concrete" is explained using the analogy of a bricoleur (idem p16f), or professional odd-job man. The bricoleur uses whatever is at hand, he keeps whatever may come in handy. The craftsman, the exponent of modern science, however, keeps only that which is pertinent to the job, excluding everything he perceives to be redundant to his aim. That is, the *science du concret* is additive and aggregative. It involves classifying *all* phenomena so that everything has its place. Lévi-Strauss goes on to argue that according to the science du concret every eventuality can be explained *totally*, both the why and the how. Modern science is, in contrast, subordinating and analytic. It attempts to reduce as many phenomena as possible to as few a number of "laws" as possible. It deals in concepts, and answers only the question "how?". For Godelier this means seeing modern science as:

> "erasing from the surface of things the network of intentions which man had originally ascribed to them in his own image; (having) destroyed fragment by fragment, level by level the imaginary representations of 'intentional' causes and replacing these with the representation of unintentional and inevitable relationships."[507]

---

[504] For Popper the reasons behind the emergence of new scientific theories are psychological, not logical. The parties in question may come to their hypotheses in any number of ways – none is illegitimate. Observation may be one of them, but is not necessarily primary. Once the hypotheses are formulated, only then do the necessary observations to back them up begin. See also the debate between Kirk and Popper outlined in Ch.5 of Lloyd (1991).

[505] An understanding of why scientific revolutions occur, for which Lakatos (1978) 9–10 criticises Kuhn.

[506] Idem (1966) Ch.1. The premise is that human kind has experienced two major periods of scientific activity, the "neolithic, or early historical" and the post-Renaissance European one. Each ascent corresponds to: "two distinct modes of scientific thought. They are certainly not a function of different stages of development of the human mind, but rather of two strategic levels at which nature is accessible to scientific enquiry: one roughly adapted to that of perception and the imagination: the other at a remove from it" (idem p15). The term "natural philosophy" used by Lindberg (1992) and French (1994) is often meant in a manner similar to that of the *science du concret*.

[507] Quoted in Larsen (1987) 208.

Does the science du concret idea characterise the hypotheses lying behind celestial divination in Mesopotamia? For Larsen (1987) 216:

> "Lévi-Strauss has in fact analysed fundamental features of a conceptual universe which is in many ways comparable to the one found in Mesopotamian traditions."

Larsen argues that the lexical tradition, for example, *is* additive, and not subordinating, and that the omen literature treats *everything* in the universe as a sign, and is similarly aggregative. Certainly, in the case of celestial divination the variable-reducing categorisation did make available for interpretation *all* heavenly phenomena, though as argued in Ch. 3.1.1 signs and events were not linked causally (as implied by both Godelier and Larsen). Celestial divination did, however, provide a total explanation of observed events. They were generated by the gods, and their aim was to inform humanity of their opinions.[508]

I noted above, however, that much in EAE was subordinating and analytic, and thus corresponding not to a *science du concret*, but to modern science. In contrast, in the case of the PCP Paradigm, while only those aspects necessary for accurate prediction were incorporated into the MAATs and NMAATs, it cannot thereby be argued that a "network of intentions" was erased from these predicted events. On the contrary, the predictions were made in order to facilitate divination in its efforts to elicit the meaning of the forthcoming signs. The PCP Paradigm was thus fully a science du concret, in spite of the superficial similarity of the MAATs with their elaborate mathematical character to modern sciences.[509]

However, it *is* likely that the new ability to predict ominous phenomena, and thereby predict the gods' actions in some small way, did have some impact on the perception of the rôle played by those gods in the behaviour of the universe. If modern science has in large part eliminated the "why?" behind phenomena, in successfully accounting for the "how?" of their occurrence, perhaps the PCP Paradigm was the first well-documented stage in this elimination.[510] For example, NB scribes composing planetary ephemerides continued, I suggest, to use the principles of the EAE Paradigm to provide an answer to the question "to what end did Jupiter rise on the 22nd?", even while accurately predicting that same event. Did not the prediction have the effect of reducing the *arbitrary* nature of the gods' behaviour, though? Did not accurate prediction show that that gods had established, at least in terms of celestial movement, an order, knowable by man, and with which they no longer interfered? Did this not thereby increase by a little more the distance of the gods from mankind? In modern cosmology, the gods, if present, are so distant that if they had any rôle in creating

508 Oppenheim (1978) 641: "In a way that is never explicitly stated or even hinted at, Mesopotamian man assumed the existence of an unknown, unnamed, and unapproachable power or will that intentionally provided him with "signs"."

509 It should not be surprising to us that a *science du concret* approach may still incorporate advanced mathematics, though this was not perhaps recognised by Lévi-Strauss himself. Goody (1977) 2f, 50, and 150, in particular, criticises the dichotomy such as that in Lévi-Strauss (1966) between so called 'savage' minds and modern minds, between 'us' and 'them'. He stresses instead the importance of the technologies of communication in accounting for any transition from 'them' to 'us'. He writes p16:

> "I have tried to take certain of the characteristics that Lévi-Strauss and others have regarded as marking the distinctions between primitive and advanced, between wild and domesticated thinking, and to suggest that many of the valid aspects of these somewhat vague dichotomies can be related to changes in the mode of communication, especially the various forms of writing."

510 Compare Lévi-Strauss's model of the two "ascents" in n506. My model would suggest one, more gradual ascent.

the current behaviour of the universe, it was in formulating fundamental laws back at the beginning of creation.

The distancing of the gods in Mesopotamia has been considered before. Bottéro (1992, Ch.7) argued that the invention of omens through what he calls "deductive divination" occurs only at a time when the gods have already begun to be removed to a great distance from the human domain. He suggests that earlier forms of divination (for which the only evidence is found in the myths) were modelled on direct or inspired discourse with the gods, as when Enki speaks to Atraḫasis in the myth of that same name. He calls this "intuitive divination", when presumably the knowledge of the "network of intentions" of all things was considered to be directly accessible. Gradually, his argument goes, the gods withdrew from direct communication and informed humanity of their decisions only through signs.[511]

The myths express the same belief in the priority of orality over literacy, as a means of accessing the truth, that Derrida notes in Plato and elsewhere (see n286, above). However, it is perhaps also the case that the emergence of a literature that attempts to justify the distance of the gods from the human sphere may well be part of the development of a concept of causality which lies at the very heart of much of modern science. It is, for physicists, gravity that *causes* Jupiter to rise, and if a divinity were in someway involved in that, it was only to establish gravity and the initial starting conditions of the universe. For the scribes of the PCP Paradigm in Mesopotamia the gods had *caused* the heavens to move in a cyclical and predictable manner, without the intermediary of gravity, and left it to run without much further direct interference. They were, of course, still considered able to affect a predictable event in an unpredictable way by altering the concomitant weather and so forth. For the practitioners of the EAE Paradigm all celestial events were caused arbitrarily by the gods. None were predictable. Even the variations around the ideal periods of the heavens were considered to be unknowable. They were affected by the gods directly, and were signs, as I have shown. The cyclical character of celestial behaviour was not, therefore, an indication that the gods had left the heavenly bodies they had created to run unperturbed in accurately predictable patterns. The divinatory aspect of the ideal periods suggests that the gods were still considered to be *continually* affecting celestial motion, in particular.

If Bottéro is correct, and "intuitive" divination was considered to precede "deductive" divination, and that the gods were considered to be less distant in the past, then this belief can be extended, I argue, to the inevitable distancing wrought by the discovery that the actual periods between successive heavenly phenomena were not altered from the ideal *arbitrarily*, but in a consistent and predictable manner – this being the great discovery, the new core hypothesis of the PCP Paradigm, the revolutionary thought of the late NA period.

So far as both celestial divination and the later astronomy that facilitated it were concerned, the ideal periods, irrespective of the predictability or otherwise of the variation of reality about them, were a central part of the perceived construction of the universe by the gods.[512] They were as central as the core hypotheses that the gods created the heavens, assigned various bodies to various divinities, divided the heavens into night and day and into star paths, and so created "signs" for mankind to read. The three known incipits to EAE, for example, state:[513]

[511] As much was proposed by Gadd (1948). Should we perhaps see Ea talking to Ut-Napištim through the intermediary of a reed hut in Gilgameš XI: 20f as a transitional point between intuitive and deductive divination?
[512] The following is a summary of work presented more fully in Brown forthcoming (b).
[513] Also quoted with references in App.1 §21.

"When Anu, Ellil and Ea, the great gods, created heaven and earth, fixed the signs (gis-kim), established stations (*na-an-za-za*), founded positions, [appointed] the gods of the night, divided the (star)-paths, designed ([*uṣ-ṣ*]*i-ru*) the constellations, the patterns of the stars, divided night from daylight, [measured] the months and created the year" (end of EAE 22).

"When Anu, Ellil and Ea, the great gods, in their sure counsel had fixed the designs (g̃iš.ḫur.meš) of heaven and earth, they assigned to the hands of the great gods (the duty) to form the day well (and) to renew the month for mankind to behold. They saw the Sun god within the gate whence he departs (and) in between heaven and earth they took counsel faithfully" (Akkadian version of the opening to EAE 1).

"When An, Enlil and Enki, the [great] gods had established in their firm counsel the great divine powers (me gal.gal.la) and the boat of Suen (the Moon god) so that the crescent Moon should grow and give birth to the month and establish signs (giskim) in heaven and earth, the boat was sent forth shining in the heavens – it came forth into the heavens" (Sumerian version of the opening to EAE 1).

These incipits clearly show that the gods were considered by the celestial diviners who put together EAE to have created the universe *in such a way* as to produce signs (*ittum* = giskim).[514] This was believed to have been done, in part by "designing", to which I will return in a moment, but also by assigning various functions to various divinities, by establishing "stations" (*manzāzu*[515]) and positions, and by dividing the heavens. The construction and division of the universe are thus referred to in these introductory passages, but the ideal periods are not described explicitly.

In *Enūma Eliš*, the Babylonian creation epic Marduk is said to have:[516]

"set up 12 months... the designs (*uṣ-ṣu-ra-ti*) of the year... made the moon appear (saying) "on the 15th day you shall be in opposition at the midpoint of each month... on the 30th you will be in conjunction with the sun a second time. I (thereby) *defined* the sign (giskim)"."

Although this is an otherwise exceptional composition, I have no hesitation in asserting that in this passage the heavens Marduk is described as having set up are those believed to underpin celestial divination of the EAE variety. The parallels are too close. The heavens are again "designed" and divided, the gods assigned functions, and "signs" are made. As in much else, Marduk has simply appropriated from An, Enlil and Enki the same ideas of universal creation as were current in earlier times. In addition to the EAE incipits, however, this part of *Enūma Eliš* shows that the ideal year and ideal 30-day month were also considered to have been formed by Marduk. That is, he was regarded as having *created the universe in an ideal manner*, with years of 12 months, months of 30 days, lunar opposition on the 15th, and so forth. A variety of other texts[517] allude to the creation of the heavens in similar ways, so without doubt the use of the ideal periods was not unique to the Babylonian creation epic.

---

514 See the references to giskim in *Gudea Cylinder A* in App.1 §3. ^{Giš}an.ti.bal = *ṣaddum*, also means "sign" in divinatory context. ^{Giš}an.ti.bal often appears in Sumerian literature meaning "emblem". This idea was perhaps also appropriated into celestial divination. See Ch.2.1.1, Jup. Other technical terms referring to the entire omen are sometimes used to mean "sign" in this sense, e.g. *pišru* "interpretation", *dibbū* "report", and *šumu* "line".

515 "Station" probably had some calendrical function. See App.1 §19.

516 Tablet Vif. See App.1 §19 and Ch.3.1.4.

517 See App.1 §20.

Neither are these other texts divinatory in purpose, and they and *Enūma Eliš* thus attest to how widespread this conception of the universe in fact was. Importantly, it is found in texts from the MB period until the NB, and confirm what was argued before – that the core hypotheses of the EAE Paradigm did not change over time.

The text i.NAM.giš.ḫur.an.ki.a, amongst other things, elaborates on the ideal date of lunar opposition, the ideal month and year to produce an ideal period scheme. The title of the work certainly refers to the "design" (giš.ḫur = *uṣurtu*) of the sky (an) and earth[518] (ki). This, the use of "design" in the incipits to EAE, in *Enūma Eliš*, in *The Exaltation of Ištar* quoted in App.1 §20, in *KAR* 307: 33 quoted in Ch.3.1.1, and in Sennacherib's annals *OIP* 2 94: 64 quoted in Ch.3.1.3 (n315) leads me to suggest that this term described the primeval, ideal arrangement[519] by the gods of the heavens, including their establishing of its first movement in periods made up of nice, round numbers (particularly so in base 60). In addition to the other core hypotheses of the EAE Paradigm just noted, I suggest that OB and earlier celestial diviners considered the universe once to have had years lasting 360 days, and months of 30 days. The moon was considered originally to have moved in such a way that it was always "seen with the sun on the 15th", the stars to have moved in such a way that they always first appeared in the same month, and so forth. Consequently, if by chance, observed phenomena corresponded with those implied by the ideal, this meant that things were as the gods had originally intended. This meant that the gods were pleased, and so an event occurring according to an ideal period or ideal scheme derived from an ideal period boded well, and so forth.

In the Sumerian version of the incipit to EAE, rather then "ğiš.ḫur", the term "me" is used. The me, or "divine powers" are sometimes abstract, sometimes concrete entities and are, or are symbolic of, the performance or offices associated with civilised human life.[520] Me such as "being old", "heroism", "wickedness"/ "decent behaviour", "shouting"/ "whispering", lamenting"/ "rejoicing", "being on the move" / "being sedentary" are enumerated in Sumerian myth of the OB or earlier periods.[521] They formed part of a description of the *entire* universe, as those me construed in opposite pairs indicate. Repeatedly in myth the gods are described as having brought order to the world, and this included making human existence possible. The powers that made this ordering possible were the me.

An explicit connection between the me and ğiš.ḫur is found in the Sumerian composition *Ninurta and the Turtle*:

> "As I let the divine powers (me) slip from my hand, these me returned to the Abzu. As I let the divine design (ğiš.ḫur) slip from my hand, this ğiš.ḫur returned to the Abzu".[522]

and in *Enki and the World Order* it is said that:

[518] Underworld, probably. This is incidental to my argument, here. Note also that ğiš in Sumerian is rendered giš in later times.

[519] This arrangement is otherwise described as a "writing" (Ch.3.1.1), suggesting something *static*, in contrast to the ordering of movement implied by ğiš.ḫur.

[520] See G. Farber (1987–90) and Klein (1997), both of whom include references to an extensive earlier literature on this term. Me is usually rendered *parṣu* in Akkadian (loaned into Sumerian as ğarza) which means "cultic office" or similar.

[521] In particular in *Inana and Enki* in which some 110 me are enumerated.

[522] Segment B lines 3–4 in www-etcsl.orient.ox.ac.uk:
Me šu-ğá šu ba-ba-ğu$_{10}$-dè me-bi abzu-šè ba-an-gi$_4$
ğiš-ḫur šu-ğá šu ba-ba-ğu$_{10}$-dè ğiš-ḫur-bi abzu-šè ba-an-gi$_4$.

> "Father Enki... counting the days and putting the months in their houses so as to complete the years... taking decisions to regularise the days... all the divine powers (me) are placed in your hand".[523]

The creation of the universe is described along what appear to be ideal lines.

I suggest, then, that the underlying construction of the universe implied in EAE is drawn from a widespread idea[524] that the Sumerian gods ordered the universe using the me, and part of this ordering was described either separately or additionally using the term ĝiš.ḫur or "design".[525] This always did, or came to, refer to the means by which the gods ordered the heavens into ideal periods. Because I consider the elaboration of the ideal periods into ideal schemes (through the number play technique known as *arû*, and in one text whose title explicitly refers to the ĝiš.ḫur) to have been instrumental in generating an interest in the dates of planetary rising and setting, and in the lunar six, and that the continuous record of *these* in particular made possible the discovery of periodicities in the data base which led to cuneiform astronomy – because of this, I consider cuneiform astronomy to have *depended* (at least in part) on the same view as to the nature of the universe as did texts of the EAE Paradigm. Indeed, the internal evidence for the adherence of cuneiform astronomy to the form and structure of divination, and the external evidence of the continuity in the names and practices of its practitioners and of the very names of their compositions (*arû*, "wisdom" and "secret"), indicates that the cuneiform astronomers *themselves* considered this view of the nature of the universe to underpin their work.

So, while in the preceding discussion of the scientific character of cuneiform astronomy, no mention was made of whether or not it constituted an investigation into the nature[526] of the universe, I argue now that it did. Of course, being able to predict the magnitude of a forthcoming eclipse by modelling the movement of the moon in latitude appears to *us* to indicate that a more profound understanding of the nature of the universe than was possible under the EAE Paradigm was entertained by the cuneiform astronomers. However, we have no texts dating to the LB period that discuss or even allude to arguments concerning the make up of the universe in a manner implied by the astronomy, and no evidence that being able to predict planetary behaviour accurately implied that heavenly motion *in general*, say, was considered. I do not wish to argue from silence, however, for the general properties of motion or behaviour may have been discussed orally, or preserved on perishable materials. Why should not such erudite scribes have discussed the nature of the universe amongst themselves? The point is that whether they did or not is irrelevant if the evidence is lost to history. It has been suggested, for example, that the astronomy of the PCP Paradigm can be compared with those sciences that model behaviour *mathematically*, without recourse to

---

[523] Lines 17f according to www-etcsl.orient.ox.ac.uk:
(17) ud šid-e iti é-ba ku$_4$- ku$_4$ mu šu du$_7$- du$_7$-da
(19) eš-bar kíĝ ud-da si sá-sá-e-da
(65) me mu-un-ur$_4$-ur$_4$ me-ĝu$_{10}$-šè mu-un-ĝar

[524] On the positivism of this approach see already n176, above.

[525] The term ĝiš-ḫur "design", also "rules", "regulations" by context, rendered as both *uṣurtu* and *gišḫurru* in Akkadian, occupies the same semantic area as the term me, without being identical. I am not arguing that "design" is identical to, or even one of the "divine powers", just that its usage in celestial context draws on the idea of a divine ordering of the entire cosmos for the purposes of human existence implied by the me.

[526] The term nam = *šimtu* "determined order, divine decree" overlaps with what we consider to be the meaning of the word "nature", as does *šiknu* "appearance". The gods decree both the nam and the ĝiš.ḫur – references cited in *CAD* Š/3 12.1a.

any underlying laws.[527] An observational record is made, and reproduced with a mathematical model. Periodicities in that model are then exploited to make predictions. Such sciences are often highly successful, but are to be differentiated from physics, say, which attempts to describe in laws the nature of the universe, often in terms of what it is made of. Oppenheim (1978) 645, for example, outlines this position:

> "The (parameters of the planets) were used by Greek astronomers in a manner utterly alien to their Mesopotamian counterparts. What the latter regarded as a sequence of points in time (based on observations and projected into the future by computation) Greek thinkers explained by geometry in such a way that a mechanical model could be constructed to produce these "irregularities" automatically. The Greeks posited a universe functioning in time as well as in space, in a continuous and regular circular movement of the planets that, combined with the ingenious invention of secondary circles (epicycles) did what the philosopher is said to have demanded of the astronomer, that is, " to save the phenomena"."

To suggest that the MAATs are like this, however, is to argue from silence.

Instead of relying on the absence of evidence, or imagining what may *also* have been discussed about heavenly motion amongst LB astronomers, using what evidence we do have – namely, that lying behind cuneiform divination there was a discussion of the nature of the universe, and that the adherence of the PCP Paradigm to the EAE Paradigm is very close – I would argue that the MAATs are in effect the most *elaborate* expositions of that same underlying belief as to the nature of the universe.

So, if we turn to Greece, briefly, for comparison, it is relatively well known that many astronomers adhered to Platonic and Aristotelian ideas that circular motion was *appropriate* to the heavens.[528] Various mathematical procedures were adopted to ensure that the behaviour of the heavenly bodies could be modelled using combinations of circular motion. In Ptolemy's case, for example, the consequences of adhering to this idea as to the fundamental nature of the heavens meant that observation and theory sometimes failed to cohere. His model of the varying motion of the moon implied that the angular diameter of the moon should vary by a factor of 2.[529] In reality it does not, though so wedded was Ptolemy to the need to use circular motion that this discrepancy was passed over in silence. This adherence, broadly speaking, was termed "saving the phenomena",[530] as noted by Oppenheim in the above quotation.

---

[527] See, for example, my comments in n277, above. Such a position is also suggested by the comments of Theon of Smyrna in his *Expositio Rerum Mathematicarum de legendum Platonem utilium* 177.20f, cited in Lloyd (1987) p311 n95. Lloyd himself argues in 1991, 292–4, that Babylonian geometry, and astronomy are "non-theoretical", that is, so far as we know "proof" – the proceeding deductively from certain premises to the required conclusion – is lacking. However, although there are no texts of which I am aware that begin with the Sumerian "designs" of the universe and proceed deductively through to the procedure employed in the MAATs, I argue that these steps *were* made via celestial divination and the NMAATs, as shown. The difference lies not so much in the theoretical nature of "Mesopotamian" as opposed to "Greek" astronomy, but rather in the way in which each was presented in the texts that have survived.

[528] I hesitate to characterise Greek astronomy as a whole, but Lloyd (1987) 312f argues that most Greek astronomers set their discussions firmly within a framework of certain *physical* assumptions as to what constituted the universe. The "matter" that was considered to make up the heavens (*aether*) was the most homogeneous, and thus it and the bodies within it were spherical, so argued Ptolemy in *Almagest* 1.3, 1.13.21ff, cited in Lloyd (loc. cit.) p314 n98.

[529] *Almagest* 5.4 366.15f, cited in Lloyd (1987) p316 and n101.

[530] See Lloyd (1987) 293f, especially n28 for references to earlier literature on the subject.

I suggest in opposition to the prevailing notion that cuneiform astronomy was characterised by having no underlying model of the nature of the universe, that the NMAATs and MAATs in fact relied on just such a model – one in which years were made up of 360 days, months of 30 days, the length of the longest and the shortest days were in a small number ratio and intermediate lengths varied as straight line functions. It is this model of the ideal, primeval universe that explains the 360 UŠ in the zodiac, the 30 *tithis* in a month, and the daylength columns in the lunar MAATs. Perhaps, even the use of a value for lunar diameter twice reality in the lunar MAATs (see nn 247, 404 & 429 and Ch.4.1.1) should be explained as a form of "saving the phenomena" – one that resonates particularly well with that cited for Ptolemy, above.

To summarise, then, I have raised the question of the scientific nature of cuneiform astronomy-astrology using models borrowed from the philosophy of science and anthropology. Some correspondences and some differences with these models were noted, but enough has been done to show, I hope, that any future discussion of the nature of science ought to incorporate the Mesopotamian evidence. I ended with a suggestion that Mesopotamian astronomy-astrology, contrary to current opinion, drew heavily on an intellectual tradition in which the nature of the universe had been explained since "Sumerian" times in the 3rd millennium BC in terms of "divine powers", a cosmic "design" in space (broadly, celestial "writing") and time (broadly, "ideal periods") and the production of "signs" for the benefit of mankind. Ptolemy, half a millennium after Aristotle, continued to subscribe to a universe made up of five "elements". These elements, constructed in pairs of opposites "explained" movement.[531] And just as Ptolemy mathematised the problem of astronomical movement, so the LB astronomers mathematised the problem of predicting celestial signs – signs that were "explained" ultimately by the me, which were also (sometimes) construed in opposites. Explicit justifications for their astronomical methods may be more apparent in the works of Greek astronomers, but I have shown that the same form of justification was also present, albeit more obscured, in cuneiform astronomical works. By extrapolating backwards from the astronomical and divinatory texts, I have shown that the "Sumerian" model of divine powers and design, despite the evidence for it being largely literary in content, ought also to be recognised as philosophy, just as Aristotle's works on similar subjects are considered in this light. We should perhaps also reconsider the use of such terms as "esotericism" and "pre(philosophical) speculation" in Assyriology to describe those few surviving learned elaborations on the order of the universe, mostly from the 1st millennium BC.[532]

5.2 Conclusions

I have described the evolution of Mesopotamian cuneiform planetary astronomy-astrology from OB to AD times, in particular concentrating on material dating to the period c. 750–612 BC. A study of this narrow time frame has enabled me better to explain what came be-

[531] Fire was hot and dry, water cold and wet, earth cold and dry, air hot and wet – elements after Empedocles – see Aristotle *Met* A4, 985a31–3 in Kirk, Raven & Schofield (1983) 286.
[532] E.g. the texts treated by Livingstone in *MMEW*. Such descriptions of these and similar compositions are made by Beaulieu (1992) 107f, for example.

fore – the EAE Paradigm – and what came after – the PCP Paradigm – and I hope to have shown that during this period a "scientific revolution" from the EAE Paradigm to the PCP Paradigm took place. Why did this revolution take place then?

Lloyd (1979) identifies the factors behind the emergence of science in Greece, which he describes as the "significant changes or developments [which] occurred during the period from the sixth to the fourth centuries BC." He points to the developments in the techniques of argumentation, refutation, persuasion, and demonstration,[533] on the one hand, and of observation on the other, relating the former in particular to the socio-political situation in Greece at this time. I noted in §5.1.2 that the monopolising of astronomy-astrology by certain families of scribes, whilst not precluding premise-challenging intellectual activity, suggests that preservation and adherence to old norms might have been a more dominant force at play in their work, and that divergences typical of "normal science" better explain any differences between the approaches of various scribal guilds. I did note, though, that the extent of the entourage of Scholars surrounding the late NA kings probably promoted a measure of personal scientific activity, accelerating the embellishment of EAE, and leading some Scholars towards astronomical prediction. I suggested that the direct employment of these Scholars by the king created a unique situation in the history of Mesopotamian scholarship – one in which personal rivalry was able to coexist with the traditional preservation of a corpus of work, and in which innovative texts, practices, and techniques were carefully couched in traditional forms.

For example, the Scholars Balasî, Nabû-aḫḫe-eriba and Issar-šumu-ereš dispute vigorously over the supposed visibility of Mercury and Venus.[534] At one point Balasî writes of Issar-šumu-ereš that:

> "[He who] wrote to the king, my lord, "Venus is visible" is a vile man, an ignoramus, a cheat!" (x072: 6).

There appears to have been little love lost between these most senior Scholars. Importantly, this dispute centred on the Scholars' competence at predicting planetary phenomena. In *LAS* II p284 on r.10' Parpola noted that the few dated texts in the corpus of NA and NB Letters and Reports included predictions. Some at least of the Scholars wanted the king to know that they had accurately anticipated events in advance. In Report 8388 Rašil boasts:

> "(Already) when Venus became visible, I said to the king my lord, "An eclipse will take place.""

Presumably being able to predict accurately was valued by the king,[535] and consequently by the Scholars who curried his favour whenever possible. The pressure brought to bear by one's colleagues to be accurate with such predictions must have been severe. Humiliation and poverty were just around the corner, as noted in Ch.1.3.

The late NA king, thus, played a critical rôle in ensuring that accurate astronomical prediction should begin in the 8th and 7th centuries BC. In general terms, c. 750 BC marked an upturn in the economic fortunes of both Babylonia and Assyria. Celestial divination was an

533 Lloyd (1992) 574 "the debates may be rather a distinctive, or at least a distinctively prominent feature of Greek science".

534 X023, x050, x051, x072 & 8083, see Parpola *LAS* II pp68 & 78.

535 "Successful prognosticators could be sure of royal favour," notes Parpola loc. cit.

activity restricted to the kings and royal family, and the fundamental connection between celestial divination and royalty meant that the fortunes of the former followed those of the latter.[536] In particular, the Assyrian king concentrated the attentions of a large number of the most senior Scholars from both Babylonia and Assyria[537] on one thing – himself. He increased the specialisation of the Scholars,[538] put them in mutual competition, and made them beholden to him financially. His wealth ensured that the Assyrian capital was exposed to foreign culture (n101), and Assyria's direct involvement in the affairs of Babylonia meant that a wide variety of Scholars were employed to study the sky for omens. Each Scholar brought with him his guild's knowledge of the heavens and of the EAE Paradigm. This was likely to produce results different from the ruminations of one individual Scholar, I suggest. The particular interest of the Sargonid dynasty (Sargon and his descendants) in the cult of Issār, perhaps further turned those Scholars towards a consideration of the planets (n157). The Scholars' job was to legitimate him in his new rôle as ruler of the world, and protect him against supernatural attack.

The Assyrian king was a particularly hard task-master. He demanded the constant study of the heavens, as the Letters concerning omens for routine affairs of state attest to – e.g. x052. He also required expertise in astronomical prediction from his Scholars:

> "Concerning the solar eclipse about which the king wrote to me (saying): 'Will it or will it not take place? Send me definite word!' " (x170: 1),

and as noted in Ch.1.3 the Scholars' livelihoods depended on his goodwill towards them:

> "The king, my lord, must not give up on me! With deep anxiety, I have nothing to report" (x045: s.1).

As to the protective rôle played by the celestial diviners, it is apparent from our understanding of the mechanics of celestial divination itself that the continual study of often near-invisible phenomena, at night-time, and in any weather was a necessary part of the discipline. There can be little doubt that being able to anticipate forthcoming heliacal events greatly facilitated this work. Any narrowing of the time interval during which an eclipse, say, *might* occur would have proven useful, and knowledge that an ominous event was occurring despite being obscured by bad weather[539] would have assisted the diviners in ensuring that the king remained under constant surveillance. Rituals to avert the evil portended by forthcoming celestial events could also have been prepared in plenty of time. For example, knowing that an eclipse might occur permitted the Scholars to place a substitute on the throne *before* it occurred:

---

536 Astrologers and the powerful commonly seek each other, corroborating one another's activities. See Barton (1994) 211.

537 Incidentally, there is no evidence to suggest that the NA Scholars were any less competent in the application of either Paradigm than their NB colleagues during the period of revolution, even though the emergence of the EAE Paradigm undoubtedly took place in Babylonia, *and* the full-flowering of the PCP Paradigm also took place in the south. This suggests that an intimate connection existed between the Babylonian Scholars and their Assyrian rulers, a connection which undoubtedly reflected the rising power of the Chaldaeans and Aramaeans in the south. This provides yet one more clue in the reconstruction of Assyro-Babylonian relations in the 7th century BC.

538 See Ch.1. Perhaps this specialisation too was part of the transition from "bricoleur" to modern scientist – see §5.1.3.

539 See n82.

"The substitute king, who on the 14th sat on the throne in [Ninev]eh and spent the night of the 15th in the palace o[f the kin]g, and on account of whom the eclipse took place" (x351: 5f).

Similarly, by predicting planetary behaviour the Scholars ensured that the king was not caught performing important business when Mars entered Scorpius, say. In x038 Issar-šumu-ereš warns the king that he should not go out the following day. No doubt the signs were not *going to be* good. Prediction thereby enhanced the Scholars' ability to protect this most significant figure in the empire, the *šangû* of Aššur – the king.

To summarise, it was the particular requirements of the practice of celestial divination, in combination with the unique circumstances under which the Scholars worked in the courts of the last NA kings that explain why the scientific revolution occurred when it did. It was the enhanced demand for supernatural protection, and the creation of competition between experts that led to the first cuneiform astronomy.

I have considered the issue from a functional standpoint, reducing the causes of the transition from the EAE to the PCP Paradigm to the needs imposed on the Scholars by the political environment that confronted them. I do not wish to suggest, however, that the technology of cuneiform writing and enhanced communication under the Assyrians were not significant – the former afforded the means by which the records of phenomena could be preserved for centuries, the latter the means by which expertise could be pooled. Nor do I suggest that the influence of particular individuals[540] was not of some importance, but in the absence of more evidence as to the rôle of individual genius, I limit my explanation to those circumstances surrounding the *institution* of royal divination.

I have, of course, made assumptions as to the Scholars' motivations (see n161), but have been guided by what little is preserved of what the Scholars themselves thought about their discipline. Compositions such as *The Catalogue of Texts and Authors* give some idea of how the scholars saw their rôle – as the modern equivalent of that fulfilled by the mythical sages, and to preserve a corpus of learned material. I noted those descriptions of the heavens as "celestial writing" and "designed", for these show us that the Scholars, too, thought of celestial divination as a decoding of the sky. Those examples in Ch.3.2.2 of the abstracting of the code suggest this also.[541] I picked out those elements of the practice and compositions of the PCP Paradigm that adhered to the forms and practice of celestial divination, and suggested that these too indicated that the scribes *themselves* undertook astronomical prediction in accordance with a particular tradition.

This study has tried to establish that prior to the mid-8th century or so no astronomical (n294) texts were composed in Mesopotamia, and at the same time that the invention of cuneiform mathematical astronomy can be pushed back some centuries to the late NA period, and thereby be located firmly within an Akkadian milieu (nn488 & 499). I have attempted to raise the question of science in Mesopotamian to a level whereby serious comparison with other ancient science can be made. I believe that the particular circumstances out of which predictive astronomy in Mesopotamia emerged offers a challenge to certain cur-

540 See I.4, *sub* Kalḫu.

541 See also n27. Scrutinising not just the translations, but the original cuneiform, has revealed more clearly the metaphoric and syntagmatic relationships between omens. Previous translations *hid* the invented aspect of omens, and enhanced their empirical one. See n93. Clearly, keeping close to the original sources is advice well heeded, but the discovery that an empiricist *agenda* lay behind previous studies of omens is important, and Ch.3 has been a corrective to that.

rent assumptions that "intellectual interest" and individuals play central rôles in scientific endeavour. Cuneiform astronomy emerged out of the demands of a well established divination industry, and was constructed over a period longer than the lifetime of any one person. I argued that the zodiac, amongst other technologies, was created in order to facilitate astronomical prediction, but that its form depended heavily on those of EAE-type divination, the construction of which I ascribe to a few scribes in the OB period. The dependence of current Western popular astrology on the zodiac, and on some other aspects of Mesopotamian celestial divination, is profound. It is not without importance that the learned "play" of a few diviners, and not prolonged empirical observation, can now be recognised to lie behind it. Thus, I have been able to move from texts more than 2000 years old to today, and suggest that more than "study for its own sake" characterises *Mesopotamian Planetary Astronomy-Astrology*.

APPENDIX 1

A Chronological Bibliography of Cuneiform Astronomical-Astrological Texts

	Celestial Omens	*Period Schemes*	*Observational Records*	*Other*
Archaic	(1) Establishment of the basic schemata,	the 12 month year	and the discovery of the planets	
Sumerian & Akkadian evidence	(2) OAkk origin theory (3) Sumerian astrology-astronomy	(4) Sumerian and early Akkadian calendars	(3) Star tablet in Gudea Cyl A?	(2) OAkk liver models from Mari
c.1950 *OB* *c.1530*	(5) Eclipse omina (6) Non-eclipse omina (7) Proto-Enūma Anu Ellil	(8) **"Ideal" calendar** (9) "Ideal" Venus scheme	(9) Records of Venus' (dis)appearances (10) OB Sumerian star-lists	(11) *Prayer to gods of the night* and others (12) Maths
Periphery *MB/*	(13) Boghazköy material in Hittite and Akkadian (14) Emar, Qatna, Alalakh, Nuzi, Susa and Ugarit	← (13) Boghazköy star list →	(14) Solar eclipse record from Ugarit	
MA	(15) "Transitional omina"	(16) *Astrolabe B*	(17) HS 245 from MB Nippur - maths exercise	(19) Epic of Creation (20) Other relevant SB literature
	← (21) ***Enūma Anu Ellil*** →			
c.1000			(18) "Kudurrus"?	
			(22) Two early 8th century eclipse reports	
c.750	(23) 714 BC Sargon II's 8th Campaign eclipse omen etc.	(26) Zwölfmaldrei and 10-star lists (31) Seasonal hours	(32) Observational records that are not omens – **Diaries** and **Eclipse Records**	(24) Assyrian literature and the Babylonian myth of *Erra* and *Išum*. The eclipse ritual.
	← (30) Mul. Apin →			
NA/ *NB* *c.612*	(27) **Letters** and **Reports** to the Assyrian Kings from Assyrian and Babylonian Scholars (28) Commentary texts (29) Explanatory works	(35) i.NAM.giš.ḫur. an.ki.a and other series (36) Diviner's manual (37) Predictions to the day (38) BM 36731	(33) *Ziqpu* stars (34) BM 78161	
c.539	(47) LB use of omens	(39) The Saros (40) The 19-year cycle	(41) Planetary Records	
	← (42) ZODIAC →			
c.331	(48) Horoscopes	(43) Auxiliary, "early", "primitive", or "atypical" **MAATs**		
0	(49) Zodiacal astrology	(44) ***ACT* MAATs** (45) **Non MAATs**		
		(46) The latest datable texts		

(1) It is difficult to believe that "astrological" concerns appear only with the advent of writing. Folklore derived from prolonged observation similar to "red sky at night, shepherd's delight" may well have existed, though this is distinct, I suggest, from "deductive divination". The assigning of gods to the Sun, Moon, and Venus (probably also to Jupiter and Mars) appears to have been most ancient. That the morning and evening stars were known to be the same celestial body (Venus) is suggested by the early 3rd millennium BC text in Nissen ed. (1993) 17. This body was assigned to Inana. See Ch.2.1.1. The seasons were perhaps correlated early to the rising of particular stars, as in Hesiod's *Works and Days* vii 383–4. Note the reference to a star in the OB Sumerian *The Farmer's Almanac* l.38 (Civil, 1994). It is commonly suggested that correlating the risings of stars to seasonal events led to certain stars becoming associated with particular lunar months, a connection that was first recorded in writing in the so-called "astrolabes" and their precursors (see below §§13, 16 & 26). It is by no means clear, however, that the purpose of these astrolabes was to help regulate the lunar calendar against the sidereal/stellar one, though this is commonly argued. See, for example, Horowitz (1998) 162–5. For an alternative or complementary interpretation see Ch.3.2.2.

Attempts to work back on the basis of precession to a time before written records, in order to relate iconography to a situation that pertained in the sky, say, must be treated sceptically, but see Hartner (1965).

(2) Weidner (1928/9) and (1941/4a) 175–6 and n19 discusses some of the historical material to be found in the great celestial omen series *Enūma Anu Ellil* (EAE), discussing references to Ur III and Old Akkadian kings. Schaumberger (1949) and (1954–6) discusses the possibilities of dating omens from tablets 20 and 21 of EAE allegedly recording eclipses from Ur, Gutium, Babylon, and Akkad. Huber (1987) calculated that a series of eclipses did indeed occur close to the death of a number of the Old Akkadian kings (using his own, still disputed, dating of Babylon 1 and the still-controversial relative dating of the OAkk dynasty) and suggests that this may have been instrumental in spawning the eclipse omina found in EAE. Against this hypothesis is Koch-Westenholz (1995) 34–6. Bottéro (1992) 37 also discusses the Old Akkadian "origin" of divination, pointing to the murder of king Maništušu etc. See also Hirsch (1963) 7f – "Die Berichte der Omina". Given how few references to divination are found in Sumerian writings, Semitic origins have been posited (Nougayrol, 1966, 12 and note the reference to the Pleiades in a text from Ebla dating to c. 2400 BC – see Durand, 1994, 4), but attempting to draw distinctions in what is perhaps best understood to be a bi-lingual culture is dangerous (against seeking origins see Lloyd, 1992, 572). That aspects of celestial divination, in particular its "ideal schemes", derive from concepts such as the Sumerian me or "divine power" and g̃iš-ḫur or "design" attested in a number of literary texts is discussed in Ch.5.1.3. Accounts of the creation of the universe in Sumerian are discussed in Horowitz (1998) 134f. Legends concerning the Old Akkadian kings Sargon and Narām-Suen, who were believed to have regarded or disregarded celestial and other omina, were current from the OB period on, and the revival of celestial divination under Sargon II in the NA period may have formed part of a wider associating of the expanding Assyrian empire with the Old Akkadian dynasty founded by his namesake. The phenomenon of historical omens is discussed in Ch.3.1.1.

(3) For a short list of references to what may loosely be termed "astrology" in pre-OB Sumerian, see Falkenstein (1966) 64–5. He refers to the *Keš Temple Hymn* (see now Biggs,

1971, and Gragg in Sjöberg and Bergmann, 1969), and to four lines in *Gudea's Cylinder A*. These latter describe (iv 26, v 23) a star tablet, dub.mul.an, being consulted by Nisaba, and stars of the pure/clear sky, mul.(an).kù.ba (vi 1, ix 10), corresponding in some way to the temple plan. In xvii 19 the me of the temple Gudea is having (re-)built is said to be linked with heaven and earth. The connection between plans in heaven (an) and on earth/underworld (ki) is found also in the *Keš Temple Hymn* ll. 45–6, where dimensions in the former correspond to those in the latter. In *UET* I 300, a Sumerian inscription of Kudur-Mabug, it is said of Nanna the Moon god that "he gives birth to the day and night, establishes the month and keeps the year intact" (ref. courtesy W. G. Lambert). Further reference to the Sumerian star tablet, commonly said to be made of the blue stone lapis-lazuli, are gathered together by Horowitz (1998) 166–8.

Koch-Westenholz (1995) 33 writes that there is "no real trace of astrology in Sumerian sources". However, the divinatory element in *Gudea's Cylinder A* is quite strong. In line 1 (see now Edzard, 1997) destinies are assigned in heaven and earth. In ii 1, iii 26, iv 12 Nanše is described as a dream interpretress, kù.zu.me.te.na, and interprets Gudea's dream. Ningirsu accepts Gudea's invitation to take part in the lunar èš.èš festival – see (4) below. In iii 16–18 Gudea asks g̃atumdug to send him a good sign, giskim. In v 19–20 Gudea's personal god Ning̃išzida is said to rise with the Sun. Later Ning̃išzida is linked with the constellation Hydra (Mul.Apin I ii 8), the god's symbol being the horned snake. It is thus possible that the god was already associated with a constellation long before the composition of this text. In viii 19 Gudea lacks a giskim from Ning̃irsu, who responds (in a dream) by saying that he will give Gudea a sign in the stars (ix 9–10, xii 11). In x 17–18 Enlil is said by Ning̃irsu to perform great rites at the start of the lunar month and at full Moon, itu.da u_4.šakar.ra. These dates are very important to the EAE diviners, too, as a glance at the Scholars' Reports will immediately reveal. In lines xii 16–17 the well-known description of an extispicy of a white kid is to be found. Another extispicy is seemingly performed for the brick-mould in xiii 16–17, and another in xx 5 during building. In xx 6 a form of divination based on throwing grain is mentioned. Thus, in a Sumerian text prior to the OB period reference is made to extispicy, dream interpretation, grain-throwing, to plans and signs in the stars, to events connected to the lunar phases, and to a star-tablet – the last of which almost certainly constitute a form of celestial divination.

For texts with lists of star names in Sumerian see (10). For the Sumerian title of EAE in OB literary catalogues see (7). For concepts like the "cow pen" and stars=cows in Sumerian texts see Heimpel (1989) 249f and Horowitz (1998) 153 n5.

(4) Evidence for pre-OB Sumerian and Semitic use of a lunar year made up of 12 or 13 months of 29 or 30 days each is now to be found in Sallaberger (1993) and Cohen (1993), where the earlier literature is summarised. The earliest Sumerian month name attested is from Fara dating to the mid-3rd millennium BC, and a variety of pre-OAkk month names are known. A Semitic calendar is known from Ebla, Mari, Gasur, Abu-Ṣalabikh, and Ešnunna from c.2600–2200 BC. Eventually, of the various Sumerian calendars used in the Ur III period (Sallaberger, 1993, 7f), the Nippur calendar came to dominate. The Sumerograms used there for the 12 months became pretty much those used in the SB calendar until the end of the cuneiform tradition. Because some of the earliest Sumerian month names were connected to seasonal activities it has to be assumed that intercalation (adding in an extra month every three years or so, and thus regulating the lunar year with the solar) took place at least

as early as the first half of the third millennium BC.

A year of 360 days comprising 12 30-day months probably used in simplifying administrative bookkeeping was also employed early in the third millennium. See Englund (1988) 136–64. The administrative year underpins the "ideal" year (8 & 11), the elaborations of which form an important part of celestial divination. See Ch.3.1.2.

Lunar festivals such as the èš.èš or *eššešu* were connected to the lunar phases. For those attested in the Ur III period see Sallaberger (1993) p37f and pp306–7. The days upon which these festivals usually took place were the 1st, 7th and $14^{th}/15^{th}$ of the month, that is when the Moon was new, half-full or full, though the day when the Moon was not visible, and perhaps also day 21 were important. The *eššešu* festival is attested until NB times. See also *CAD* E p373. These days recur in the divinatory material and the Moon's behaviour on them was considered ominous by the Scholars.

(5) Collections of lunar eclipse omens are attested from the OB period. Four tablets are discussed in Rochberg-Halton *ABCD* p19ff. Perhaps "forerunners" to EAE 15–22, they constitute a small schematised collection in their own right. There are references to eclipse omina in OB Mari; a fragmentary lunar eclipse text discussed by Dossin (1939) p101 (cf. Durand, 1994, 5); a reference in a letter to king Zimrilim of the fortune implied by a lunar eclipse (idem 1951 46ff, Parpola *LAS* II 486, and *CAD* A/II 507 b); and a menological lunar eclipse text (Koch-Westenholz, 1995, 37 n2). A 28-line late OB text held in private hands dealing with omens connected to the "darkening of the sun" was published by Dietrich (1996). The Boghazköy texts *KUB* 4 63 and 64 (*RA* 50 p11) record omens concerning solar eclipses in an apparently OB style (see Koch-Westenholz, 1990, 235). Omen 38 of text 1 of the OB oil omina (Pettinato, 1966) includes the prediction of an eclipse of the Moon *nam-ta-al-le* ^{d}*Sin* in the apodosis. It is clearly meant to imply something bad, for the client is also predicted to die.

(6) Only two OB texts are known to me which contain planetary omina that do not deal with eclipses. They are BM 97210 (unpublished, ref. in *ABCD* Ch.1 n5), and the text published by Šilejko (1927) and discussed by Bauer (1936). Walker (1982) 22–3 discusses the fragmentary text BM 26472 which he suspects may include Jupiter omens written in the OB period describing events from the time of the Ur III king Šulgi, though the text itself is much later in date. Note also the reference to a Mars omen in the apodosis of a liver omen – perhaps also OB in date (*CAD* N 266 and Reiner, 1995, n19). We await Rochberg's publication on second millennium BC celestial divination.

(7) Collections of celestial omens were being put together in the OB period, as the discussion in *ABCD* pp19–22 reveals. There it is suggested that the lunar eclipse omens were the first celestial omens to be systematically incorporated into omen collections. However, the presence of weather and solar eclipse omens at that time suggests a wider scope of interest, more like that found in the later Enūma Anu Ellil (EAE). This is what I have termed "proto-EAE" – collections of celestial omens similar to those in EAE, that show evidence of the same "categorising logic" (see Ch.3.2.1 and 3.2.2) seen in the canonical series (21). The title of this proto-EAE is attested in two OB Sumerian literary catalogues (nos. 3 & 7 in Hallo, 1963, 169), one at least from Ur. The Ur example, first published in Kramer (1961) has the title in both Sumerian and Akkadian, where they appear to have been counted as

separate compositions – loc. cit. 76 n6). These titles show beyond doubt that a form of EAE existed in the OB period, though this fact is not widely stated in the secondary literature.

(8) BM 17175+17284 was copied by Walker and published by Hunger & Pingree *Mul.Apin* 163–4. It is a purchased OB text probably from Sippar or Tell ed-Dēr, and gives values in unnamed units of time or weight (of water pouring from a water-clock – see Brown, Fermor & Walker, 1999) measuring the length of watches of the night. It presupposes a 12-month year, a ratio of 2: 1 *in time* for the longest to the shortest night, and a spring equinox on the 15th of month XII. It is without doubt that each month was considered to last 30 days, and the year 360 (see also 11). This 360-day year of 12 30-day months is identical to the administrative year known throughout the third millennium BC (see 4), and with the addition of the 2: 1 ratio and the location of the equinox in the middle of month XII constitutes what I term the "ideal" year. The "ideal" year is fundamental to celestial divination, as I show in Ch.3.

The "ideal" year underpins some of the values in the so-called OB coefficient lists. One value corresponds to the daily change in the length of the night (based on the 2: 1 ratio and a 360-day year), another to the daily change in the length of time the Moon is visible during the month that contains the equinox (based on what I term the "ideal lunar visibility scheme" in Ch.3.1.2), and the last to the period of time for which the Moon is visible on the first and last days of the equinoctial month (which becomes the Nippur tradition of EAE 14 – an alternative "ideal lunar visibility scheme"). See now Robson (1999) §8.2 for the texts and Brown, Fermor & Walker (1999) for a detailed study of the units involved.

(9) The 63rd tablet of EAE is known as the *Venus Tablet of Ammiṣaduqa* by modern commentators. All previous publications have now been superseded by Reiner & Pingree *BPO*1 and Walker's (1984b) corrections. All of the exemplars of this composition date to the late NA period, or later. The canonical version comprises 4 sections and 59 omens. The first 21 omens follow a standard pattern. Each protasis records Venus's (Ninsianna) date of disappearance and its date of reappearance. Each apodosis is general and deals with the king, land, or state. The protases seem to record actual observations over 16 years. After omen 10 come the words : mu giš.dur$_2$.gar ku$_3$.sig$_{17}$.ga.kam – the golden throne year name of Ammiṣaduqa 8. It is largely on the basis of this statement that attempts have been made to date the observations absolutely, and thus date the 8th full year of this OB king's reign. The most comprehensive attempt has been made by Huber (1982). It appears plausible that this section of EAE 63 *does* indeed record observations made *to the day* of Venus in the OB period.

However, it is not clear that the observations were originally recorded in the form of omens. Reiner & Pingree state p9, *BPO*1 that they believe omen 10 to have been originally a report of an observation without an apodosis. As I demonstrate in App.3 most, if not all of the apodoses were added later. Section III of EAE 63 contains observations of Venus, but it is not possible to determine when they date from. Section IV is simply a re-statement of the omens in sections I and III, ordered by month. Section II stands apart from the rest of the text, describing the rising and setting dates of Venus according to a *periodic* scheme. The visibility periods in both the east and west are 8 months and 5 days, and the invisibility periods are 3 months and 7 days respectively. I have termed this the "ideal Venus scheme " in Ch.3.1.2. A total period in this model is thus 19 months and 17 days, which is 587 days if the schematic 30d month is used, and about 578 days if we take an average figure for the synodic month. In fact Venus's synodic period (phase to phase) is about 584 days, or about 8/5 of a year. It is not known when this model was first used. Presumably, it dates at least to the canonising period of EAE – the second half of the 2nd millennium BC (see 21) – perhaps even to the OB period. It remains unclear, then, if this model of Venus's behaviour can be included amongst the known examples of OB astronomical-astrological period schemes.

(10) Several mono-lingual OB star-lists in Sumerian are known. Some are classed "forerunners" of Ur_5-ra XX-XXIV and its commentary texts in *MSL* 11. The relevant ones are AO 6447 viii 38–44 and ix 9–10 (*MSL* 11 p129f), *OECT* 4 157 rev ii' 1'-8' (*MSL* 11 p136f), *OECT* 4 161 x 13–28 (*MSL* 11 140f), BM 78206 (= *CT* 44 47) iii 1–9. The Nippur forerunner to Ur_5-ra XX-XXII is published in *MSL* 11 pp93–109. Lines 387–410 (pp107–8) comprise a star-list, known also as the *Nippur Star List*. Although these lists imply that the observation, categorising, and naming of the stars and constellations had taken place, there is no evidence that the order in which the stars were listed was based on observation.

(11) Two of the oldest versions of the 24-line *The Prayer to the Gods of the Night* are compared in Dossin (1935). AO 6769 is published and compared to the text published by Šilejko in 1921. Horowitz (1993) argues that only AO 6769 is OB, the other being MB. Horowitz & Wasserman (1996) publish the OB student text CBS 574 which also appears to be a copy of the prayer. *KUB* 4 47: 39f from Boghazköy is a version of the prayer in Hittite, and includes what may be an OB or MB star-list (see 13). Several NA fragments of the prayer exist, including K3507, *OECT* 6 74–75 pl.XII (with šul.pa.è written in l.13 – references from Horowitz, 1993, 158) and those in Oppenheim (1959). See also von Soden (1936). The text refers to Ištar, Sîn, Šamaš, and Adad making judgements, which has overtones of EAE, for these are the names of the sections into which the series is divided. It describes various constellations coming forth and establishing the truth of an extispicy.

An OB prayer to Venus is attested in two copies from Tell ed-Dēr, IM 80213 and 80214. See Meyer (1982). It includes a statement that the ("ideal") year lasts 6 times 60 days and nights, from the 20th of month I to the 20th of month I. Some elements of the Sun as an 'astrological' body are to be found in the OB *Šamaš Hymn* – see for example Reiner (1985) Ch.IV ll.151f.

Other OB Akkadian literary texts show the influence of the concerns of celestial divination in so far as they include references to the ideal year and to celestial omens. OB *Atraḫasīs* refers in II.13 to the middle watch, and in IV to the key lunar days of the 1st, 7th, and 15th, for example. See also Walker (1983) on the fragmentary remains of the OB *Myth of Girra and Elamatum*, especially lines 36f. Some other OB texts contain references to the scale and construction of the universe, for example the *Etana* myth and *Gilgameš* IX iv 45' where the hero travels 12 *bēru* along the path of the Sun. This last may derive from the 12 *bēru* division of the nychthemeron, and anticipates that of the ecliptic. For details see also Horowitz (1998) Chs.3 and 5.

(12) It is asserted here, and by all other commentators, that the mathematical methods employed in the mathematical astronomical-astrological texts (MAATs) dating to the last centuries BC (44) are all present in the extant OB mathematical texts, and thus known to some scholars at least by the first centuries of the second millennium BC. Very few of those who

composed the OB mathematical texts are known to us by name or title, and none are known from later catalogues, to my knowledge. There is no Mesopotamian term "mathematician", and the OB mathematical texts formed, no doubt, part of the corpus of other scholarly professions.

A few OB mathematical texts deal with issues connected with astrology-astronomy. Some of the constants listed in the OB coefficient lists bear on celestial timing (8). These coefficient lists formed part of the reference works used by those engaged in calculating or writing mathematical exercises. The OB water clock texts (Thureau-Dangin, 1932) are mathematical exercises, but tell us something about the devices that may have been used in the timing of celestial events. It is not until the late NA period that we have evidence that the accurate timing of heavenly phenomena was attempted. Prior to this the only attested timings recorded were extrapolated mathematically from "ideals" (see Ch.3.1.2) and, I argue, appear to have had a largely divinatory significance. For details see now Brown, Fermor & Walker (1999). See also (17).

(13) See Koch-Westenholz (1993). The attested texts include lunar eclipse omens in Hittite and Akkadian (*ABCD* 33f), a solar-eclipse tablet in what appears to be OB script (*KUB* 4 63), some omens concerning comets and/or meteors, and omens concerning other lunar phenomena in Hurrian, Akkadian, and Hittite. There also exists a fragment of a Hittite translation of the introduction to EAE (*KUB* 34 12). There are no non-solar or non-lunar planetary omens attested, even though "astrological" texts comprise half of all the divinatory texts found there. The texts from Boghazköy date from the late OB period, at the earliest, to about the 12th century BC. They probably reflect Babylonian originals from this period, perhaps with some Hurrian influence. See Koch-Westenholz op.cit. 231f, Wilhelm (1989) 68–71.

KUB 4 47: 39f contains a Hittite version of *The Prayer to the Gods of the Night* (see 11), but in lines 43–46 it gives a list in Akkadian (with Hittite influence) of 17 stars belonging to the star-path of Ea, the first 5 of which are likely to be the planets other than the Sun and Moon. This gives the earliest *terminus ante quem* for the discovery of Mercury and Saturn. The following 12 are perhaps stars allocated to each month, in which case this text would also represent the first attested "astrolabe" (26). If so, then the star rising in month 1 is mul.mul – the Pleiades. This is one month earlier than is the case in all subsequent astrolabes. See Weidner *Hdb.* 60f, *BPO* 2 p2, Lambert (1987) 93–6, Horowitz (1998) 158. The existence of an OB version of *The Prayer to the Gods of the Night* does not unfortunately prove that the star-paths of Ea (and of Anu and Ellil in the following line of *KUB* 4 47) were established in the OB period, since *KUB* 4 47 could be of MB date *and* represent a local innovation. Nevertheless, I suggest here and in (17) that the 3 star paths were designated, all the planets were observed and named, and the basic form of the so-called "zwölfmaldrei" or "astrolabes" was established by the end of the OB period, or by the early MB period at the latest. Further evidence to support the original Mesopotamian, as opposed to Hittite, association of the star list and the prayer comes from the Kuyunjik text published by Oppenheim (1959), in which both are found together, and from the MB text HS 1897 which appears to be a forerunner of section B of *Astrolabe B* (see 16).

(14) Dating from the 15th to the 13th centuries BC there exist celestial omen texts from **Emar** (Arnaud, Emar IV/4 no.652: 80–82 parallels the Sumerian version of the title of the

series), many still unpublished; **Qatna** (AO 12960 in Bottéro, 1950 117 – part of EAE 22); **Alalakh** (Wiseman, Alalakh 451–2 – eclipse omens using the OB spelling an.ta.lú for eclipse); **Nuzi** (Lacheman, 1937 – earthquakes, Adad section of EAE); and **Susa** (MDP 18 258 in *RA* 14 139–42 – a tiny fragment of EAE 22 in Akkadian and *RA* 14 pp29–59 – a text in Elamite concerning atmospheric phenomena). From slightly later, texts concerned with celestial omens are attested in Ugaritic from **Ugarit**, as well as the 14th century solar eclipse report. For more details see now Koch-Westenholz (1995) 44–51.

(15) Very few celestial omen texts have been recovered from within the Mesopotamian heartland for the period dating from the end of the OB dynasty to the late NA period. What the few texts that are known from before 1000 BC seem to show is that, although EAE existed, its form was different from its NA redactions, that distinctions existed between Assyrian and Babylonian recensions, and that texts transitional between the known OB omen collections and the later EAE were still being written (*ABCD* p23). From MB Nippur we have the text Clay, *PBS* II/2,123 (an *imgiddu* one-column tablet – perhaps an early Report) which reproduces part of the Adad section of EAE, and the unpublished text Ni 1856 mentioned in *ABCD* pp 19 &25 which concerns lunar eclipses. Rochberg-Halton loc. cit. argues that the Ninevite text BM 121034 is MA on the basis of internal criteria. It also deals with lunar eclipse omina, but is not the same as Ni 1856. She describes the two lunar eclipse texts as "transitional" between the OB compilations of eclipse omina and the NA canonical EAE. I know of four MA celestial omen texts from Assur (not including the "astrological" section of *Astrolabe B*). Three (VAT 9803, 9740+11670, and Assur 10145 – in AfO 17 pp71, 80, and pl.II respectively) reproduce EAE 15 and 20 more or less exactly, and are discussed in *ABCD* p25f. VAT 9740+11670 appears to show traces of the title of EAE. Rochberg-Halton suggests loc. cit. 26 and n46 that the new terminology found in these three MA texts might stem from actual observations done at that time. The evidence is too scanty in both the OB and MA periods to be sure, but references to the Isin II king Adad-apla-iddina and Kassite king Burnaburiaš (see 21) do seem to indicate that the MB compilers of EAE did more than simply bring together already formed omens. See also Horowitz (1998) 158–9. Finally, the MA text *KAR* 366 appears to contain celestial omens, although the protases are fragmentary.

(16) The best preserved text of *Astrolabe B* is Schroeder *KAV* 218 from the "Tiglath-Pileser I library" at Assur and is dated to c.1100 BC. (The library concept was suggested by Weidner 1952/3 201 and was criticised by Lambert (1976) 85 n2. It is now referred to as Library N1 in Pedersén's *ALCA*.) It attests to an early Assyrian "royal" or at least "institutional" interest in things astronomical-astrological. See also (15). There are many duplicates (see *BPO*2 61–63 and 81–82 and Horowitz, 1998, p155 n10) which suggests that the text was considered by the Scholars to have been of some importance. Its relationship to EAE is very close in that it shares the same "ideal year" attested in EAE 14 and omens derived from the rising of stars in certain months form the basis of much of EAE 51. Also, in Mul.Apin (30) Iii36-iii12, 36 stars are listed in order of their heliacal risings in astrolabe-style. I discuss the use to which the "ideal astrolabe" was put in the context of celestial divination in Ch.3.1.2. I consider it to be unlikely that it had a practical use for farmers as Horowitz (1998) 164 suggests.

Horowitz (1993) 158 and (1998) 159 refers to the MB tablet *VAS* 24 120 which includes a Sumerian-only version of the menological section of this text, and would thus demonstrate what has sometimes been suspected – that *Astrolabe B* is a Babylonian original. This also appears to be the case from the colophon of *KAV* 218, see Horowitz (1998) 159 n17. He dates its composition (1998, 157–61) to the MB period. *Astrolabe B* comprises four sections numbered A-D. Part C is a list-astrolabe similar to the Pinches-type (see 26), part B is another astrolabe and includes a commentary, and part A is the menology. Part D lists stars rising and setting simultaneously. Horowitz (1993) 159 suggests that the Pinches-type astrolabes and the so-called Hilprecht text HS 245 (see 17) derive from a common OB tradition that leads ultimately to the compendium we know as *Astrolabe B*.

(17) The Hilprecht text (HS 245, formerly HS 229) from MB Nippur, but seemingly in OB script, appears to describe distances in the sky between stars, though in the form of a mathematical exercise using star names in a particular order and relationship to the Moon (see Rochberg-Halton, 1983, Høyrup, 1993, and Horowitz, 1998, 179–82). A NA parallel, Sm.1113, is also attested – see Horowitz (1993) 151. The order of the stars appears to bear some relationship to the astrolabes, and *CT* 33,11 = Sm.162, a NA circular astrolabe (see 27) on the obverse, has on its reverse a text similar to HS 245, as Horowitz loc. cit. shows. Thus, on the same tablet are found the HS 245-type text, which dates back to a MB exemplar in OB script, and the circular Pinches-type astrolabe, which has long been suspected of being older than *Astrolabe B* (16).

(18) Tuman has made various attempts to date the Kassite period "kudurrus" on the basis of the planets and constellations that are often inscribed thereon; e.g. 1986 & 1987. See also Koch-Westenholz et. al. (1990). I remain to be convinced that the kudurrus, even in some instances, constitute an observational record of sorts. The symbols could equally well reflect "astrological" concerns different from that which pertained in the sky on the occasion of the kudurru's manufacture, and/or the stylisation of the situation could be such as to make dating hopeless. A study by K. Slanski of the kudurru genre is forthcoming. It indicates amongst other things that these objects should now be referred to as "entitlement *narûs*".

(19) *Enūma Eliš*, the creation myth celebrating Marduk, has been dated to the late OB or Kassite period. Parts of it have been heavily influenced by celestial considerations. The relevant sections are IV: 19–26 (creating a constellation by the power of his word), V: 1–25 (setting up the sky – *lumaši* constellations, three star paths with a star for each of the 12 months; the designs of the year; Nēbiru, the "station" of Anu, who fixes the paths; creating the "stations" of Ellil and Ea; entrusting the night to the crescent Moon in order to mark out the days – the lunar scheme days 1, 7, 15, and 30; a "sign"), VI: 87–91 (Bowstar), and VII: 124–31 (Nēbiru is defined as the one who holds the turning point *kunsaggû* of the heavens). "Station" appears to refer to the positioning of the star in month XII in Astrolabe B, thus in some way designating the start of the new year – see Horowitz (1998) 116 and 115 n12 for *kunsaggû*. The 50 names of Marduk (his special number) also provide examples of the form of word-play discussed in Chs. 2 & 3. The universe as constructed by Marduk is very much the one envisaged in the "ideal" schemes of the astrolabes, of Mul.Apin, and of EAE, which are outlined in Ch.3.1.2.

(20) A number of other literary texts describe the organisation of the heavens in ways which parallel those described in *Enūma Eliš* (19) and in EAE (21). These include the MB

The Exaltation of Ištar (*TCL* 6 51 pl.97, see Horowitz, 1998, 144 for references to earlier literature), a Sumerian/Akkadian bilingual account of Ištar's elevation to the upper echelons of the gods. It includes a brief description of night and day being assigned to the Moon and Sun gods respectively, and to their gathering of the stars into furrows (r.3–10), which is reminiscent of the part played by Nēbiru in *Enūma Eliš*, and of the paths of Ea, Anu and Ellil which permeate the astrolabes, Mul.Apin, and EAE.

One section of the diorite statue from the Kassite capital Dur-Kurigalzu (fragment BbII) deals with the 30 days in a complete month (iti.ur.a). See Kramer (1948).

K 7067 (*CT* 13 31 see Horowitz, 1998, 147–8), K10817+11118 and K2313 (refs. courtesy W.G. Lambert) are SB tablets from the Kuyunjik collection that describe the gods dividing up and measuring the heavens, appointing "stations", assigning "watches" and indicating the lengths of watches. They are fragmentary, but clearly allude to ideas that find a fuller, if less technical, expression in *Enūma Eliš*. *CT* 46: 55 (Horowitz, 1998, 178) deals with the measurements of the "circle of the heavens" [*kip*]*pat šamê* and may be related to HS 245 (17) and the *ziqpu* texts (33).

(21) **EAE** – The title is listed in OB Sumerian and Akkadian (see 7), in a Hittite copy (see 13), in Sumerian from Emar (see 14), and traces are found in source Z of EAE 20 = VAT 9740+11670 from MA Assur. EAE 22 shows OB orthography (*ABCD* p251 and Farber, 1993). EAE 14 uses the OB-style calendar with the vernal equinox on XII 15, and a lunar visibility scheme already known in the OB period as the mathematical coefficient lists indicate (see 8), and similar to that in Mul.Apin IIii41f. EAE 63 seems to contain observations from the OB period. Eclipse, lunar, and weather omens are attested in the OB period which show the same "omen logic" as EAE, but which are not in the same order, or use the same orthography as EAE. Solar eclipse omens are attested in Boghazköy, as is the first of the astrolabes which are themselves subsequently reflected in EAE 51 (see also 17). Mul.Apin Iiiv3–8 (*ziqpu* section) corresponds to EAE 55, and IIi53–59 and 64–67 to EAE 56. It is pretty certain that EAE is the source (see Hunger & Pingree, *Mul.Apin* p10), and Mul.Apin in its final form is dated to c.1200 by Hunger & Pingree (elements of it are undoubtedly older, see 30).

The NB source S to EAE 20 RecB has a subscript "from a tablet of the 11th year of Adad-apla-iddina (c. 1154-)". Perhaps the tablet referred to here is a tablet of the series EAE (see *ABCD* p174f). In part of the Šamaš section of EAE the text Sm.2189 r.21 refers to the 14th century BC Kassite king Burnaburiaš – see Weidner (1941/4a) 176. For comparison there exist MA canonical versions of *Iqqur Ipuš* (*Labat Calendrier* 19–20), MA canonical versions of *Šumma izbu* (*Leichty Izbu* 20), and MA/MB canonical versions of *Šumma alu* (BM 108874 – see *ABCD* p25 n44).

To the best of my knowledge, then, it appears that EAE drew heavily on OB material, some of which was already known by that title (proto-EAE), and was put into its standard form during the Kassite period. It was probably transmitted to Assyria from Babylonia at the turn of the millennium along with *Astrolabe B* (16). The lunar section (Tablets 1–22) may initially have been separate from the solar, weather and planetary sections, and only added together later, since even the late NA copies of Tablet 22 end with lines that are very similar to the opening paragraph of the series. The EAE 22 version reads: "When Anu, Ellil and Ea, the great gods, created heaven and the earth, fixed the signs, established stations, founded positions, [appointed] the gods of the night, divided the (star)-paths, designed the constellations, the patterns of the stars, divided night from daylight, [measured] the month and created the year; for Moon and Sun...they determined the decisions of heaven and earth." (e.g. K5981 and K11867 – also attested in some texts from Assur, see Rochberg 1989a, 270–1, and Horowitz, 1998, 147). It is close to both the canonical Akkadian version of the opening lines of Tablet 1 (*ACh.* Sîn 1, STC II 49, VAT 7827): "When Anu, Ellil and Ea, the great gods, in their sure counsel had fixed the designs of heaven and earth, they assigned to the hands of the great gods (the duty) to form the day well (and) to renew the month for mankind to behold. They saw the Sun god within the gate whence he departs (and) in between heaven and earth they took counsel faithfully", and to the Sumerian version which reads: "When An, Enlil and Enki, the [great] gods had established in their firm counsel the great divine powers (me) and the boat of Suen (the Moon god) so that the crescent Moon should grow and give birth to the month and establish signs in heaven and earth, the boat was sent forth shining in the heavens – it came forth into the heavens".

The series comprises 68 or 70 tablets, depending on the recension. Weidner (1941/4a) 181 suggests that 5 versions were probably current in the NA period, from the schools of Uruk, Babylon+Borsippa, Kalḫu, Nineveh, and Assur, though this notion has been criticised by Koch-Westenholz (1995) 80f. EAE was used extensively in the NA period by Scholars attached to the Assyrian court. In the LB period it continued to be written. Pingree (1982) suggests that a transmission of EAE 1–49 to India into the Pāli Dīghanikāya took place in the 3rd or 4th centuries BC, and the astral omens into the Gargaṣamhitā slightly before the new millennium (see also idem 1987a). VAT 7814, a copy of EAE, is dated to 194 BC. See also *LBAT* 1521–1577. This gives an indication of EAE's importance still in the late period.

The series has not yet been fully reconstructed, though work is in progress. Weidner in *AfO* 14, 17 and 22 provided a bibliography to tablets 1–49 of the series, and Reiner and Pingree have described the probable contents of tablets 50–70 in *BPO*2 and in Pingree (1993). Virolleaud in *ACh.* and Weidner and Virolleaud in Babyloniaca 3,4 & 6 in particular published many fragments of EAE. See also Borger HKL III § 91 and for texts and fragments in the BM see now Reiner (1998). This is how the publication of the series looks at the moment:

1–13	Moon omens – a sub-series igi.du$_8$.a.me *šá Sîn* "visibilities of the Moon" – mostly unpublished. For its commentary (*mukallimtu*) – see Weidner (1912) and Koch-Westenholz (1995) 84, and idem (1998). See also n4 p52 of Al-Rawi & George (1991/2). For a LB *ṣâtu* of EAE 8, *TU* 17, see Hunger (1995).
14–	Frames the first 13. Lunar visibility tables according to an ideal scheme. See Kugler *SSB*2 p50, and Al-Rawi & George (1991/2). For an edition of the Kalḫu version, *CTN* 4 10, see Hunger (1998).
15–22	Lunar eclipse omens. See Rochberg *ABCD*.
23(24)–29(30)	Non-eclipse solar omens. See van Soldt (1995).
29(30)–40	Solar eclipse omens – unpublished.
40–c.49	Adad omens – unpublished.
50–c.51	Star omens. See *BPO*2.
52–	*Ikû* "field" omens – unpublished.
53–	Mul.Mul "stars" Pleiades tablet – unpublished.
54–?	Mars, Mercury, Saturn, Scorpius, Fish? – unpublished.

55–(K2342+)	MUL.ŠUDUN omens – unpublished.
56–	The 5 planets – see provisionally Largement (1957).
57–	Attested in K2330, but contents unknown.
58–?	Mars, Mercury, Saturn, Scorpius, Fish?
59–63	Venus tablets. 59–60 published in *BPO*3 Group C. The bulk of tablet 61 is also published in *BPO*3 Group A, and appears to group together the oldest Venus omens aside from those in EAE 63. Tablet 63 has been published often – see (9). Tablet 64 is related to material published as part of group F of *BPO*3. For a description of these tablets see also Pingree (1993).
64–c.65	Jupiter tablets – unpublished.
66–?	TE= *iṭḫi* tablet? – unpublished.
67(69)–	*mišḫu imšuḫ* tablet ? – unpublished.
68(70)–	*adir* kenning tablet ? – unpublished.

Some 24 tablets have been published, or about one-third of the series. See also Oppenheim (1975) and Koch-Westenholz (1995) 76–82.

(22) The lunar eclipse of October 23 798 BC is recorded in *ABL* 1406. The 763 BC, 15th June solar eclipse is recorded in a comment on a rebellion in Assur. See *RlA* 2 430 r8, *CAD* AII p505, Roaf (1990) 175. It is an important chronological marker.

(23) *Sargon II's 8th Campaign*; an extispicer confirms a lunar eclipse omen dated to Oct. 24th, 714 BC. Noticed by Oppenheim (1960) 136. He shows that the interpretation of the omen is based on an eclipse shadow, and the identification of the Guti in the apodosis with the real enemy, the Urarteans. See also Grayson (1991a) 96.

(24) *Erra and Išum* was probably composed during the 8th century BC – see Dalley (1989) 282 – and may reflect the arrival to the throne of Babylon of Nabonassar (though Parpola and Neumann, 1987, 180 date it to the 9th century). It contains allusions to celestial concerns in the names of the characters, and perhaps in their activities. I do not interpret this text solely as an allegory of events in the sky, but an awareness of how the myth might *also* be mirrored in the sky seems to me to be present. Erra, also known as Nergal, attempts to seize control of Babylon from Marduk, but disaster is averted by Išum. If Erra is identified with Mars, Marduk with Jupiter, and the Sebetti with the Pleiades (as we find in the late NA period), the myth can be interpreted (partly) as a description of the behaviour of these heavenly bodies. There are a large number of small pieces of evidence which cannot be listed here, but in lines I: 21 and I: 117 Erra is referred to as an en.gi$_6$.du.du and a *bibbu* respectively, both of which probably describe heavenly bodies. Marduk's decision to descend to the *apsû* and have his garments cleaned by *ummânu*s may in part be seen to be describing Jupiter's conjunction with the Sun. Al-Rawi and Black (1989), who published more of Tablet 2 of the text, note that ominous celestial phenomena play some part in the myth. The Fox constellation (a Mars E-name, like the Pleiades, see Ch.2.1.1) *ummulma* "sparkles" in IIr.6'. In IIr.10' the Star of Erra sparkles and is said to be "bearing radiance" (see n229), and all the people, it is said, will be ruined. I suggest that the absence of Marduk, and the temporary take-over of Erra is being paralleled by Jupiter's absence behind the Sun and the brightness of Mars, both ill-boding events. I suggest that Kabtî-ilānī-Marduk, the author, was aware of celestial

behaviour and possessed some knowledge of EAE.

Similarly, original Assyrian 'literary" compositions contain allusions and references to celestial bodies. See for example SAA3 text 1: "those of heaven...like the writing on the celestial firmament, does not miss its appointed time"; text 2: "your shining name is Jupiter...who makes at his rising a sign"; text 25: "in your days...Jupiter has taken a course of truth in the heavens, while Mars, your star, is *clothed with a glitter* (repeatedly darkened?) in the heavens; text 39: 30f: "the upper heaven is of luludanitu stone...lower heaven of jasper, he drew the constellations on it...40 *bēru* is the disc of the Moon, 60 *bēru* is the disc of the Sun – the inside of the Sun is Marduk, the inside of the Moon is Nabû." This latter, *KAR* 307, is also treated in Livingstone *MMEW* and Horowitz (1998) 3–19.

(25) The Nabonassar Era had significance in later antiquity, but it remains debatable to what extent its significance was realised at the time. See I.2.

(26) Astrolabes, or Zwölfmaldrei, mul.meš 3.ta.àm, come in list and circular form. They list for each of the twelve months, three stars that are supposed to rise (heliacally) in those months. Each of the three stars is allocated to a path of Anu, Ea, or Ellil in the sky. These paths appear to be bands stretching up from the Eastern horizon; the path of Ellil is to the north-east, that of Anu to the west, and that of Ea to the south-east (see *BPO2* 2.2.1.2, Koch, 1989, but see Lambert's 1987 comments). In some of the astrolabes' numbers are assigned to each month or to each group of three stars that correspond to the length of daylight in that month, or when those stars rise, much as in (8). The attested astrolabes are listed in Hunger & Walker (1977) 34 and Horowitz (1998) 155. The alleged development of the astrolabes from the *Boghazköy Star-List*, via the so-called "Pinches-type" (after the text *LBAT* 1499, first published by Pinches, 1900), and section B of *Astrolabe B* (16), to the Mul.Apin (30) lists is discussed in *BPO2* 72–8 and in Hunger & Pingree *Mul.Apin* 11. The circular-type astrolabes are thought to pre-date the list-type, since in the former the vernal equinox is assigned to month XII (as in 8), and in *LBAT* 1499 to month I, as in Mul.Apin. It is argued that during the MB period the ideal vernal equinox date changed to I 15. Note, however, that in LB text BM 82923 (Hunger & Walker, 1977) the vernal equinox numerical values, although not specified against a month name, are assigned to stars which in the circular astrolabes ideally rise in month XII.

The recognition that some stars in the astrolabes switch star-paths has led some commentators to hypothesise that the stars were originally grouped not according to over which part of the eastern horizon they rose, but for other "astrological" reasons. Van der Waerden (1949) 17 suggests that the paths of Ea, Anu, and Ellil derive ultimately from the "Stars of Elam, Akkad, and Amurru" which are found listed later in Kuyunjik texts such as *GSL* (see 29, against this see Horowitz & Oelsner, 1997/8 n59). Hunger & Pingree *Mul.Apin* 139 indicate how often the constellations assigned to the three gods do not actually fall into the respective star-paths, suggesting that the paths came later than the assignations. They note also that, generally speaking, the stars associated with the death gods lie to the north, those associated with the vegetation gods lie to the middle, and those connected to the storm gods lie to the south. Horowitz (1998) 175–7 comments on the relationship between the "Stars of Elam, Akkad and Amurru" and the astrolabes, pointing to the fragmentary *mukallimtu* commentary (28) to EAE, *ACh.* Išt. 39 = 81–7-27,81, in which the association between the two appears to have been formalised. Certainly, the "Stars of Elam, Akkad and Amurru" form part of the divinatory background to the astrolabes, and no doubt permitted the editors of EAE to interpret the phenomena of particular stars as those pertaining to one of those three countries.

That the astrolabes were in use by the NA Scholars is suggested by SAA8 19: 6. It is also known that NA astrolabes, based on Mul.Apin rather than the *Astrolabe B* or the list or circular-types, were being compiled in Nineveh, and no doubt elsewhere. Examples include *CT* 33 9 (Horowitz, 1998, 174) and NV 10 (Donbaz & Koch, 1995).

Recently, it has become apparent that a 30-star catalogue tradition, comprising three groups of 10 stars each designated as "those of Ea, Anu and Ellil" respectively, existed alongside the 36-star one. This catalogue may be related to *The Prayer to the Gods of the Night* (11), for in that text 10 stars are also listed, and is perhaps also reflected in the Sumerian menology that accompanies *Astrolabe B* (16). See Horowitz & Oelsner (1997/8). Presumably the stars of the 30-star catalogue tradition were not in the first instance related to the 12 months of the "ideal year", but since they are listed in an order that is repeated in the Zwölfmaldrei, one may presume that they were also part of the background upon which the authors of the 36-star astrolabes drew. We await Horowitz's forthcoming publication on the astrolabes.

(27) Many Letters and Reports sent to the NA kings from Babylonian and Assyrian Scholars have been preserved. Many of them contain celestial information that was felt to concern the king, state or land. Often they include the interpretations of celestial phenomena, either in the form of comments, or in extracts from EAE or other series. Sometimes, however, more explicitly predictive material was sent (see 37). The datable Letters and Reports are attested from 680 to 648 BC, but the majority are datable to between 675 and 666. These texts are the major source material for this work.

(28) From the NA and later archives, commentaries (*mukallimtu* and *maš'altu*) and explanatory lists (*ṣâtu* – see also Ch.3.2.1), excerpt collections (*rikis girri*, *liqtu* and *liginnu*), on various parts of the canonical series *iškaru* EAE, and on the "alternative" *ahû* and "oral" *ša pī ummâni* omen collections are frequently attested. For descriptions of the types see Bauer (1936), Weidner (1941/4a) 182f, Elman (1975) and Koch-Westenholz (1995) 82–92. Many examples are mentioned in the Letters and Reports, and the 29th tablet of the *ahû* series has been reconstructed by Rochberg-Halton (1987a). See also idem 1984b. For the *mukallimtu* and a *ṣâtu* to tablets 1–14 of EAE see (21). See also n30.

(29) A variety of texts which seem to be collections of the schemata more or less underpinning EAE are known from the NA periods and later. Some of the schemata may of course pre-date EAE, and some may be derived from it. Some of these texts were discussed by Weidner in the *Hdb.*, but are in need of updating. K250 and its duplicates are the best known, for which see now Koch-Westenholz (1995) App. B – the *GSL*.

Also of interest are (A) K11151 in Weidner *GD* Taf.17 & p39 and Livingstone *MMEW* 73 which appears to be a forerunner of Seleucid period texts connecting the zodiacal signs to various stones, trees and so forth in the style of later Greek astrology. Some doubt must remain as to K11151's Ninevite origins, despite the museum number; (B) The series DT 72+ which includes omens written in code, and which contains at the end of DT 78 periods of years associated with the planets similar to those found in the later Goal Year texts. See

Weidner (1957/71b) p187: 7, Gadd (1967), Hunger (1967) and idem (1975), and Ch.4.2.2 here. (C) *STT* 300, dated to 619 BC, which parallels (the non-zodiacal part of) the Persian-Seleucid text *BRM* IV 19 (Ungnad, 1944 & Neugebauer & Sachs, 1952–3), which contains the first attested "dodekatemoria".

(30) Mul.Apin "Plough Star" is a two (possibly three)-tablet Babylonian series that has usually been thought to represent a "scientific" innovation from amidst the morass of omenology of the OB and MB period. The latest edition is by Hunger & Pingree (1989). In 1991, George published BM 77054, a NB "copy of a fore-runner" to Mul.Apin I iv 19–24, which used the OB-style equinox dates. Hunger & Pingree date Mul.Apin to c.1200 BC, although the earliest extant version dates from -686. I argue in Ch.3.1.2 that most aspects of Mul.Apin date to the OB period, and that the text is by no means distinct in its aims from EAE.

(31) Seasonal hours (1/12th of the day) are attested explicitly in the Kuyunjik NB report in Reiner & Pingree (1974/7). This helped confirm that they are also present in the Nimrūd Ivory Prism (see Langdon, 1935 and Smith, 1969 and photo p78 of SAA8). They are also perhaps found in the Mul.Apin gnomon section. Rochberg-Halton (1989c and 1998, 37–8) has argued that seasonal hours are used in the later horoscopes under the name *simanu.*

(32) Observational records without attached omens are known from Assyria and Babylonia at this time. Reiner & Pingree (1975a) describe 3 Kuyunjik NA-script and NB-script texts and 1 (later?) Babylonian NB-script text which record the positions of Mercury. Walker (1982 and 1999) discusses and edits a NB-script copy of a 7th century BC set of observations of Saturn from Babylon/Borsippa. HSM 1490 = 899.2.112 is an unpublished record of Mars observations dating from c. 681 to c. 648 BC (to be published by J. Britton). Some of the Reports record phenomena without observation – see particularly the reports of Nabû'a, SAA8 p80f, and x149, x134 and x133. From Babylonia the earliest attested Diary dates to 652 BC (Hunger & Sachs *Diaries* I). Late NB-script copies of detailed records of eclipses dating back to c. 747 BC have been found in Babylonia. *LBAT* 1413 is a short collection of consecutive eclipses, *LBAT* 1414, 1415+1416+1417, 1419 is the *Saros Canon* – see (39) and Huber (1973). See Ch.2.2.3 and Ch.4.2.1 for details.

(33) *Ziqpu* or culminating stars are first listed in K.9794, for which a late, nearly complete copy AO 6478 = *TCL* 6 21 exists. See Weidner *Hdb.* 131f, Schaumberger (1952), Horowitz (1994, n10 and 1998, 182–8) for discussion and references. The intervals between the times of the stars' culminations are noted in AO 6478 in minas, UŠ *ina qaqqari*, and *bēru ina šamê*. The minas no doubt described weights of water flowing from a water clock, and it is known that times at night, particularly of celestial phenomena, were noted using *ziqpu* stars – e.g. x134: 8 and x149: r.1 and SAA2 249: 12'f. These water clock weights were correlated to the UŠ, or "degrees of right ascension", between the stars and to the "*bēru* of the sky". One full circuit of the stars in AO 6478, one nychthemeron, amounted to 60 2/3 minas, 364 UŠ, and 655,200 *bēru*, the latter measurement apparently giving the perceived *actual* length of the parallel of declination upon which these stars were located. One *bēru* measured about 11 kms. The ratio of UŠ to *bēru* in the text is 1: 30,0;0 or 1: 1800 where ordinarily it is 1: 30. *Ina qaqqari* probably also refers to the sky, and not to the earth as Horowitz (1998) 185 argues.

In the NA text *STT* 340, however, one circuit of *ziqpu* stars is said to cover only 12 *bēru* (see Horowitz, 1994, 97). In this case, the celestial distances between the stars have been described in terms of degrees of right ascension using *bēru* in the familiar sense as equal to 30 UŠ. A smaller list of *ziqpu* stars, without the time element, is also attested in Mul.Apin Iiv1–30. It is not without significance, I argue, that their potential for timing events at night was realised, perhaps *first* realised, in the late NA period.

A second group of *ziqpu* texts related the crossings of the meridian (the north-south line which passes overhead) by the stars to the risings of segments of the ecliptic. See Schaumberger (1955), Rochberg (1988) 58–9 and (1996), and Horowitz (1994) for BM 77242. These texts clearly post-date the invention of the zodiac (42). For a further study of *ziqpu* texts, and what they demonstrate in terms of evidence for the use of particular units in Mesopotamia, see Brown *CAJ* forthcoming.

(34) BM 78161 was first discussed by Walker & Pingree (1988). They saw in it an early, pre-zodiacal way of locating planetary positions against strings (gu) of stars in the sky, each string with its own *ziqpu* star. Koch (1992) sees in it a record of a particular celestial configuration, and dates it to -650. This does not explain the purpose of the duplicates. See also Koch (1999).

(35) I.NAM.giš.ḫur.an.ki.a is a learned explanatory text, and has some cross-over with EAE. It demonstrates the kind of word-play that exemplifies some of EAE and contains the "ideal lunar visibility scheme" (see Ch.3.1.2 and 3.2.1). The title of this series provides evidence for a connection between OB and earlier ideas of the "design" of the universe and the achievements of the latest MAATs – see Ch.5.1.3. For an edition of what remains, and related material, see Livingstone *MMEW*. The earliest exemplar dates to -683, but Livingstone pushes the series' composition back to the end of the 2nd millennium.

Iqqur īpuš (see *Labat Calendrier* and *RlA* 4 319–23) is a series in which the omens found in various other series have been ordered according to the month in which they occurred. It is not clear, however, whether the omens were first constructed for *Iqqur īpuš* or for the other series in which they are found. It is broadly divided into a first part concerned with various human activities, and a second that dealt with various celestial and meteorological phenomena. Thus §§ 67–102, according to Labat's nomenclature, include many celestial omens found in EAE. This second section was perhaps entitled *biblāni* – "new Moons" according to Labat, loc. cit. 6, perhaps better "lunar disappearances". Indeed §§67–8 begin with omens that concern the appearance of the Moon on the 30th and lunar "opposition" on the 14th and 15th. This bears on the "ideal month" discussed in Ch.3.1.2. See also Reiner & Pingree *BPO*3 Group D omens and Reiner (1995) 88f.

Some further celestial omen series are named in (36) and in the library record published by Parpola (1983b).

(36) This unique text combines terrestrial and celestial divination in an otherwise unattested scheme. It is published by Oppenheim (1974) who dates it to the Sargonid period. See Reiner (1995) 94f, who notes the text's connection with the interests of *Iqqur īpuš*, and Ch.3.1.2 here for a discussion of this text in the context of ideal period schemes.

(37) Examples of Letters and Reports containing or referring to predictions are:

Letters: Eclipse SAAX 26, 45, 46, 71, 78, 114, 147, 170, 216, 240, 351.
Other 8, 23, 31, 42, 48, 50, 51, 72, 74, 362.

Reports: Non-occurrence of an eclipse SAA8 42, 46, 67, 87, 321, 344, 447.
Occurrence 250, 251, 279, 320, 346, 382, 388, 487, 502.
Other 52, 60, 246, 257, 266, 293, 516.

For details see Ch.4.2.

(38) Neugebauer & Sachs (1967)183 published BM 36731 which (probably) describes the period 616–588 BC. It demonstrates the earliest known use of the abstract day, or "*tithi*" (1/30th of a month), the longest: shortest day ratio of 3: 2, and the use of fixed intervals between the equinoctial and solsticial points, and the last visibility (šú) of Sirius (implying that the sidereal and equinoctial years were equated). The text uses (though not explicitly) an average *epact* (the length of the year less 12 months) of 11;16 days, which is a dramatic improvement on all previous attested estimates of the length of the year, which were, I argue, only "idealised". This text is soon to be re-edited by J. Britton. See Ch.4.2.2 for other material demonstrating a 6th or 7th century BC knowledge of the periods between recurring celestial phenomena. See also *HAMA* 542f.

(39) The Saros, the 18-year, 223-month scheme for predicting eclipses is only directly attested in the last centuries BC, but it may well have been used much earlier. Eclipse records from as early as the mid-8th century BC were arranged into what modern students refer to as the "*Saros Canon*" – a compilation involving 24 18-year cycles incorporating when complete some 932 lunar eclipse possibilities in a 432 year span, separated one from another by 6 or 5 months. A solar eclipse version also exists. See Walker (1997) 19–22 for details and references. The existence of the Canon does not mean that the fundamental 223-month period between eclipses of the same type was known this early. The 6 or 5 month period was perhaps all that was used by the NA and NB Scholars – see Ch.4.2 and 4.2.4.3. It is evidence, however, that the dates and times of eclipses were being recorded this early, even if only later they were incorporated into the scheme. The Canon demonstrates that the 18-year Saros was known at least by the 4th century BC. Britton (1993) describes how a reform, or adjustment, in the *Saros Canon* seems to have taken place between 532 BC and 491 BC, which suggests that the 223-month interval may have been known then. John Steele is currently working on the evidence for other such reforms. The Saros is incorporated into the functions used in Text S (see 43), dating to the mid-5th century BC. In general the times of the first contacts of eclipses were predicted using the *Saros Canon*. This is not true of the eclipses predicted using the lunar MAATs. The Exeligmos (triple Saros) is attested late in Thureau-Dangin *TU* no14.

(40) From 626 BC, the record of intercalated years indicates that some form of systematic scheme was in use by the astronomer-astrologers. From 503 the 19y cycle was thought to have been attested in the record of intercalations, but as yet this remains to be proven (Walker, 1997, 22–23). See also Sachs (1952b), *HAMA* 357f, and Slotsky (1993) for references to related cycles. There is a close relationship between the 19-year cycle and the Saros (39) – see Ch.4.1.

(41) From c. -680 (HSM 1490) Mars observations – unpublished, Britton forthcoming.
From -646 (BM 76738+) Saturn observations – see Walker (1999).
From -586 (*LBAT* 1386) Mercury observations
From -586 (*LBAT* 1386) Venus observations
From -525 (*LBAT* 1393) Jupiter observations arranged in 12y groups
From -422 (*LBAT* 1411) Saturn observations
From -422 (*LBAT* 1411) Mars observations
In SH 81–7–6,135 (*SSB* 1 p45 and *BA* p107) we have what is probably a pre-4th century BC text giving planetary periods in years and days. See n412.

(42) The Zodiac first appears in the Diaries between -463 and -453. See also Rochberg-Halton (1991b) p112f and *HAMA* 593f.

(43) These are texts that use mathematical methods in order to predict the time and/or location of celestial phenomena using methods that are different from those found in the *ACT* material (44). Some predate the *ACT* texts, others appear to use more primitive techniques although they too date to the Hellenistic or later periods. **Lunar** examples include:

- Diary -567: 11 gives a calculated value for the time interval Moonrise to Sunrise. The scheme used is unknown, but the methods outlined in Brack-Bernsen (1997) 115f and attested in the Hellenistic period text *TU* 11: 29f (loc. cit. 123f) seem to have been the most likely. See Ch.4.1.2, here. These methods are simpler than those used in the lunar *ACT* texts (44).
- Cambyses 400 (-522/1) also gives calculations of these luni-solar time intervals known today as the "lunar 6". See *BA* p100. The methods used were again perhaps those outlined in *TU* 11: 29f.
- Text K (Neugebauer & Sachs, 1969, 96–111) includes another primitive, though zodiac based, scheme for determining lunar visibilities.
- Text S – Aaboe & Sachs (1969), Britton (1989), Moessgaard (1980) 78–9, *HAMA* 525–8 and augmented by Aaboe, Sachs, Henderson, Britton & Neugebauer (1991).
- Text G = BM 36580 (pp69–71) – gives eclipse possibilities from -474 to -456. It has a large empirical input, but nevertheless utilises the same (or very similar) functions used in the Hellenistic period *ACT* texts for determining month length, luni-solar visibilities and predicting eclipses. It also uses zodiacal longitudes (see 42).
- Text A – Aaboe-Sachs (1969) and Britton (1990), another early lunar MAAT uses zodiacal longitude from -397 as an argument.
- Text L (Aaboe, Sachs, Henderson, Britton & Neugebauer, 1991) -416 to -380 is a System A lunar MAAT, but with the solar anomaly unsolved.

Non-Lunar examples include:

- 4th century BC Venus MAAT BM 33552 in Britton & Walker (1991).
- Aaboe & Huber (1977) BM 3715 (Venus).
- Neugebauer & Sachs (1967 and 1969) texts C (Mars, Venus); F(Saturn); H(Mars).
- Aaboe et al. (1991) text M (Mercury).
- Cf. *ACT* ii 362–444 and Pingree (1998) n91.

(44) The earliest known full lunar system A text, BM 40094, dates from -318 to -315 (Aaboe, 1969), the earliest full lunar system B, BM 34162, from -257 to -244. Non-lunar, non-solar planetary ACT-type texts all date to the Seleucid period. According to my terminology, mathematical astronomical-astrological texts (MAATs) include the so-called "*ACT* texts", named after Neugebauer (1955), 3 vols., who published and discussed all the then known examples of the most *complicated* texts designed to predict celestial phenomena. These texts were not necessarily the most *successful* at doing this, but they conformed to a particular style of presentation and employed similar mathematical techniques. They can be differentiated from the NMAATs (45) and the "atypical" or "more primitive", perhaps "earlier" MAATs noted in (43). They include tables of numbers pertaining to celestial phenomena known as "ephemerides", and "procedure texts" that are poorly understood, but which appear to describe the methods underlying the ephemerides. For more discussion on these issues see Ch.4.1 and Ch.5.1.3. *ACT* included about 300 of the c.330 texts now known. See the bibliography offered in Britton (1993), and the recent works of the modern students named in n265 above. Some recent work has shown that some of the *ACT* functions and parameters were known as early to the 5th century BC (see 43).

(45) Sachs (1948) identified amongst the so-called LB non-mathematical astronomical-astrological texts (NMAATs) **G**oal-Year texts (GYTs), **A**lmanacs (and Normal-Star Almanacs), **D**iaries, (and **Ex**ceptions). They are sometimes referred to as GADEx texts. His description remains good, but see as well Sachs (1974) and Hunger's introduction to the Diaries. The texts are mostly copied in *LBAT* and they contain the records of celestial phenomena recorded to differing degrees of temporal and spatial accuracy. Those apart from the Diaries arrange these data in such a way as to make the prediction of future phenomena straightforward. It would appear that the Almanacs and GYTs mainly used the data recorded in the Diaries. Their relationship to the Horoscopes (48) is also close.

The Diaries are attested from –651 – see Hunger & Sachs (1988, 1989, 1996);
the Normal-Star Almanacs from –300;
the Almanacs from –281 (for two examples see Sachs & Walker, 1984, and 46 below);
the Goal-Year texts from –255.

One of the exceptional texts is published in Hunger (1988). It dates to the beginning of the 2nd century BC.

NMAATs were composed at the same time as MAATs and were concerned with predicting the same celestial phenomena. Many texts, including the Diaries, contained not only the record of observed phenomena but occasionally calculated data. This occurred presumably when observation was impossible, and the methods employed may have included those found in the MAATs. Nevertheless, in general a NMAATs required the Scholar to look up the record of an observation dating to some particular time in the past in order to predict a future phenomenon. In the case of an MAAT the Scholar looked up a *calculated* time and location.

(46) The latest datable cuneiform text (AD 74/5) is an Almanac. It, and some other very late texts (*LBAT* 1197–1201), are published by Sachs (1976).

(47) For some LB reports and letters see *LAS* II p503–5. References to this scholarly activity, and to state-craft based on omens are found in Nabonidus's dream (*VAB* 4 270), the

report *YOS* 1 39 dated to 548 BC, and the later text 83–1-18,2434. See Oppenheim (1969) 121–2 and nn49–50. See also Gadd (1966) and Reiner (1995) 76f. See under (21) for late EAE texts.

(48) The earliest of the 30 attested Babylonian "horoscopes" dates to -409, the last to -68. See Sachs (1952a) and Rochberg-Halton (1989b and especially 1998). Note also the reference in Cicero *De Divinatione* II 87 to Eudoxus (c.-370) criticising Babylonian horoscopy (or birth omens, which have much earlier parallels – Rochberg-Halton, 1989b, 109). These horoscopes indicate the times of birth of certain, rarely named, individuals and contain data on the locations and behaviour of the heavenly bodies at or near that time. Occasionally prognostications were given. The horoscopes depended on the zodiac, and required data on the signs in which the planets were located at arbitrary times, data not readily available from the *ACT* texts). Rochberg-Halton, 1989b, 120f and 1998 Ch.1 stresses their relationship to the NMAATs, particularly to the Almanacs.

(49) A further series of texts have quite recently come to be understood as examples of Babylonian astrology similar in many details to Greek and later ideas of astrology. They depend on the zodiac and including the concepts of trine, dodekatemoria etc. They include:

- Iatromathematical texts *BRM* IV 19–20 and *LBAT* 1596. Leibovichi (1956c). Reiner (1995) Ch.3.
- Zodiac + *izbu*, Biggs (1968).
- Zodiac + extispicy, Reiner (1985) 592 and (1995) 78f = *SpTU* 4 159.
- Zodiacal signs having characteristics, Koch-Westenholz (1995) 165.
- Cryptographic texts – e.g. *LBAT* 1604/5.
- Melothesia – Reiner (1993 and 1995 58f).
- Calendar texts – e.g. VAT 7815/6 in Weidner *GD* p41f, and *LBAT* 1586/7 in Hunger (1975). See now BM 47851 in Hunger (1996) who lists other related texts. See also Reiner (1995) 114f.
- Micro-zodiac texts VAT 7851, 7847+AO 6448 (= *TCL* 6: 12), *LBAT* 1580 and 1578/9 in Weidner *GD* p12f and AO 6483 (= *TCL* 6: 14) in Sachs (1952a) 65, and *BRM* IV 19&20 in Ungnad (1944) and Neugebauer & Sachs (1952–3).
- Related texts BM 34567 (Sp II 39) in Sachs (1952a) p65, *TCL* 6: 13 in Rochberg-Halton (1987b) and (1988b) and BM 36746 in Rochberg-Halton (1984a).
- Weather omen series in *TCL* 6 9, 19, and 20 in Hunger (1976b).

For a brief survey see Koch-Westenholz (1995) Ch.8. Other sources of late astrological-astronomical material include Hunger *Uruk*, *TU* 11–21 and the new Sippar texts, for which see n1 p149 *Iraq* 52 (1990), but much remains unpublished. For much on the survival of Mesopotamia celestial divinatory and astral magical traditions see now Reiner (1995).

APPENDIX 2

Comments on the Dating of the Letters and Reports

The following is the result of a re-analysis of all the texts dated by Parpola in LAS II, by Hunger in SAA8, and by de Meis and Hunger in ABABR, which has superseded the work of Neumann (1995). On the basis of the improved readings of some of the texts, particularly of the Reports in Hunger's SAA8 volume, but also of some of the Letters in Parpola's SAAX volume, redating has been necessary. In addition some errors have been spotted in LAS II Appendices I and J, wherein Parpola dated the Reports and Letters. Finally, some of the Babylonian Letters, edited for the first time in SAAX, have not previously been dated. The dating of some of these has been attempted. Where no comment is offered on the dating of a text, Parpola's LAS or Hunger's SAA8 dating has been accepted. The dated texts are listed in Table 1 of the Introduction. The texts discussed are listed below in order of the years BC assigned them in LAS, SAA8 or SAAX, and if redated, according to the year assigned them in Table 1. Within each year they are listed according to their SAA8 or SAAX numbers, Reports preceding Letters. Months are given Roman numerals. Note that the Mesopotamian year commenced around the vernal equinox, so that months X-XII in the Mesopotamian year that began in 670 BC, *for example, correspond broadly to Jan-March 669 BC. The discussions are of necessity brief and must be read in conjunction with the texts themselves, Parpola's commentaries and appendices in LAS II, and ABABR:*

Year	**SAA**	**RMA**	**Comment (all dates are BC)**
680	8244	162	Redated to 671 in *ABABR*.
680	8323	187	Datable either to the beginning of Esarhaddon's or to Assurbanipal's reign. See now *ABABR*.
680	x109		Datable to the second month of 680, or subsequent years on the basis of line 16'. Later years would not make sense from the Letter's context. This dating agrees with that suggested by Labat (1959) 115.
—			
679	8288	195	Too many Jupiter-Mars conjunctions in the period of concern to make dating possible. See *ABABR*.
679	8364	104	Hunger SAA8 also proposes the date 677, and both are considered plausible by *ABABR*.
679	x149	*LAS* 105	See year 621.
—			
678	8489	145	Any possible dates are unsure. See *ABABR*.
678	8535	*ABCD* 280f.	See year 659.
—			
677	8503	184	In *LAS* II App. J 664 is also posited as a possible date for this text. Still more are possible – see *ABABR*.
677	x113		Bēl-ušezib's Letter is datable to late 677 or early 676 on the basis that Saturn and Cancer are said to have been standing in the halo of the Moon. See *LAS* II 397.
—			

676	8100, 8438 & x084		= *RMA* 192, 193 & *LAS* 14, were not dated in *LAS* II, but in *AfO* 29/30 48 to April 27, 676.
—			
675	8245	199A	Parpola (App. J) interprets this Report as describing the rising of Mercury near Regulus. There is, however, no evidence that the planet's rising is meant. Mercury is frequently near to Regulus, so many dates are possible. See also *ABABR*.
675	8246	208	*LAS* dating rejected in *ABABR* and several alternatives offered.
675	8317	115C	The restoration of r.1f in SAA8 has made this text datable to within a year. In *LAS* II App. J Parpola only dated it to the period 677–675.
675	8324	216	Redated to 5/8/675 in *ABABR*.
675	8337	219	Two other possible dates fit the evidence. See *ABABR*.
675	8338	210	Various possible datings are listed in *ABABR*.
675	8370	95	Insufficient of this Report remains to be sure that it describes Jupiter, Venus, and Scorpius in a lunar halo in my opinion, though the dating is not rejected in *ABABR*.
675	8380	244A	Hunger's dating is uncertain, he admits, as it is based on a restoration of the text, though acceptable astronomically according to *ABABR*.
675	8469	235A	*LAS* dating excluded in *ABABR*, and there are problems with any of the possible alternative dates.
675	8500	67	Redated to 678 in *ABABR*.
675	x111–2		These Letters from Bēl-ušezib concern the 675 conflict with Mannea. They post date x113, which is wrongly positioned in SAAX.
—			
674	8040	180	See *ABABR*. Datable to periods separated by c.15 years, but insufficient data exist to exclude a possible dating to 15 or 30 years later than 674.
674	8251	30	Parpola and *ABABR* have dated this text, which predicts an eclipse in *elūlu* (VI), on the basis that an eclipse actually took place in that month. The text makes it clear, however, that a prediction is intended, probably on the basis only of the omen in line 8. Any *elūlu* eclipse concerns Elam (r.6) – see Ch.3.2.2. Were a predictive system being used, then the text would probably date to a period six months after an *addāru* (XII) eclipse. These occurred in a number of years during the period of concern.
674	8248	195A	Jupiter set in the west on the 10th of *kislīmu*, which dates it to Nov. 26th 674, which slightly corrects *LAS* II and SAA8. Venus, however, is nowhere near its morning or evening rising. This implies that kur (r.1) *napāḫu* is being used by Nergal-eṭir to describe not a *heliacal* rising, but merely Venus's daily appearance out of the gloom. Although queried in *ABABR*, the dating is not rejected. Nor is it here.
674	8249	224	Many possible dates exist, even assuming udu.idim means Mercury. See also *ABABR*
674	8403	174A	*ABABR* propose 674 or 666.
674	8451	203	*LAS* date rejected in *ABABR* and several alternatives proposed.
674	8454	218	Dating rejected in *ABABR*.
674	8548	243C	*ABABR* links this text with 8351 and 8455 and dates it to 28/11/674, even though the commentary to 8351 indicates that the preferred date (of several) is 6/6/673. I consider these texts to be undatable.
674	x001–3 = *LAS* 31–2		The optional date of 679 is offered in SAAX, based on the correction in *LAS* II p516. The arguments in *LAS* II p35 seem sound for all years

			after 678, but in 679 a fractional lunar eclipse and then a nearly full solar eclipse took place (reported in 8502). Parpola, *LAS* II p516 favours the 679 date, based on his observation that the term lú.engar *ikkāru* "farmer", used for the king during the "substitute king ritual", is only attested after solar eclipses. See *LAS* II App. L. However, Parpola's argument is circular. He uses lú.engar to date texts (e.g. *LAS* 137, 138, 167 & 162 = x210–2 & x216) to periods immediately following solar eclipses. Of those texts that refer to the king in this way, only *LAS* texts 25, 77, and 334 (x026, x128, and x381) are unambiguously datable to periods shortly after a solar eclipse. In fact x216 shows clearly that the king referred to himself as lú.engar after a *lunar* eclipse, since in that text a solar eclipse is being sought. Doing so only makes sense following a lunar eclipse. Parpola has to argue that the Scholars were looking for *yet another* solar eclipse, following a solar eclipse/lunar eclipse pair in 669. This seems unnecessarily complicated, and assumes unjustified levels of observational ignorance on the part of the Scholars. Unfortunately, this means that a large number of the Letters dated to 669 by Parpola must remain undatable. See below. This follows inevitably from Parpola's own correction in *LAS* II p516, but is not noted in SAAX.
674	x303	*LAS* 33	Clearly undatable – must be an error in *LAS* II App. I.
—			
673	8041	98	Datable only to the period 674–671. See *ABABR* and 8181.
673	8042	271A	This Report predicts that an eclipse will *not* happen in *elūlu* (VI). Even if this is determined by the observations of the Moon on the 13th, it may have occurred in any year in which an *addāru* (XII) eclipse took place. Since an eclipse *did* occur in VI/673, Parpola is also assuming that the methods used to rule out eclipses based on observations just prior to their possible occurrence had failed! See also *ABABR*.
673	8181	100	Datable only to the period 674–671. See *ABABR* and 8041.
673	8250	272B	Four possible datings pertain; see *ABABR*.
673	8357	211	Dating impossible due to contradictory data. See *ABABR*.
673	8371	151	It is unclear that Mercury's heliacal rising is meant in this Report.
673	8504	223	Mercury is said to have risen in *araḫsamnu* (VIII) into Scorpius. Possible datings other than 673 exist.
673	x114	274	Parpola has tentatively dated this text to the end of *addāru* (XII) by assuming that the "signs of the eclipse" which appeared "in *addāru* and *nisannu* (I)" (r.1f) were other eclipses. This does not provide evidence for a secure dating.
—			
672	8069	178	Also *LAS* 328. The month in question is *a-da-ri* in line 3 corresponding to month I and not XII as Parpola has it. Parpola's method of dating the text is speculative. He notes (n637) that the flood prediction was essentially linked with Jupiter's rising in the path of Anu, (e.g. 8115: r6, *ACh.* Išt. 16: 13). However, this is not strictly true. The corresponding omen in EAE for Jupiter rising in the path of Ea is broken (*ACh.* Išt. 16 15), and not attested elsewhere, to my knowledge. Also, two alternative omens concerning Jupiter's rising in the path of Anu are attested, neither of which seem to concern floods – x362: 14' and *ACh.* 1Supp. 53: 10, 2Supp. 62: 14. This Report remains undatable. In *ABABR* alternative dates are also proposed.

672	8080	256B	Not datable for reasons outlined in *ABABR*. See 8101.
672	8101	235	Insufficient data remain for a secure dating. See *ABABR* and 8080.
672	8182, 8001 & 8043		= *RMA* 256C-258 Because in *RMA* 256B=8080 thunder in *ābu* (V) is also recorded, this hardly provides grounds for dating these texts.
672	825	237	Dated using the intercalary *addāru* (XII_2). This is uncertain, as Hunger notes in SAA8. See *LAS* II p381.
672	8253	225	Also datable to 664 according to *ABABR*, but the reference to the intercalary *addāru* (XII_2) makes Parpola's dating secure, in my opinion. Parpola erroneously leaves it out of *his* chronological list on *LAS* II p423.
672	8381	227	Dating rejected in *ABABR*. Based on the reference in rev.5 to the son of the king (= Esarhaddon) the date 26/XII/669 seems secure, and I have plotted the text accordingly.
672	8452	236A	Parpola has dated this and 8462 on the basis of their similarity to 8101. However, 8101 is not datable. See above. Also, note that in 8452: 6 Mars in Cancer elicits an omen which concerns the king, where 8462's scribe sees nothing ominous in the situation, nor does the author of 8101. This suggests that different occasions were being recorded in 8462 and 8452.
672	8462	236	See 8452 above.
672	8477	163	Dated in *LAS* II p420, but not listed on p423. Several dates comply with the surviving data, however.
672	x005-x007 & 273		= *LAS* 1–3 & 211 are dated by Parpola on the basis that the *adê* in x006: 19 is the famous oath of allegiance to Esarhaddon (Wiseman 1958 etc.). This may be plausible (see his discussion *LAS* II pp1–6), but is not sufficiently certain for our purposes.
672	x009	*LAS* 4	Date is uncertain, even as to the identity of the king.
672	x013, x193 & x191		= *LAS* 7, 140 & 144 These can only be dated to the period 672–669, as noted in SAAX.
672	x187	*LAS* 171	Parpola dates this Letter by connecting it to the securely dated x185 = *LAS* 129, but although probably datable to the period 672–669 any more accurate dating is purely speculative.
672	x188	*LAS* 132	Although written after Assurbanipal had become crown prince and after Esarhaddon's main wife had died, we are less confident than Parpola (*LAS* II p120) in dating this Letter to the period *shortly* thereafter.
672	x233–5	*LAS* 195, 197–8	These Letters may concern the funeral of Esarhaddon's Queen. However, they may not, and cannot therefore be dated.
672	x236	*LAS* 182	Parpola attempts to associate this Letter with *RMA* 257 = 8001 which itself cannot be securely dated (see above). The king is also likely to have had more than one *murṣi šatti* "seasonal illness", which forms the basis of Parpola's association.
672	x238	*LAS* 172	Dated to 672 or 671. Plotted in Table 1 under 672.
—			
671	8045	237	Many datings are possible, as indicated in *ABABR*. Cf. 8081.
671	8081	256B	See *ABABR* for a number of possible datings. Cf. 8045.
671	8140–2 = *LAS* 344–5 & 351		The table provided by Parpola p360 does strongly suggest that the Reports reflect observations made in one of a few groups of years. I feel unable to decide which group to plump for, noting only that another Nabû'a Report, *ABL* 817 = 8130, dates to the time when Assurbanipal was crown prince, as noted by Parpola in *LAS* II p469.
671	8244	162	See year 680.

671	8340	220	As noted in *LAS* II App. J and SAA8 the events described also occurred in 670. The dates are refined in *ABABR*. I have plotted it under year 671.
671	x008	*LAS* 13	I agree with Parpola's long and complex argument in *LAS* II pp16–18, except at the last. In choosing between non-ominous retrogradations of Mars, whilst Saturn is about to retrograde, in the years following the death of the scribal "servant of the king", I cannot see why Parpola has dismissed the situation that pertained in 669 BC. His own figures on p17 show that while Mars was retrograde on Feb.20 (in non-ominous Virgo) Saturn was indeed about to "push itself" *ramanšu ida'ip*. I cannot choose between 671 and 669.
671	x039, x068 & x347		= *LAS* 34, 59 & 278 are all dated by Parpola to early *tamūzu* (IV) 671. X039 and x068 are clearly connected, are dated to the period of the crown prince (672–669), and were written after Esarhaddon's absence from Nineveh and arrival back in *tamūzu*. Parpola maintains that this absence from the capital was that caused by the Egyptian campaign in 671, even though it requires Esarhaddon to have left Egypt long before the main battles, all of which took place in that month – see *LAS* II p64. I find this hard to believe, especially because Esarhaddon's absence from Nineveh could have been prompted by any number of things that did not find their way into the chronicles. Similarly, in order to date x347 to the first eclipse in 671, Parpola has to insist on Esarhaddon's early return from the Egyptian campaign. Perhaps, his first instinct, to date the Letter to the second eclipse in 671, was better (ibid. p267). The year is still 671 for x347, of course.
671	x040, x041, x348 & x353		= *LAS* 57, 58, 276 & 281 all concern a tiara for Nabû. The first three have been dated to 671, and the last to 670 by Parpola *LAS* II p264. The defining limits to the period are set by the mention of Egyptian (booty) in x353: 13 and the reference to the crown prince in x041: 13. Because this offers such a small flexibility of only a year I have decided to consider these texts dated, after Parpola.
671	x045	237	As for x081, except that Nabû-aḫḫe-eriba *is* attested before 674, and it is not clear to me that this Report *does* describe Mars being stationary in Leo.
671	x081	234A	Mars was stationary in Leo in 674 and 671 as *LAS* II App.C3b shows. November 674 predates the otherwise earliest attestation of Balasî by fewer than two years, and cannot be excluded.
671	x168		Datable by the 15/X middle watch eclipse to Dec. 27th 671. This date is missing in SAAX.
671	x359	*LAS* 275	See year 670.
—			
670	8046	274E	Parpola's dating is rejected in *ABABR*.
670	8047	274G	Parpola dates this Report by assuming that certain periods of time between eclipses were known and used by the Scholars. See Ch.4.2.4.3. In this Report the fact that a watch was being made for a solar eclipse at the end of *araḫsamnu* (VIII) suggests that one of these periods of time must have elapsed since an earlier eclipse. None of this is in dispute, but using the simplest cycle, which predicts lunar eclipses mostly every 6, occasionally every 5 months, a lunar eclipse would have been predicted to fall in *kislīmu* (IX) in 679, 678, 670, 669, 661, and 660, as the tables

			prepared in Britton (1993) 65 and *LAS* II App. F readily show. A solar eclipse would then have been predicted for the end of that month and for the end of the preceding one, namely month VIII. Clearly, it is not possible to say for which year the *predictions* were being made. Assuming the (still unknown) methods of predicting eclipse occurrences at this time anticipated every possible eclipse, solar eclipse possibilities occurred in 670, 669, 668 and 667 during the 'known' active life of this Scholar.
670	8141	*LAS* 345	See 8140–2 in year 671.
670	8143	244D	See *ABABR* for why the dating in *LAS* II App. C is unsafe.
670	8162	*LAS* 332	Clearly this Report cannot be dated securely.
670	8254	186	Jupiter rose in May 658, too, which was perhaps month II of the local calendar. I cannot eliminate this possible dating. See 8326 and *ABABR*.
670	8255	86	Venus was standing in front of Orion at the start of *ābu* (V) 678, as well. I cannot see how Parpola can eliminate this possibility so confidently. See also *ABABR*.
670	8326	187A	Datable to a period 12 years later than that proposed by Parpola in *LAS* II, namely to 658. See *ABABR*.
670	8382	85	Apart from the fact that other *kislīmu* (IX) eclipses are attested, this Report is only a prediction based on an EAE omen in line 6. No details of the eclipse are given, so it cannot be seen to be a report. The prediction could, of course, have proved to be false.
670	8486	226	*ABABR* states that other dates are possible for this text.
670	8487	274F	Parpola lists this in *LAS* II App.J.2, giving it a date in 670, when in App.J.1 he tentatively dates it to 667. Hunger suggests it might date to 649 BC. This Report is unusually exact in its prediction; *aiāru* (II) 14th, morning, even that the Moon set while still eclipsed, and so points strongly to May 1st, 649. Hunger's translation phrases the Report as a prediction. However, it appears to me that Nadinu says in line 1 that he had written to the king, before the eclipse had happened, stating as follows – lines 2–3: an.mi gar-*an en-na* [*a-du-ú*] *ul it-te-iq* gar-*an* "an eclipse will take place, [now] it will not pass by, it will take place". I have included line 3 within the inverted commas. This makes better sense, since the rest of the Report can now be seen to record the details of the eclipse *after* it had happened. Nadinu is reminding Assurbanipal that his earlier prediction had been correct. I accept Hunger's dating, as do *ABABR*.
670		*LAS* 317	Deleted.
670	x010	*LAS* 16	As Parpola states, *LAS* II p24, many earthquakes occurred and occur in the geographical region of concern. Dating on the basis of a morning earthquake remains speculative.
670	x042 & x069		= *LAS* 38 & 349 discuss the same matter and must date either to 670 or to 668, as Parpola shows in *LAS* II pp44–5. 670 BC is the likelier of the two choices, based on harvesting considerations, but the later date cannot be excluded. Parpola asserts that *LAS* 326 = 8095 shows that the year 668 can be excluded. This must be a mistake. Perhaps Parpola means *LAS* 324 = 8084, which dates to 668 and includes Jupiter observations?
670	x045 & x071		= *LAS* 41 & 62. Parpola dates these texts (*LAS* II p51) on the basis that solar eclipses were watched for 2 weeks either side of a *predicted* lunar eclipse. The evidence from the Letters implies that such a lunar eclipse

			must have been predicted for *kislīmu* (IX). See the discussion under 8047, year 670.
670	x070	*LAS* 72	Dated by Parpola, but without confidence.
670	x136	*LAS* 94	Only datable to the period of the crown prince.
670	x147	*LAS* 101	Parpola attempts to limit the possible dates of this unassigned Assyrian Letter (*LAS* II p89) by asserting that the presence of clouds (*urpu*) suggested winter. This is hardly reliable. The Letter states very clearly that during the morning watch of the 14th day an eclipse took place. From *LAS* II App. F the possibilities are 14/III/669, 14/VII/667, 14/V/664, and perhaps others after 662. I cannot understand the choices made by Parpola in *LAS* or in SAAX, and note the error in the latter – the December 27th eclipse was in 671 BC.
670	x194–201		= *LAS* 151, 130, 143, 123–4, 133 & 159–60 concern Esarhaddon's illness, and are all datable to the crown prince period of 669–672 BC Parpola, *LAS* II p137 connects this illness to the one related in x315–6 = *LAS* 246–7, the latter being datable on the basis of the plot referred to in lines 19ff – see *LAS* II p238. The connection is a little tenuous, I believe, but because the texts are all connected, and because they can be confidently dated to a narrow band of time, I have accepted Parpola's dating.
670	x202–4		= *LAS* 147–8 & 158. X202–3 are perhaps connected by reference to an earthquake, the last showing that it dates from the period of the crown prince, 672–669. More than this is mere speculation, and Parpola recognises this in *LAS* II p133. X204 is dated to 670 by Parpola on the basis of the spelling of the author's name, see *LAS* II p133. This is insufficiently precise for my purposes.
670	x205	*LAS* 170	Parpola dates this Letter only on the basis of the rest of Adad-šumu-uṣur's correspondence. It cannot be dated from internal evidence. Note that *LAS* 170 appears twice in the list of datable Letters in *LAS* II App. I.
670	x207	*LAS* 145	Parpola's grounds for dating this text more accurately than to the period 672–669 are simply speculative.
670	x241–61		= *LAS* 180–1, 183–4, 186–94, 196, 173–7 & 163–4 – all by Marduk-šakin-šumi. X241–53 concern Esarhaddon illness(es). X253 can be dated to 670 on the basis of the intercalary *elūlu* (VI) in that year (*LAS* II p186). X252: r.9'f describes the completing of statues in month XI and is perhaps connected to x258: r.5f. X254 concerns the entry into Arbela and Cimmerians, and is probably to be dated to 670 (*LAS* II pp192–3). X255–61 all concern rituals connected (Parpola argues *LAS* II p164) to the eclipse of 15/IX/670. Some of the associations between the texts seem tenuous, but in general I agree with the dating. X250 and x251 are not dated in SAAX, and x249 is too fragmentary to associate with the other texts, so I have not plotted them in Table 1.
670	x274	*LAS* 208	This Letter is dated to V/670 during the period of the Queen Mother's illness, as the lines r.7f indicate. The letter is contemporary with x200, and precedes x201, x297, x244, whose dates should consequently be pushed forward a month.
670	x284	*LAS* 213	Parpola expresses doubts as to the date. I agree.
670	x296	*LAS* 218	Only "possibly" 670, as stated in *LAS* II.
670	x306	*LAS* 220	The date of 670 follows from x301–2 and 305 and x196 etc. It is missing in *LAS* II App. I.

670	x317	*LAS* 256	This Letter is too fragmentary to date with confidence to 670. It may well have been written during one of Esarhaddon's illnesses, but also concerns the illness of a female personage, which does *not* connect it with *LAS* 246 = x315, as Parpola suggests in *LAS* II p254.
670	x323–4 = *LAS* 248 & 253		Parpola erroneously dates x324 in *LAS* II App. I to 670, when in the commentary to the Letter he dates it only to the period 672–669, as he does with x323.
670	x338 & 343 = *LAS* 272–3		They are connected through x343: r.1'ff, and Parpola recognises that the dating of the year is uncertain for x338; see *LAS* II p262.
670	x355	*LAS* 282	This Letter was erroneously left out of the list in *LAS* II App. I, since Parpola dated it on p275. Nevertheless, I feel the text is too fragmentary at lines r.2f to be sure of the dating.
670	x358	*LAS* 286	See year 667.
670	x359	*LAS* 275	This Letter concerns Akkad and must date to between V/671 and VIII/669, as Parpola shows in *LAS* II pp275–6. It probably predates x352 = *LAS* 280, and I have chosen to redate it to the end of 671 BC. Note that this text is not listed in *LAS* II App. I.
670	x360	*LAS* 295	This fragmentary Letter may well be connected with BM 135588 (see discussion *LAS* II p303), but the dating of the latter is by no means secure.

—

669	8003	274B	See year 657.
669	8005	206	This Report describes the el of Venus, not the ml as stated in *LAS* II App. I and in SAA8. I am also not sure that the text has been correctly restored. See *ABABR*
669	8048	236H	Mars and Saturn pass by each other approximately every two years and a bit, which means the 669 date proposed by Parpola is uncertain, without the additional information provided in 8049, 8082 or 8102, for example. Cf. 8168–9, 8327, 8383, 8416 & 8491.
669	8050	246G	*ABABR* opt for 669, without excluding a 673 date. I have plotted accordingly.
669	8384	269	For a redating to 657 see *ABABR*.
669	x016–7 *LAS* 21 & 230		I cannot understand how Parpola can argue that these Letters to Esarhaddon's mother can be dated to the period shortly after Esarhaddon death when it states in x016: r.2 "may Mullissu best[ow...] and give long-lasting days, happin[ess] and joy [...] to the king, [the crown prince], and..." Perhaps, this is why he no longer opts to date them in SAAX.
669	x022	*LAS* 19	Parpola does not now offer a date for this text in SAAX.
669	x056	*LAS* 35	The earthquake correlation to x055 is tentative.
669	x137	*LAS* 93	In *LAS* II App. I Parpola lists this text with an asterisk implying that the date of 669–06–10 = 669/III/14 was considered certain. In the commentary on the text, however, he opts for 670/IX/15 as the most likely date for a morning eclipse that occurred during the crown-prince period (based on x136, which shares the same introductory formula – a means of dating which does not inspire excessive confidence). The two possibilities are given in SAAX. Clearly, the date was incorrectly listed as secure in App. I. It is not even clear that x137 does refer to the crown prince period (672–669), and other morning eclipses which occurred on the 14th (App. F) remain possibilities.

669 x190 *LAS* 146 Parpola cannot decide between XI/671 and XI/669 in the commentary to the Letter in *LAS* II p132f. Why does he in App. I?

669 x206 & 262 = *LAS* 233 & 149. Parpola argues that these Letters refer to rituals born of the same celestial event – Mars' brightness. This hypothesis is by no means assured, since Mars inspired šu.íl.la "hand lifting prayers" in many ways (e.g. *LAS* 334, and Koch-Westenholz, 1995 129). This casts doubt on the possibility of dating x262. In addition Parpola asserts that the brightness of Mars referred to in x206: r.1 implies its retrogradation. Certainly, this is when Mars is brightest, but how bright is bright? Perhaps, the king's concern over Mars related not to its being at its brightest, but to it being bright*ish* near Virgo (Spica is not an assured identification, despite *LAS* p222: 10'), which occurred in 673, 671, 669, 668 & 666/665. The dating of both texts is therefore unsure.

669 x209 *LAS* 139 Dates in 669 or 671 are equally possible according to Parpola *LAS* II p126 r.8f.

669 x210-x223 = *LAS* 126, 135–8, 142, 150, 152, 154–6, 162, 166–7 are Letters written by Adad-šumu-uṣur, some with the help of Marduk-šakin-šumi (x216 & x221) and Urad-Ea (x212). Parpola has dated them all to 669 BC, x210-x219 to *simānu* (III) 669. I feel these dates to be far from secure. The attested dates for the three Scholars cover the period 672–667. X219: 7f contains, in a fragmentary state, a reference to the substitute *p*[*u-u-ḫi*] and to a date in *elūlu* (VI). Parpola has read this as "[concerning the su]bstitute about whom the king [my lord, wrote] to me: "I was told [that he should sit (on the throne) until the] xth of *elūlu*". Parpola works back 100 days from this to *aiāru* (II) or *simānu* (III) to find the month of the eclipse that required the installation of a substitute. This allows him to date the Letter to 669. However, it is possible that lines 7f of x219 read differently, perhaps "[that he should sit (on the throne) "from", ta, *is-su,* the] xth of *elūlu*", in which case the Letter should be dated to 674. Letters x213–214 and x217–219 belong together. They all refer to the crown prince's baby. Adad-šumu-uṣur alone wrote them, so it is possible that these lines refer to the *simānu* eclipses of 679/678.

Letters x210–212 all concern the *qirsu*, refer to the king as lú.engar, and no doubt date to the same period. X216 may also be connected through the use of lú.engar, as may x220 and x221 through the solar eclipse that was sought and not found. These six Letters were authored by the three Scholars together. X221 provided Parpola with what he thought was clear evidence that the six should be dated to 669. In x221: 15f it reads: lú.engar *ma-la* 2-*šú e-pu-šú-ni* "which the 'farmer' has (already) done twice" and Parpola argued, p154, that this dated the current substitution to the period after 671. However, as his own note on p516 makes clear the first substitution during Esarhaddon's reign took place in 679, the second in 677, and the third in 674. I see no reason not to date all six Letters to 674 BC. This undermines Parpola's attempted dating on the basis of the introductory blessing (p143), and so makes x223 and x215 undatable.

669 x222 *LAS* 142 This Letter is only datable to the crown prince period (672–669); noted by Parpola in his commentary in *LAS* II and in SAAX.

669	x257–261 = *LAS* 174–7 & 163		depend for their dating on x255 = *LAS* 173. Parpola's argument (see *LAS* II pp163–5) relies on identifying a lunar eclipse as the cause of the rituals alluded to in this group of Letters. I am unable competently to disagree with this interpretation, and since the group of Letters clearly relate to the same subject, and date to the period of the crown prince (672–669), I have dated them all to 670 BC, see above!
669	x304	*LAS* 235	Parpola dates this Letter in *LAS* II App. I and in the commentary p224, but does not do so in SAAX. He notes in *LAS* II p224 that his dating is uncertain, which in any case relies on a connection to x217 etc., whose dating I feel to be insecure.
669	x325	*LAS* 257	This Letter dates to one of the periods when the king was addressed as the lú.engar; *LAS* II pxxiii. It cannot be dated more precisely.
669	x326	*LAS* 153	The dating of this Letter relies on its similarity to x217, the dating of which to 669 I am sceptical.
669	x340	*LAS* 269	The performance of the kettledrum ritual before [Ne]rg[al] and Saturn on the 25th night is no clear demonstration that the ritual corresponded to Mars' and Saturn's conjunction. Parpola's dating is speculative (as he admits).
669	x361	*LAS* 294	Parpola dates this text to 669 on the basis that it refers to a strong flood in r.14, which mirrors the rise of waters described in x364: 12'f, which does definitely date to 669. Parpola admits to the dating being only conjectural.
669	x365	*LAS* 288	This Letter cannot be dated and should not have been assigned a definite year in *LAS* II App. I.
669	x371		The eclipse of IV/671 is mentioned in line 7. Kudurru describes how he dispatched the plant of life of the month IV eclipse in *nisannu* (I) *last year*. This dates the Letter to the year after I/670, i.e. 669 (and before Esarhaddon's death).
669	x378	*LAS* 242	This appears in Parpola's list of dated Letters (App. I), even though no comment is made on its dating in his commentary. None appears in SAAX.
—			
668	8051	209	Dating adjusted only slightly in *ABABR*.
668	8084	188	Several dating possibilities exist. See *ABABR*.
668	8185	211F	Too little remains to date, despite a possible connection with 8051.
668	8323	187	See year 680.
668	8437	222	In 681, 674, 669, 656 and 654 Mercury also rose in *elūlu* (VI) in Leo.
668	x064	*LAS* 330	Too fragmentary.
668	x172		The information provided in lines 4'f makes this Letter datable. Mars rising in V (668, 653), then near Libra (c. 188°–197° = c. 17/X/668), Mercury in Capricorn (c. 264°–288° = c.17/X/668).
—			
667	8054	191	This Report describes Jupiter rising, probably in Leo (an Ellil constellation) in an unknown month. The apodosis in line 2: "the land will experience joy" suggests we can eliminate some months by using what remains of the relevant sections of EAE. This can only be done by accepting a consistency in the versions of EAE used which is, as yet, unjustified. Nevertheless, *ACh.* Išt. 16 & 2Supp. LX, 8254: r1, 8326: 1, x362: 10', 8084: 1, 8289: 1, 8184: 3, 8369: 1 & 8356: 1 perhaps allow us to elimi-

			nate months II, III, IV, VI, VII & VIII, since different apodoses are attested for them. Jupiter rises in Leo in 679 in month V, in 678 in month VI, and in 667 at the very end of month IV (see *LAS* II pp394–5). The 679 date seems the most likely, though this pushes the date of the first Letter or Report sent by Nabû-aḫḫe-eriba back by some 6 or 7 years. Clearly this is not impossible. *ABABR* accept the restoration of month V, but reject 667, though not for very good reasons, and referring to the "sign" of Leo seems unwise. I have not plotted this text in Table 1.
667	8098	251	= *LAS* 325. Parpola's dating in *LAS* II p325–6 and fig.5 relies on the identification of the stars of the head of scorpio, sag.du *ša* mul.gír.tab as β and δ Sorpii. However, lines r.4f clearly state that the Moon stood in front of ^{mul}*ṣur-ru* and ^{mul}li$_9$.si$_4$. The latter has been convincingly identified as Antares, α-Scorpii, (*BPO* II p13 etc.) which does not undermine Parpola's calculations, but does prevent him readily from choosing between 672 and 667. Note also the different translation given by Hunger in SAA8 for lines 7f, from that offered by Parpola in *LAS* II: "tonight a star at the head of Scorpio stood before the Moon". Hunger's translation, "this night a star stood [in] the head of Scorpius in front of the Moon" is suggestive of a planet. It just so happens that on *šabāṭu* (XI) 21, 667 Mars was located next to Antares. Perhaps, this is what is meant in lines 7f. I favour the date 667, but do not feel confident enough to plot it in Table 1. I do not know why this text was not treated in *ABABR*.
667	8144	266B	Jupiter rises in Leo in 679, 655, *and* 667. See *ABABR*.
667	8146	228	This Report only describes Mercury's proximity to Regulus, not Mercury's evening last, as Parpola asserts in *LAS* II App.J. Consequently, the Report is undatable.
667	8385, 8417 & 8328		= *RMA* 250A, 249A & 250 The dating of these texts depends on identifying the fog they record in *šabāṭu* (XI) with that described in 8098. This is in itself dubious, but in any case, I do not feel able confidently to date 8098.
667	8386	247A	The name of the planet in line 1 of this Report is broken. I cannot see how Parpola and Hunger can identify it as Saturn, as also noted in *ABABR*.
667	8388	273	Insufficient information is presented in this Report to date it with confidence, even assuming the prediction to be accurate. An eclipse is predicted for the 14th. The predicted apodoses concern Elam (probably because the eclipse was due to occur on the 14th), and Amurru (which perhaps means the eclipse was due to take place in months III, VII, or XI) – see *LAS* II App.F4.4. Suitable eclipses occurred in 679, 678, 669, 667, 666, 662, and possibly later (see *LAS* II App. F1). If the prediction included the watch during which the eclipse was due to take place, the absence of references to Akkad or Subartu might indicate that the eclipse was due to take place in the morning. This would eliminate all possibilities except the eclipses on 14/III/669 and 14/VII/667. Also, Venus rose heliacally some time before the predicted eclipse (lines 7–8). In 667 this indeed happened in *ābu* (V), and in 669 in *nisannu* (I). Thus, even by using every possible (and suspect) means to try and identify which year is most probable, there is still no way of differentiating between 669 and 667. I hardly think it possible that we can assume the Scholars were capable of predicting which quadrants of the Moon were to be eclipsed in order to date this text, as seems to be implied in *ABABR*.

667	8418	22	Dated by Parpola *LAS* II p420, but not included in App. J.
667	x078	*LAS* 63	Assuming this prediction to have anticipated an actual eclipse, which is in itself debatable, more than one morning eclipse occurred during the possible period of this scribe's activity, some of which fell multiples of 6 or 5 months after other eclipses.
667	x225	*LAS* 119	Only datable to Assurbanipal's reign. The use of the repeated, emphasising adverbial form in the introductory blessing *adanniš adanniš* "very, greatly" may reflect the petitioning nature of this Letter rather than the time at which it was written.
667	x358	*LAS* 286	Parpola's discussion in *LAS* II p282–3 makes it clear that this Letter was written either in 667 or in 670. Parpola prefers the former alternative. I disagree. In r.7'f the eclipse of the Moon was predicted for month VII, rather than for intercalary *elūlu* (VI_2). This fixes the year to either 667 or 670. An eclipse predicted for 670/VII would have fallen 10 months after the previous eclipse in 671/X. An eclipse predicted for 667/VII would have fallen 7 months after the last attested eclipse of 667/I. Only in the first case could the Scholar have confidently stated that no eclipse would occur in VI_2, since 9 months after 671/X was not a multiple of 5 and/or 6 months. I have consequently redated this Letter to 670, with the result that the only piece of evidence that Mar-Issar continued working under Assurbanipal has now evaporated.
—			
666	8007	21	Mars set close to the end or beginning of a month in 681, 675, 668 and 666, and possibly thereafter (see *LAS* II App. C3a). Parpola's choice may be the likeliest, but the other possibilities cannot be excluded. See also *ABABR*.
666	8056	204	Venus disappeared in the morning in *nisannu* (I) in 666 and, of course, eight years earlier in 674 (showing how well the lunar and solar years were synchronised at this time), which is a possible year for this Scholar.
666	8145	205	Venus disappears in month I every 8 years. See *ABABR*.
666	8146	228	There is no evidence that this Report describes the el of Mercury near Regulus, as implied in *LAS* II App. J.
666	8175	207	Dates spaced by 8 years either side of Parpola's proposed date are possible. See *ABABR*. The text is possibly related to 8247, dated to 674, and to the undated 8349, but this is not sufficient evidence upon which to base a secure dating.
666	8419	233	Mars appears in the east just before Sunrise in *elūlu* (VI) in 681 and 666, and possibly thereafter – see *ABABR*.
666	x226–8	*LAS* 121–2 & 125	all concern the same subject and must closely follow x224 = *LAS* 120. Although the date cannot be pinned down precisely, the temporal parameters are closely defined and dating has been accepted.
666	x089	*LAS* 289	It is *not* clear that this Letter is the earlier Letter referred to in x090: r26e.
666	x088	212A	= *LAS* 111. Venus approached Cancer on many occasions during the period Akkullanu was corresponding.
666	x091–2	*LAS* 301 & 308	are only dated by Parpola on orthographic and stylistic grounds to the years around 666. They cannot be dated more precisely.
—			
665	x094	*LAS* 302	The year assigned to this Letter by Parpola is based on stylistic similarities with x090, and on the *absence* of omens. Dating on this basis must remain speculative.

—

664 8093 91 Dated by Neumann (1995), but this dating is criticised in *ABABR*.

664 8147 94 Jupiter was in Scorpio during the time of full Moon on a number of occasions during the period of concern – see *ABABR*.

664 x063 Tentatively dated by Parpola to IX/664.

—

660 x160 This long Babylonian Letter is not dated by Parpola in SAAX. Hunger, who first published the text (1987) suggested that the most likely date was June 660 BC (op.cit. p162). He argued that the observational evidence (lines 11f) pointed to Jupiter's first station in kun.meš = *zibbāti* "tails", which occurred in June 672, 660, and 648 (cf. Jupiter's 12-year period). In fact gub in line 14 does not necessarily refer to a stationary point, and 8254 shows that to "become steady in the morning" (line 11) can mean to rise heliacally. We can therefore interpret this section of the Letter to mean Jupiter's heliacal rising into "the tails". This occurred in March 671, 659, and April 647. This last date is after the Šamaš-šumu-ukīn rebellion, and therefore can be excluded (as Hunger does the 648 date, op.cit. p162). The Babylonian Kudurru in line r.31 is perhaps the same as the sender of x371 (dated to 669) or, perhaps less likely, the sender of x179. Perhaps, the 671 date is favourite?

—

659 8535 *ABCD* 280f. This Report is undated in SAA8. It describes an evening eclipse (with the Moon rising partially eclipsed, l.10) in month III on the 15th. Rev.5 suggests that Jupiter was visible. As with the month III eclipse reported in 8300, 8316 and 8336 (which occurred on the 14th, in the region of Sagittarius, during the evening watch and with Jupiter visible), that of May 22nd 678 seems to be the only possibility. The date given as the 15th in 8535 presents a problem if the 678 eclipse were meant. The 15 is clearly written in line 9, as the photo plate XIV SAA8 shows. Loath as everyone is to resort to the explanation of a scribal slip, this is my only suggestion, and I plot this text in year 678. I do not understand why *ABABR* notes that Jupiter was not visible during the May 678 eclipse. It is noted in 8316: r.4 and 8300: r.12, and the table of retrocalculations in *ABABR* p55 clearly shows a positive altitude for Jupiter at that time.

—

657 8003 274B Also *LAS* 329. In *LAS* Appendices I & O Parpola dates this Report (previously thought to be a Letter) to 657–04–15, despite the fact that in his discussion of this text on page 346 he writes that the eclipse of 669 was much the most likely to have been the one described. Hunger has noted this error in SAA8 and opted for the earlier date, but *ABABR* indicates that 657 is the more likely possibility. Despite a possible connection with 8384, I do not consider this text to be datable.

657 8384 269 See year 669.

657 x101–2 *LAS* 331 & 358 These two texts from the hand of Akkullanu may well date to the 650s, but cannot be dated more precisely.

657 x159 An eclipse predicted for month VIII passed by (l.4f). This suggests that an eclipse occurred $5\frac{1}{2}$ to $6\frac{1}{2}$ months earlier. Possibilities are the lunar eclipses of II/650 or II/649, and the solar eclipses of II/669 or I/657 (*LAS* II Apps. E-F). A solar eclipse is stated to have occurred in line 3, on

the 28th of an unknown month. This unknown month must lie between I and VIII, if the simplest predictive scheme were being used (Ch.4.2.4.3). Suitable solar eclipses were 29/II/669, 28/I/657 and ?/VIII/650. In 650 a lunar eclipse did occur in month VIII, which contradicts lines 4f. In 669 the date of the solar eclipse was the 29th, where as the "8" in line 3 is apparently clear. 657 appears to be the year, and a lunar eclipse did occur in VII. Consequently lines 2 and 3 must be reconstructed differently from SAAX. In line 3 the month must be *nisannu* (I). In line 2 the month referred to is, perhaps, that of the lunar eclipse following, as suggested by line 5. $5^{1}/_{2}$ to $6^{1}/_{2}$ months after VII/657 stretches from the beginning of intercalary *addāru* (XII_2) to the start of month I. I am still not fully confident in dating this fragmentary text to 657, but have plotted it nevertheless.

—

651 x131 *LAS* 83 As Parpola shows this Letter must date to one of three midday solar eclipses in the period 657–648. Note the times in Steele & Stephenson (1997/8) 196. I do not believe that *muṣlalu* "siesta time" < *ṣalālu* "to be at rest/asleep" is sufficiently accurately defined to choose between the three possibilities.

—

650 x105, x132–5, x142 & x345–6 = *LAS* 80–82, 84, 99, 108 & 368–9 have all been tentatively dated to 650 BC by Parpola, since none provide adequate data for greater precision. W.r.t. x134, even accepting the somewhat dubious dating by style to the period 660–640 BC (*LAS* II p84), the record of an evening eclipse does not prove sufficiently accurate to reduce the number of possible dates to one. W.r.t. x142, x136 shows us that Issar-nadin-apli was writing Letters at least as far back as 672–669. Similarly, the author of x105, Akkullanu, is attested long before 650. This makes any precision in attempted dating impossible.

—

649 8487 274F See year 670.

649 x139 *LAS* 97 *Šabāṭu* (XI) 650 is early 649 BC!

—

621 x149 *LAS* 105 I have redated this text to June 2 679 BC. See n65, above. This same eclipse is reported in 8502: r.8.

APPENDIX 3

An Analysis of the Published EAE Planetary Omens

The intention here is to demonstrate that most of the omens found in the official series of EAE – the iškaru – were, at least in part, invented. I have adopted a discursive approach, since a line by line analysis of each omen is beyond the scope of this work. I give some examples of invented omens, and have limited myself to those parts of the series that have been published recently, and which concern the 7 seven planets visible to the naked eye. This constitutes less than one-third of the entire series. For details on these publications see App.1 §21. Ideally this analysis needs to be read in conjunction with these editions, Ch.3.2.1 and Ch.3.2.2.

Tablets 15–22 *(ABCD) = Lunar eclipse omens:*
The schematic nature of the omens in these tablets of EAE is well known. The structure of the protases is outlined in detail by Rochberg in *ABCD* 27–29, and does not need to be repeated here. In EAE 15, for example, omens in which the direction of eclipse impact and clearing are noted are elaborated for each of the four cardinal directions. The associated winds are also given for each of these four directions. In EAE 16 eclipses are described for each month of the year and for days 14, 15, 19 and 20. On the last two dates eclipses *cannot* occur, indicating that the omens were wholly invented. Those omens for eclipses on day 14 follow a standard pattern, including protases for eclipses with the basic colours of white, black, red, and yellow and lastly multi-coloured. Again these protases were invented through metaphoric textual play. In EAE 17, part I, an eclipse occurring in month I, day 14, was assigned the colours red, yellow, dark, and black in order that still more omens could be generated. In EAE 17 part II eclipses occurring on days 14–21 for months I to VI are interpreted. Those on days 16–21 *cannot* occur. EAE 18 continues the patterns of EAE 17 into months VII to intercalary XII. EAE 19 part III utilises days 14–16, 20–21 in months I-IX upon which to interpret a setting, eclipsed Moon. EAE 20 stands alone in this section of the celestial divination series in providing one long omen for day 14 of each month. The apodoses include short "historiettes" that probably contain the remnants of an empirical record. The protases are presented in a systematic manner, but are sufficiently detailed to suggest that they record observations of actual eclipses. It is, however, quite clear that the events described in the apodoses and protases were *not* observed simultaneously, but were simply brought together because the interpretations of the celestial phenomena were *already known* – that is already encoded. This is most clearly indicated in omens §IV and §V wherein the protases and apodoses are separated by an additional line "if in month IV/V the eclipse does not occur according to its count: there will be famine/flood…", which incidentally obeys the encoding of Ch.3.2.2 (xix)[542]. In EAE 20 those apodoses which *cohered* with the decipherment of the protases were appended for the purposes of embellishment, and probably in order to add the illusion that these omens had empirical origins. The decoding of the protases relied on only a small amount of the data presented in each example. The other information presented merely added richness to the text. This is typical of many EAE tablets – see for

[542] See also Rochberg's comments in *ABCD* 176 and p42–43 for some comments on the technical use of *ina la adannišu* and *ina la minātišu* in EAE 15–22.

example EAE 63 below. EAE 21 provides a long omen for day 14 of each month (a "quasi-observational" protasis, according to Rochberg *ABCD* 231), and then includes omens for days 15, 16, 20 and 21. Tablet 22 completes this part of EAE with lunar eclipse omens for each month of the year on days 14–16 and 20–21, and in part II for any day from the 1st to the 30th. They were clearly all invented with little of no empirical input.

In all cases, except for the omens in EAE 20, the apodoses in this part of EAE are general or stock: "downfall of the king of Elam in battle"; "the cattle will perish"; "grain will decrease" etc., and aside from when the textual play rules interact with the code, the eclipses bode ill, particularly for the king of the country designated by the relevant eclipsed quadrant. See Ch.3.2.2 and *ABCD* Ch.4.

Tablets 23–29 *(Van Soldt, 1995)*

23 *Sunrise omens arranged by month:*

The protases describe the Sun's rising on the 1st of each month. Too few apodoses are preserved to try and reconstruct any schemata of the code, but the repetition of "black cloud", "yellow cloud" etc suggests that there was one.

24 *Disk (aš.me) omens (only some describe the Sun):*

Omens I2 – I13 repeat the same protasis changing only the wind in question, and various aspects of the luminosity. They were elaborated *metaphorically*. Omen I5 reverses the prognosis of I4 as the east wind changes to the west in the respective protases. Omen I14 is too simple to have been observed, and was extended to omen I21. Omen IIa is also too simple to have been observed. Omen III8 "if the disk when it rises stands next to the Moon and Venus is visible in front of it at noon: a well known important person will rebel against his lord" was probably derived from associating the disk or Moon with the king and Venus with the rebel. See also omens III5–7, III9. Omen III10 appears to contain a play on words with *kabta* and *qibītu*. Omens III12–18 all concern the presence of varying numbers of stars when the disk rises, and all bode ill in similar ways. Clearly, this interpretation was the result of encoding and not observation. Omens III19–26 pertain to the disk rising during various watches of the night. Most can be explained by the code in Ch.3.2.2 (ix). Omens III28a-44 interpret "normal disks", and then the presence of increasing numbers of disks up to the number 7. Clearly, many of these protases were invented. Omens III45–64 record similar protases for various days of months I to III. Again, their protases were invented according to the rules of textual play. Omen III65 probably records an observation. The remainder is fragmentary, except for omen IV6 "if the Sun rises at an unexpected time and swallows a star: a well-known ruler will be captured", which interprets non-correspondence with the ideal as expected from code Ch.3.2.2 (xix), and puts it in a form in which the physical swallowing of star is syntagmatically (pictorially) linked to the capturing of a ruler.

25 *'Sun' (man) omens, mainly concerned with early rising:*

As with the previous tablet the Sun's (if this is what is meant by man) rising at unusual times bodes ill (Ch.3.2.2. (xix)). This is extended to omens I4–10 in which the 'Sun' rises during one of the three watches of the night. Omens I11-III45 are fragmentary or have no apodoses preserved. The protases in part III1–21differ from each other only in terms of the colour and direction of the Sun's radiance. Those in III22–40 vary the protasis "if the Sun rises early" by the day of the month and by the addition of accompanying meteorological effects.

They were elaborated through the rules of metaphoric textual play. The omens in III45–68 predict eclipses depending on the day on which the Sun rises and winds (with a few other sundry happenings such as halo, stars etc.) Again, non-correspondence with the ideal time of rising bodes ill, but too few apodoses are preserved to determine the precise schemata used to generate them. I suspect that the days in the protases determined the countries to which the eclipses pertained, much as in Ch.3.2.2 (iv).

26 'Normal Sun' (man sag.uš) omens:
The prognosis of the first omen, "if the normal Sun flares up when it rises: there will be obedience and peace in the land", is reversed in the subsequent omens in which the normal Sun rises "red" (no.2), or is "dark" (no.3). Omens I4–41, many of whose protases were invented by analogy, all begin "if the Sun rises...", and describe the Sun being surrounded or accompanied by clouds, stars, and rain. Significance is attached to its turning yellow, red, white and black (the standard colours). Its right side, then its left, are able to turn dark (I25–26), and the behaviour of clouds are also interpreted. Where preserved the apodoses are all brief and formulaic, derive not from observation, but from the standard repertoire of phrases designed to express good or ill fortune. The remaining omens in the tablet also concern Sunrise and the accompanying colours, stars, winds, dust storms, and clouds. Too few apodoses are preserved to determine whether or not any underlying schemata were at work, but many of the protases were clearly invented through metaphoric textual play. Omen IV25 also contains the word play on *Kabta* and *qibit*, noted above and IV7 draws from the Sun's wearing of a crown just like the Moon, the prognosis that the king (the crown wearer) will conquer an enemy's country (just like his own).

27 Atmospheric phenomena at Sunrise and Sunset:
"If the Sun rises and its light is strong: one not of royal descent will be appointed king" was hardly derived from observation of simultaneous events on earth and in heaven, but the metaphor of a new bright day for a change in dynasty seems clear enough. The first omen was invented. The following omens in the first part of this tablet are too broken to analyse. Omen II2 interprets the Sun burning like a flame as indicative of rebellion in the land. Fire caused by civil unrest probably explains the syntagmatic connection between protasis and apodosis. Omen II4 describes the Sun appearing in the afternoon and setting within 2 hours, the Moon being surrounded by a halo which breaks to the east. An eclipse is predicted for the following noon. This (type I – see Ch.3.2.1) omen may well have recorded two observations. As is apparent from tablets 24 and 25, the Sun's appearance at unusual times boded ill, so the occurrence of an ill-boding eclipse would have conformed perfectly with the existing interpretation, and was no doubt included in the series for that reason. Omens II5–7 interpret the Sun's darkness as indicative of a forthcoming eclipse – that is when the Sun is also darkened. Omen II8 links a blood-sprinkled Sunrise with hostilities (no doubt with associated blood-spattering). In II11 the burning Sun again evokes destruction. All these omens demonstrate a link between protasis and apodosis and were invented at least in part – probably completely – without any recourse to specific observation. The remainder of the tablet is fragmentary.

28 *Solar colours, glows (anqullu), haloes arranged by month:*
Omens I1–26 almost all concern the Sun rising and looking as if it has been sprinkled with blood on various days in month I. The apodoses all bode ill and frequently refer to battle, flesh, rebellion, mourning and other horrors which resonate with blood, thereby linking them syntagmatically to the protases. The omens in part II (28–46) repeat the protasis "if the Sunrise is dark in Nisannu" replacing month I with the other months of the year. In this scheme the protasis bodes ill for the king in months I, II, III, IV, VII, VIII, IX, X, XI, XII and XII/2 and well for months V and VI. Omens 47–58 repeat the protasis "if the Sun stands in a glow" for each month of the year, but the apodoses are not sufficiently well-preserved to determine whether or not a scheme similar to the one above pertained. Omens 59–65 are also fragmentary, but those following describe the glow in terms of its looking like canebrake, wild animals, sheep, sediment etc. They invariably bode ill, often in terms of a predicted epidemic. Omens 73–77 concern the significance of a GU – perhaps *qû* "web"- with which the Sun is seen. The apodoses depend on the colour of this GU. The four standard colours and gùn "multi-coloured" are used to generate 5 omens. White, black and yellow GU bode ill, where red and multicoloured ones bode well. Omen 77 is "if the Moon and Sun are looking at each other: the king of the country will increase in wisdom". This of course refers to the day of lunar "opposition" occurring on the ideal, and thus bodes well – see Ch.3.2.2 (xix). Omens 80–83 repeat "if with (the halo) a cloudbank (*nīdu*) lies to the right of the Sun", replacing "right" with "left", "in front of" and "behind", which bode respectively well, ill, well, ill. Clearly binary rules are at play here, or perhaps the *pars hostilis*, *pars familiaris* code noted in §3.2.2 (C). In omens 85–88 the left/right = bad/good code is also used, but "in front of" and "behind" have become encoded with the values Elam and Guti respectively. In the following 4 omens "left" is now assigned to Subartu and "right" to Ešnunna, but "in front of" and "behind" signify well and ill respectively. The four relative orientations have been encoded to signify one of four countries, and either good or bad. This is the simple code outlined in Ch.3.2.2. The remaining omens (94–108) follow the same pattern. In each section of 5 omens an initial omen is embellished by adding to its protasis the line referring to a cloudbank, and locating that cloudbank in each of the four relative orientations, each signifying either a country or good or bad. The tablet ends with the line "If the Sun falls behind in the count of the month; an enemy will devour Akkad[543]", which shows again how non-correspondence with the ideal has been encoded with malefic significance.

29 *Cloudbank phenomena:*
This tablet begins with "if at Sunrise the cloudbanks are normal; the destruction of the country will be hastened", which suggests immediately that the *nīdu* was encoded as ill-boding. This is also borne out by all the following omens in which the cloudbank is systematically described as red, yellow, white and black and is situated to the right, left, in front of, behind, in the path of, and beside the Sun. In omens III45f the number of cloudbanks is extrapolated up to the number 7, and the curvature of the meteorological phenomenon in all four directions is interpreted. The co-occurrence of the cloudbank with a halo still bodes ill, except in so far as the rules interact with the code in order to reverse the expected prognostication simply because one aspect of the description in the protasis has been inverted (see n348). For example omen III66 reads: "if the Sun is surrounded by a halo and a red cloudbank stands

543 I man *ana* šid.meš iti *muṭ-ṭi u* kúr kur uri.ki kú.

to the right: Adad will beat down the crop of the country", where omen III67 reads: "if the Sun is surrounded by a halo and a red cloudbank stands to the *left*: Adad will beat down the crop of the *enemy* country". Omens III68–75 follow the same pattern. The remaining omens (III76–106) all demonstrate the use of comparable metaphoric rules of construction. Frequently, red and white in the protases bode well, whereas black and yellow colours bode ill, the last often described in terms of natural events such as epidemics, famine, Adad's destruction of the crops and so forth – see Ib8, Ib14,III29, III68. This may point to a colour encoding. Certainly, much if not all of this tablet was constructed using the rules of textual play adapted to the original malefic encoding of the *nīdu*.

Tablets 59–62 & 64? *(BPO3) = Venus omens:*[544]

EAE 59–60 (op. cit. 110f) list Venus omens by month. Those for month II contain protases which describe celestial events that cannot happen. These are described in *BPO*3 p24 and do not need to be repeated here. They simply show that those omens were wholly invented. VAT 10218 (p40f – a version of EAE 61 according to K148) omens 6–7 are parallel, by which I mean that the protasis "in front of" in omen 6 is replaced with "behind" in omen 7, and the latter's apodosis is reversed. In omens 13–17 Venus at its rising is described as red, black, white, yellow and red/yellow respectively. When red, it bodes ill, otherwise it bodes ill, though when yellow, famine in Amurru is predicted. Compare the colour scheme in EAE 29, described above. Omens 18–25 all concern "crowns". It is unclear whether or not a code is at work here or if some observational component lies behind the omens. Omens 26–47 also concern Venus's occultation and they all bode ill, obeying code Ch.3.2.2 (xvi). Omens 40–41 are parallel. Omens 51–59 deal with the close approach of Venus and Jupiter. They bode ill. Omens 64 and 65 are parallel, as are 67 and 68, 70 and 71, 73 and 74. Omens 80–87 elaborate on Venus and a *mešḫu* by altering the colour of the latter in the standard manner. Omens 90–111 largely concern the unexpected appearances of Venus, all of which bode ill in line with non-coherence with the ideal boding ill. Omens 124–128 parallel each other with constellations interchanged.

It is possible to analyse the rest of the material published in *BPO*3 in similar fashion. Many of the omen protases were invented by paralleling others in the manner described in Ch.3.2.1, and many apodoses show syntagmatic links with their accompanying protases. Encoding such as that which accompanies planetary absences (Ch.3.2.2 (xvi)) are attested in BM 40111: 15' (p73). Omens such as K.3111: 9 (p91): "If Venus enters into the Moon: Elam will destroy a border city of mine…" rely on the identification of Venus with Elam (code Ch.3.2.2 (xiv)) and interpreting the celestial phenomenon *pictorially* in order to derive a syntagmatic link between protasis and apodosis. Another such link is found in omen K.2226: 11 (p93): "if Venus enters the Sun: the king's son, his father will kill him", based on identifying the Sun with the king (see Ch.2.1 B-names). Lunar absence at the neomenia predicts an eclipse in K.10688: ii1 (p98) – another lunar absence. In the commentary K.35: 9, Venus exceeding her "appointed time" explains an ill-boding prognosis and in line 15 her position being "complete" (*gummura*) bodes well, as expected from the code which interprets correspondence and non-correspondence with the ideal. See also omen VII2 (p135) of K.2097+. A dim Venus bodes ill in EAE 59–60: I1 (p110) and IV16–22 (p128) – see code Ch.3.2.2 (xiii). The basic interpretations of Venus's heliacal rising in the 12 months of the

[544] This material is not presented in *BPO*3 in separate tablets for reasons explained in loc. cit. p1.

year are discussed in the context of EAE 63 omens below. The same apodoses apply and were invented, not observed. This also applies to many of the omens attested in Group D and E manuscripts published in *BPO*3 pp143–97. Venus is referred to both by the name Ištar and Delebat in the protases, and those apodoses which concern women and sex (e.g. K.229: r3, r10, r32; K2153: 10) no doubt reflect Ištar's particular interests. Another impossible protasis is attested in K.7936: 5 (p211). Omens 7–9 of the same text reproduce omens 1–3 with igi "becoming visible" replacing kur "rising", and the apodoses predicting the same consequences albeit written slightly differently (also found in the parallel text K.3601 on pp220f). Again this demonstrates that the decoding preceded the particular way in which the apodosis was expressed. Observation of terrestrial phenomena played no rôle in the creation of these omens. The interpretations of the celestial events, whether seen or not, were "read" directly. Omen K.3601: 23 has an impossible protasis and the reverse of the tablet reveals some of the code underpinning the Venus omens. For example: "if Venus rises in the east; she is female: favourable – if Venus rises in the west; she is male; unfavourable" (r.31–2) and so forth. See also 81–2-4,239: 2f (p253).

Tablet 63 *(BPO1) Venus omens in four parts:*
All the apodoses in this text are formulaic. Only 7 basic types are attested. They are listed in *BPO*1 pp13–14. In part 2 the protases are also invented (according to the scheme outlined in Ch.3.1.2). Their accompanying apodoses rearranged by month are as follows: (Venus appears in the West = (ef) or East = (mf) in month N day M, followed by the apodoses)[545]

I	mf – there will be mourning in the land
II	ef/mf – hostilities in the land
III	mf – downfall of large army
IV	ef/mf – hostilities in the land, land's harvest will prosper
V	mf – rains from the sky, ?
VI	ef/mf – land's harvest will prosper, land will be happy
VII	mf – hostilities in the land, land's harvest will prosper
VIII	ef/mf – hard times in the land/ (VIII 15 ef – land's harvest will prosper – §30)
IX	mf – scarcity of barley and straw in the land
X	mf – land's harvest will prosper, land will be happy; ef – land's harvest will prosper (§§31 & 32)
XI	mf – land's harvest will prosper
XII	ef/mf – king will send king messages of hostility

Consistently (aside from one case in month VIII, perhaps due to the use of day 15) the same apodoses appear *regardless* of the day of the month in which Venus rises. Note also that the apodoses for Venus's appearances in months IV & VII and VI & X are identical, and that the good boding apodoses are found in half of all months, and concentrated in months IV-VII and X-XI. Clearly there was a desire on the part of the redactors to have equal numbers of

545 This arrangement is implicit also in Labat (1965) §104A, p200f who reconstructed what he believed to be a section of *Iqqur īpuš* using an EAE 63 text (K160) since *some* of the above omens appear in the remains of the "Séries Mensuelles" of *Iqqur īpuš* (op.cit. 210f). It is not clear, however, that *all* of this part of EAE 63 was excerpted into *Iqqur īpuš* – see Reiner's comments in *BPO*1 p10. Indeed the borrowing may have been the other way.

good-boding and ill-boding prognoses, and to arrange them more or less evenly into four parts.

In part 1 of EAE 63 (omens 1–21) *every* omen apodosis corresponds to those used in part 2 for Venus's heliacal risings, except for omens 1, 5, 8, 15, 17 & 21. In omens numbers 9 and 11 the apodoses are only very slightly different from those listed above. In number 21 the sending of messages of "peace", rather than "hostility" are predicted, but they are still messages. This apodosis repeats that in omen 1, and thus frames the first part of EAE 63 – a form of literary *metaphoric* learned apodosis invention. In omen 1 the prognosis is positive, just as required by the scheme used in part 2. Omens 8 and 17 both concern Venus rising in month VIII, the apodosis of which is identical to that for month V. Omens 5 and 15 are both month II omens. Thus, it is without doubt that the apodoses for the "observational" part 1 of EAE 63 were added to an empirical record, according to an already established scheme in which the only significant ominous fact was the month in which the planet rose.

In part 3 a new omen for intercalary month VI is attested, the others correspond to those in part 2 except that the contradiction in apodoses for month II in part 1 and 2 is resolved, with both apodoses being given (omen 36). In part 4 the omens are listed again, this time by the month in which Venus *disappears* (*itbal*). In part 2 they were listed by the month in which Venus *appears*. As noted, only the latter determines the prognosis. Hence not one omen in EAE 63 can be explained as being the result of empiricism (see Ch.3.1.1), even though the text does includes what are believed to be the records of a few observations.

I hope to have demonstrated during the course of this brief study of the published planetary omens of EAE that not only were *many* celestial omens either wholly or partially invented, but that the *vast majority* were. It is, in fact, quite hard to find omens which describe events in the human sphere that are specific enough to suggest that they were records of particular one-off events. The bulk of the apodoses describe either ill or propitious happenings in terms of the stock motifs of warfare, weather, harvest, royal death or success, and disease. Often the prognosis is simply described in terms of "evil" or "good" (e.g. EAE 59–60: II10f) and any further elaboration is dispensed with. Sometimes even the most detailed of apodoses are little more than extensive and *invented* elaborations on an existing decipherment, again with no observational component. However, I have also indicated that those events in the human milieu which corroborated the existing decipherment were included in order to add colour, and (I suggest) an illusion of empirical background. A surprisingly large number of the protases were also clearly invented, shown either by their obvious paralleling of other protases, or by the fact that they describe events which cannot happen in nature. Many of the attested protases were, however, the records of observed phenomena. Many of their details indicate that the skies had been studied and their phenomena committed to writing, albeit in broad categories. Not all these details, however, were ominous, and only those for which a decoding existed determined which of the stock apodoses were to be attached in order to form the complete omen. This is perhaps best exemplified by EAE 63 part I.

Bibliography

Aaboe A.H. 1958 "On Babylonian Planetary Theories" *Centaurus* 5 209–77

1964 "On Period Relations in Babylonian Astronomy" *Centaurus* 10 213–31

Aaboe A.H. and D.J. Solla-Price 1964 "Qualitative Measurement in Antiquity. The Derivation of Accurate Parameters from Crude but Crucial Observations" *L'aventure de la science, mélanges Alexandre Koyré* 1–20

Aaboe A.H. and A. Sachs 1966b "Some Dateless Computed Lists of Longitudes of Characteristic Planetary Phenomena from the Late Babylonian Period" *JCS* 20 1–33

Aaboe A.H. and A. Sachs 1969b "Two Lunar Texts of the Achaemenid Period from Babylon" *Centaurus* 14 1–22

Aaboe A.H. 1972 "Remarks on the Theoretical Treatment of Eclipses in Antiquity" *JHA* 3 105–18

1974 "Scientific Astronomy in Antiquity" *Phil.Tran.R.S.Lon.* A276 21–42

Aaboe A.H. and P.J. Huber 1977 "A Text Concerning Subdivision of the Synodic Motion of Venus from Babylon: BM 37151" *Fs.Finkelstein* 1–4

Aaboe A.H. 1978 "Review of *HAMA*" *Isis* 69 441–5

1980 "Observation and Theory in Babylonian Astronomy" *Centaurus* 24 14–35

1991 "Babylonian Mathematics, Astrology, and Astronomy" *CAH 2nd ed.* III/2 276–92

Aaboe A.H., O. Neugebauer, A. Sachs, J. Henderson and J.P. Britton 1991 "Saros Cycle Dates and Related Babylonian Astronomical Texts" *TAPS* 81/6

Al-Rawi F.H. 1989 "The Second Tablet of "Išum and Erra"" *Iraq* 51 111–22

Al-Rawi F.H., J. Black and J. Friberg 1990 "Seeds and Reeds, A Metro-Mathematical Topic Text from Late Babylonian Uruk" *Bagh.Mitt.* 21 481–560

Al-Rawi F.H., H. Hunger and A.R. George 1990 "Tablets from the Sippar Library" *Iraq* 52 149 n.1

Al-Rawi F.H., H. Hunger and A.R. George 1991/2 "Enūma Anu Enlil XIV and Other Early Astronomical Tables" *AfO* 38/9 52–73

Alster B. 1975 "On the Interpretation of "Inanna and Enki"" *ZA* 64 33 n33

Aro J. 1966 "Remarks on the Practice of Extispicy in the Time of Esarhaddon" *CRRA* 14 109–17

Baigent M. 1994 *From the Omens of Babylon. Astrology and Ancient Mesopotamia*

Baolin L. 1991 "On the Length of the Synodic Month" *The Observatory* Vol. 111 No.1100

Baolin L. and F.R. Stephenson 21–22

Barton T. 1994 *Ancient Astrology*

Bauer T. 1936 "Mondlaufprognosen aus der Zeit der ersten babylonischen Dynastie" *ZA* 43 308–14

Beaulieu P-A. 1992 "New Light on Secret Knowledge in Late Babylonian Culture" *ZA* 82 98–111

1993 "The Impact of Month-lengths on the Neo-Babylonian Cultic Calendar" *ZA* 83 66–87

1994 "Rituals for an Eclipse Possibility in the 8th Year of Cyrus" *JCS* 46 73–86

1995 "An Excerpt from a Menology with Reverse Writing" *Sumer* 1–14

Berger P-R. 1993 "Imaginäire Astrologie in spätbabylonische Propaganda" *Graz Vol.* 275–89

Bergmann E. 1969 See Å.W.Sjöberg

Bezold C. 1916 "Die Angaben der babylonishe-assyrische Keilinschriften über farbiger Sterne" *Antike Beobachtungen farbiger Sterne* ed. F.Boll 97–147

Biggs R.D. 1968 "An Esoteric Babylonian Commentary" *RA* 62 51–8

1971 "An Archaic Version of the Kesh Temple Hymn from Tell Abū Ṣalābīkh *ZA* 61 193–207

1985 "The Babylonian Prophecies and the Astrological Traditions of Mesopotamia" *JCS* 37 86–90

1987 "Babylonian Prophecies, Astrology, and a New Source for "Prophecy Text B" " *Fs.Reiner* 1–14
Black J.A. 1984 *Sumerian Grammar in Babylonian Theory*
1989 See Al-Rawi
Black J.A. and A. Green 1992 *Gods, Demons and Symbols of Ancient Mesopotamia*
Black J.A. 1996 See Wiseman
1998 *Reading Sumerian Poetry*
Bobrova L. 1993 From Mesopotamia to Greece: to the Origin of Semitic and Greek Star
Bobrova L. and A. Militarev "Names" *Graz Vol.* 307–27
Boll F. 1911 "Zur babylonischen Planetenordnung" *ZA* 23 372–77
1913 "Neues zur babylonischen Planetenordnung" *ZA* 25 340–51
Borger R. 1956 *Die Inschriften Asarhaddons, Königs von Assyrien AfO* Bh.*9*
1957–8 "Die Inschriften Asarhaddons, Nachträge und Verbesserungen" *AfO* 18 113–8
1959 "Mesopotamien in der Jahren 629–621 v. Chr." *WZKM* 55 62–76
1963 *Babylonisch-assyrische Lesestücke* (2nd ed. 1979)
1967–75 *Handbuch der Keilschriftliteratur* I-III
1971 "Gott Marduk und Gott-König Šulgi als Propheten: Zwei prophetische Texte" *Bi.Or.* 28 3–24
1973 "Keilschrifttexte Verschiedenen Inhalts" *Fs.Böhl* 38–55
1996 *Beiträge zum Inschriftenwerk Assurbanipals: die Prismenklassen A, B, C = K, D, E, F, G, H , J und T sowie andere Inschriften*
Bottéro J. 1950 "Autres textes de Qatna" *RA* 44 105–12
1974 "Symptomes, signes, écritures" *Divination et Rationalité* ed. J.P. Vernant 70–197
1977 "Les noms de marduk, l'écriture et la "logique" en mésopotamie ancienne" *Fs.Finkelstein* 5–28
1978 "Le substitut royal et son sort en Mésopotamie ancienne" *Akkadica* 9 2–24, = Ch.9 of Bottéro (1992)
1985 "Le manuel de l'exorciste et son calendrier" 65–112 *Mythes et rites de Babylone*
1992 *Mesopotamia: Writing, reasoning, and the gods* translated by Z. Bahrani and M. Van de Mieroop from *Mésopotamie: L'écriture, la raison et les dieux* (1987)
Bouché-Leclerq A. 1899 *L'astrologie greque* (repr. 1963) 35–71
Bowen A.C. 1988 "Meton of Athens and Astronomy in the Late Fifth Century BC" *Fs.Sachs*
Bowen A.C. and B.R. Goldstein 39–81
Brack-Bernsen L. 1969 "On the Construction of Column B in System A of the Astronomical Cuneiform Texts" *Centaurus* 14 23–38
1980 "Some Investigations on the Ephemerides of the Babylonian Moon Texts, System A" *Centaurus* 24 36–50
1990 "On the Babylonian Lunar Theory: A Construction of Column Ø from Horizontal Observations" *Centaurus* 33 39–56
1993 "Babylonische Mondtexte: Beobachtung und Theorie" *Graz Vol.* 331–57
Brack-Bernsen L. and O. Schmidt 1994 "On the Foundations of the Babylonian Column Ø: Astronomical Significance of Partial Sums of the Lunar Four" *Centaurus* 37 183–209
Brack-Bernsen L. 1997 *Zur Entstehung der babylonischen Mondtheorie*
Bremner R.W. 1993 "The Shadow Length Table in MUL.APIN" *Graz Vol.* 367–82
Brinkman J.A. 1973 "Sennacherib's Babylonian Problem: An Interpretation" *JCS* 25 89–95
1980–3 "Kudurru" *RlA* VI 268–74
1984 *Prelude to Empire* 747–626 BC
1991 "Babylonia in the Shadow of Assyria (747–626 BC)" *CAH 2nd ed.* III/2 1–70
Britton J.P. 1987 "The Structure and Parameters of Column Ø" *Fs.Aaboe* 23–36
1989 "An Early Function for Eclipse Magnitudes in Babylonian Astronomy" *Centaurus* 32 1–52

1990 "A Tale of Two Cycles: Remarks on Column Ø" *Centaurus* 33 57–69
Britton J.P. and C.B.F. Walker 1991 "A 4th Century Babylonian model for Venus. BM 33552" *Centaurus* 34 97–118
Britton J.P. 1991 See Aaboe
1993 "Scientific Astronomy in Pre-Seleucid Babylon" *Graz Vol.*61–76
1994 See Beaulieu
1996 See Walker
forthcoming "Babylonian Theories of Lunar Anomaly"
Brown D.R. and M. Linssen 1997 "BM 134701 = 1965–10–14,1 and the Hellenistic Period Eclipse Ritual from Uruk" *RA* 91 147–66
Brown D.R., J. Fermor and C.B.F. Walker 1999 *AfO* 46 (forthcoming) "The Water Clock in Mesopotamia"
Brown D.R. forthcoming (a) *Cambridge Archaeological Journal* "The Cuneiform Conception of Celestial Time and Space"
forthcoming (b) "Mesopotamian Contributions to the Philosophy of the Exact Sciences"
Burstein S. 1978 *SANE* 1/5 *the* babyloniaca *of Berossus*
Cagni L. 1969 *L'epopea di Erra*
1977 *The Poem of Erra SANE* 1/3
Campbell-Thompson R. 1900 *The Reports of the Magicians and Astrologers of Nineveh and Babylon in the British Museum* I & II
Chadwick R. 1992 *BSMS* 24 7–24 "Calendars, Ziggurats, and the Stars"
1993 "Identifying Comets and Meteors in Celestial Observation Literature" *Graz Vol.* 161–84
Civil M. and E. Reiner 1974 *MSL* XI *The Series HAR-ra = ḫubullu: Tablets XX-XXIV*
Civil M. 1976 "Lexicography" *Fs.Jacobsen* 123–57
1987 "Feeding Dumuzi's Sheep: The Lexicon as a Source of Literary Inspiration" *Fs.Reiner* 37–55
1994 *The Farmers' Instructions: a Sumerian Agricultural Manual*
Clay A.T. 1912 *PBS* 2/II n123 (see Ungnad, *OLZ* 15 446f)
1923 *BRM* 4 nn.6, 19 & 20 = MLC 1872, 1886 & 1859 (SE - see Ebeling, 1931:24 and Ungnad, 1944)
Cogan M. 1983 "Omens and Ideology in the Babylonian inscription of Esarhaddon" *History, Historiography, and Interpretation: Studies in Biblical and Cuneiform Literature* eds. Tadmor and Weinfeld 76–87
Cohen M.E. 1993 *The Cultic Calendars of the Ancient Near East*
Contenau G. 1940 *La Divination chez les Assyriens et les Babyloniens*
Cooper J.S. 1980 "Apodotic death and the Historicity of "Historical Omens" " *CRRA* 26 99–106
Cornelius F. 1966 "Die Mondfinsternis von Akkad" *CRRA* 14 125–9
Craig J.A. 1899 *Astrological-Astronomical Texts*
Cramer F.H. 1954 *Astrology in Roman Law and Politics*
Dalley S. and J.N.Postgate 1984 *The Tablets from Fort Shalmaneser = CTN* 3
Dalley S. 1989 *Myths from Mesopotamia*
De Meis S. and H. Hunger 1998 *Astronomical Dating of Assyrian and Babylonian Reports = ABABR*
Deller K. 1969 "Die Briefe des Adad-šumu-uṣur" *AOAT* 1 45–64
Denyer N. 1985 "The Case against Divination: An Examination of Cicero's *De Divinatione*" *Proc.Camb.Phil.Soc.* 211 (NS 31)
Derrida J. 1972 *Dissemination* = Derrida 1981
1981 *Dissemination* (trans. B. Johnson)
Dhorme E. 1949 *Les Religions de Babylonie et d'Assyrie* (Chs. IV & X)
Dietrich M. 1967–8 "Neue Quellen zur Geschichte Babyloniens" (I) & (II) *WO* 4 61–103, 183–251
1969–70 "Neue Quellen zur Geschichte Babyloniens" (III) *WO* 5 176–90, Bēl-ibni letters

1970 *Die Aramäer Südbabyloniens in der Sargonidenzeit (700–648) AOAT* 7
1970–1 "Neue Quellen zur Geschichte Babyloniens" (IV) *WO* 6 157–62, Nabû-ušallim letters
1979 *Neo-Babylonian Letters from the Kuyunjik Collection CT* 54
Dietrich M. and O. Loretz 1986 "Astrologische Omina" *Deutung der Zukunft in Briefen, Orakeln und Omina* eds. M. Dietrich, K. Hecker, et al. 94–5
Dietrich M. 1996 "Altbabylonische Omina zur Sonnenfinsternis" *WZKM* 86 99–106
Dijk J. van 1976 *Texts in the Iraq Museum* IX Pl.LXVII (IM 62257 EAE)
Donbaz V. and J. Koch 1995 "Ein Astrolab der Dritten Generation: N.V. 10" *JCS* 47 63–84
Dossin G. 1935 "Prières aux "Dieux de la nuit" " *RA* 32 179–87
1951 "Lettre du devin Asqudum au roi Zimrilim au sujet d'une éclipse de lune" *CRRA* 2 46–8
Driel G. van 1969 *The Cult of Aššur*
Dubberstein W.H. and R.A. Parker 1956 *Babylonian Chronology, 626 B.C - A.D. 75*
Durand J-M. 1994 "Les cieux, premier livre de lecture..." *Les Dossiers d'Archéologie* 191 = *Astrologie en Mésopotamie* 2–7
Dvorak R. and H. Hunger 1981 *Ephemeriden von Sonne, Mond und hellen Planeten von -1000 bis -601*
Ebeling E. 1931 *Tod und Leben nach der Vorstellungen der Babylonier*
1948 "Ein gebet an einen "verfinsterten Gott" aus neuassyrischer Zeit" *Or.NS* 17 416f & Tf.49
Edzard D.O. 1997 *Gudea and His Dynasty* = RIM 3,I 68–101
Eichler B.L. 1987 "Literary Structure in the Laws of Eshnunna" *Fs.Reiner* 71–84
Ellis M. de J. 1989 "Observations on Mesopotamian Oracles and Prophetic Texts" *JCS* 41 127–86
Elman Y. 1975 "Authoritative Oral tradition in Neo-Assyrian Scribal Circles" *JANES* 7 19–32
Englund R. K. 1988 "Administrative Timekeeping in Ancient Mesopotamia" *JESHO* 31 121–85
Eph'al I. 1982 *The Ancient Arabs*
Epping J. Strassmaier 1889 *Astronomisches aus Babylon oder das Wissen der Chaldäer über den gestirten Himmel*
Fales F.M. and J.N. Postgate 1992 SAA 7 *Imperial Administrative Records, Part I*
Fales F.M. and J.N. Postgate 1995 SAA 11 *Imperial Administrative Records, Part II*
Fales F.M. 1975 "L'ideologo Adad-šumu-uṣur" *ANL* 23 453f
Falkenstein A. 1966 " "Wahrsagung" in der sumerische Überlieferung 2. Astrologie" *CRRA* 14 64–5
Farber G. 1987–90 "me" *RlA* 7 610–613
Farber W. 1993 "Zur Orthographie von *EAE* 22: Neue Lesungen und Versuch einer Deutung" *Graz Vol.* 247–57
Fatoohi L.J. and F.R Stephenson 1994 "The Babylonian unit of time" *JHA* 25 99–110
1997/8 "Angular Measurements in Babylonian Astronomy" *AfO* 44/45 210–14
Fermor J. 1999 See Brown
Ferry D. 1990 "Prayer to the Gods of the Night" *Fs.Moran*
Finet A. 1966 "Le place du devin dans la société de Mari" *CRRA* 14 87–93
Finkel I.L. 1996 "Assyrian Hieroglyphs" *ZA* 86 244–68
Finkelstein J.J. and O. Gurney 1957 *The Sultantepe Tablets* 1 = *STT* 1
Finkelstein J.J. 1963 "Mesopotamian Historiography" *PAPS* 107 461–72
Foster B. 1974 "Wisdom and Gods in Ancient Mesopotamia" *Or.NS* 43 344f
Fotheringham J. 1928, S. Langdon and C. Schoch *The Venus tablets of Ammizaduqa*
Foucault M. 1961 *Histoire de la folie à l'âge classique* = Foucault 1967
1967 *Madness and Civilization. A History of Insanity in the Age of Reason* (transl. R. Howard)
1969 *L'archéologie de savoir* = Foucault 1972
1972 *The Archaeology of Knowledge* (transl. A. Sheridan Smith)
Frame G. 1992 *Babylonia 689–627 BC A Political History*
French R. 1994 *Ancient Natural History*
Friberg J. 1987 *Mathematics and Metrology in Near Eastern Texts*

1990 See Al-Rawi
Gadd C.J. 1948 "Ideas of Divine Rule in the Ancient Near East" *The Schweich Lectures of the British Academy*
1966 "Some Babylonian Divinatory Methods and their Interrelations" *CRRA* 14 21–34
1967 "Omens Expressed in Numbers" *JCS* 21 52–63
George A.R. 1986 "Sennacherib and the Tablet of Destinies" *Iraq* 48 133–45
1990 See Al-Rawi
1991 Review of Hunger and Pingree *Mul.Apin* *ZA* 81 301–6
1991/2 See Al-Rawi
Gingerich O. 1963 and W.D. Stahlman *Solar and planetary longitudes for years −2500 to +2000*
Ginzel F.K. 1899 *Spezieller kanon der Sonnen- und Mondfinsternisse für das Ländergebiet der klassischen Altertumwissenschaften und den Zeitraum von 900 vor Chr. bis 600 nach Chr.*
Goldstein B.R. 1988 See Bowen
Goldstine H.H. 1973 *MAPS 94 New and Full Moons from 1001 BC to A.D. 1651*
Goody J. 1977 *The Domestication of the Savage Mind*
1986 *The Logic of Writing and the Organization of Society*
Gössmann F. 1950 *Šumerisches Lexicon 4/2 Planetarium Babylonicum, oder die sumerisch-babylonischen Stern-namen*
Gragg G. 1969 "The Keš Temple Hymn" in Sjöberg and Bergmann (1969) 157f.
Grasshoff G. 1993 "The Babylonian Tradition of Celestial Phenomena and Ptolemy's Fixed Star Catalogue" *Graz Vol.* 95–137
Grayson A.K. 1964 "Akkadian Prophecies" *JCS* 18 7–30
Grayson A.K. and W.G. Lambert 1966 "Divination and the Babylonian Chronicles: A Study of the Rôle which Divination Plays in Ancient Mesopotamian Chronography" *CRRA* 14 69–76
Grayson A.K. 1974–7 "The Empire of Sargon of Akkad" *AfO* 25 56–64
1975a *Assyrian and Babylonian Chronicles TCS* 5
1975b *Babylonian Historical-Literary Texts*
1980–3 "Königslisten und Chroniken" *RlA* 6 86–135
1991a "Assyria: Tiglath-Pileser III to Sargon II (744–705 BC)" *CAH* 2nd *ed.* III/2 71–102
1991b "Assyria: Sennacherib and Esarhaddon (704–669 BC)" *CAH* 2nd *ed.* III/2 103–41
1991c "Assyria 668–635 BC: The Reign of Assurbanipal" *CAH* 2nd *ed.* III/2 142–61
1991d "Assyrian Civilization" *CAH* 2nd *ed.* III/2 194–228
Greenfield J.C. 1982 "Babylonian-Aramaic Relationship" *Nachbarn* 471–82
Greenfield J.C. and M. Sokoloff 1989 "Astrological and Related Omen Texts in Jewish Palestinian Aramaic" *JNES* 48 201–14
Gurney O. 1952 "The Sultantepe Tablets" *Anat.Stud.* 2 25–35
1953 "The Sultantepe Tablets" *Anat.Stud.* 3 15–25
1957 See Finkelstein
Gurney O. and P. Hulin 1964 *The Sultantepe Tablets* II = *STT* 2
Güterbock H.G. 1957/8 "A Hittite Parallel" to Köcher & Oppenheim (1957/8) *KUB* XXIX: 9 col.IV: 4ff eclipse omens, *AfO* 18 78–80
1988 "Bilingual Moon omens from Boğazköy" *Fs.Sachs* 161–73
Hallo W.W. 1957–71 "Gutium" *RlA* 3 708
1963 "On the Antiquity of Sumerian Literature" *JAOS* 83 167–76
1983 "Dating the Mesopotamian Past: The Concept of Eras from Sargon to Nabonassar" *BSMS* 6 7–18
1988 "The Nabonassar Era and other Epochs in Mesopotamian Chronology and Chronography" *Fs.Sachs* 175–90
Harper R.F. 1892–1914 *Assyrian and Babylonian Letters Belonging to the Kouyunjik Collection of the British Museum*

Hartner W. 1962 "The date of the Cimmerian threat against Ashurbanipal according to *ABL* 1391" *JNES* 21 25f

1965 "The Earliest History of the Constellations in the Near East and the Motif of the Lion-Bull Combat" *JNES* 24 1–16

1969 "Eclipse Periods and Thales' Prediction of a Solar Eclipse. Historic Truth and Modern Myth" *Centaurus* 14 60–71

Heimpel W. 1986 "The Sun at Night and the Doors of Heaven in Babylonian Texts" *JCS* 38 127–51

1989 "The Babylonian Background of the Term "Milky Way"" *Fs.Sjoberg* 249–52

Henderson J. 1991 See Aaboe

Hirsch H. 1963 "Die Inschriften der Könige von Agade" *AfO* 20 1–82

Horowitz W. 1993 "The Reverse of the Neo-Assyrian Planisphere *CT* 33 11" *Graz Vol.* 149–60

1994 "Two New *Ziqpu*-Star Texts and Stellar Circles" *JCS* 46 89–98

Horowitz W. and N. Wasserman 1996 "Another Old Babylonian prayer to the Gods of the Night" *JCS* 48 57–60

Horowitz W. and J. Oelsner 1997/8 "The 30-Star-Catalogue HS 1897 and the Late Parallel BM 55502" *AfO* 44/45 176–85

Horowitz W. 1998 *Mesopotamian Cosmic Geography*

Høyrup J. 1993 ""Remarkable Numbers" in Old Babylonian Mathematical Texts: A Note on the Psychology of Numbers" *JNES* 52 281–286

Huber P.J. 1957 "Zur täglichen Bewegung des Jupiter nach babylonischen Texten" *ZA* 52 265–303

1958 "Ueber den Nullpunkt der babylonischen Ekliptic" *Centaurus* 5 192–208

1973 *Babylonian Eclipse Observations 750 BC to 0* (privately circulated)

1974 See van der Waerden

1977 See Aaboe

1982 "Astronomical Dating of Babylon I and Ur III" *Occasional Papers on the Near East* 1/4

1987 "Dating by Lunar Eclipse Omina, with Speculations on the Birth of Omen Astrology" *Fs.Aaboe* 3–13

Hulin P. 1964 See Gurney

Hunger H. 1967 "Kryptographische astrologische Omina" *AOAT* 1 133–45

1968 *Babylonische und assyrische Kolophone AOAT* 2 = Hunger *Kolophone*

1972a "Neues von Nabû-zuqup-kēna" *ZA* 62 99–101

1972b "Die Tafeln des Iqīša" *WdO* 6 21–28

Hunger H. and E. Reiner 1975 "A scheme for intercalary months from Babylonia" *WZKM* 67 21–8

Hunger H. 1975 "Noch ein Kalendartext" *ZA* 65 40–43

1976a *Spätbabylonische Texte aus Uruk* 1 = Hunger *Uruk*

1976b "Astrologische Wettervorhersagen" *ZA* 66 234–60

Hunger H. and C.B.F.Walker 1977 "Zwölfmaldrei" *MDOG* 109 27–34

Hunger H. 1981 See Dvorak

Hunger H. and S. Parpola 1983/4 "Bedeckungen des Planeten Jupiter durch den Mond" *AfO* 29/30 46–9

Hunger H., F.R. Stephenson and K.K.C. Yau 1985 "Records of Halley's Comet on Babylonian Tablets" *Nature* 314 587–92

Hunger H. 1986 "Rechnende Astronomie in Babylonien" (review of *ACT*) *OLZ* 81 229–32

1987 "Emfehlungen an der König" (CT54:57,106) *Fs.Reiner* 157–66

1988 "Eine Sammlung von Merkurbeobachtungen" *Fs.Sachs* 201–33

Hunger H.

Hunger H. and A. Sachs 1988 *Astronomical Diaries and Related Texts from Babylonia Vol. 1 652 B.C to 262 BC = Diaries* I

Hunger H. and A. Sachs 1989 *Astronomical Diaries and Related Texts from Babylonia Vol. 2 261 B.C to 165 BC = Diaries* II

Hunger H. and D. Pingree 1989 *MUL.APIN - An Astronomical Compendium in Cuneiform AfO* Bh.24

Hunger H. 1990 See Al-Rawi

1991 "Schematische Berechnungen des Sonnenwenden" *Bagh.Mitt.*22 513–9

1992 SAA 8 *Astrological Reports to Assyrian Kings*

1993a "Meteor" *RlA* VIII 147–8

1993b "Astronomische Beobachtungen in neubabylonischer Zeit" *Graz Vol.* 139–47

1995 "Ein Kommentar zu Mond-Omina" *Fs.von Soden* 105–118

1996 "Ein astrologishes Zahlenschema" *WZKM* 86 191–6

Hunger H. and A. Sachs 1996 *Astronomical Diaries and Related Texts from Babylonia Vol. 3 164 B.C to 61 BC = Diaries* III

Hunger H. 1998 "Zur Lesung sumerische Zahlwörter" *Fs.Römer* 179–83

1998 See De Meis

Huxley G. 1964 *The Interaction of Greek and Babylonian Astronomy* (New Lecture Series No.16, The Queen's University, Belfast)

Jakobson R. 1990 *On Language*

Jastrow M. Jr. 1898 *Handbook on the History of Religion* 328–466

1912 *Die Religion babyloniens und assyriens* Band II Hälfte ÌII (esp. 415–748 and 996)

Jean C.F. 1935 "Vocabulaire du Louvre, AO 6447" *RA* 32 161–74

Jensen P. 1890 *Die Kosmologie der Babylonier*

Jeremias A. 1929 *Handbuch der altorientalischen Geisteskultur* (2. Aufl)

Jeyes U. 1991–2 "Divination as a Science in Ancient Mesopotamia" *JEOL* 32 23–41

Jones A. 1983 "The Development and Transmission of 248-Day Schemes for Lunar Motion in Ancient Astronomy" *AHES* 29 1–36

1984 "A Greek Saturn Table" *Centaurus* 27 1–36

1990 "Babylonian and Greek Astronomy in a Papyrus Concerning Mars" *Centaurus* 33 97–114

1991a "The Adaptation of Babylonian Methods in Greek Numerical Astronomy" *Isis* 82 441–53

1991b "A Second Century Greek Ephemeris for Venus" *AHES* 41 3–12

1991c "Hipparchus's Computations of Solar Longitudes" *JHA* 22 77–101

1993 "Evidence for Babylonian Schemes in Greek Astronomy" *Graz Vol.* 77–94

Kataja L. and R Whiting 1995 SAA XII *Grants, Decrees and Gifts of the Neo-Assyrian Period*

Kilmer A.D. 1960 "Two New Lists of Key Numbers for Mathematical Operations" *Or.NS* 29 273–308

1987 "The Symbolism of the Flies in the Mesopotamian Flood Myth and some Further Implications" *Fs.Reiner* 175–80

King L.W. 1913 "A Neo-Babylonian Astronomical Treatise in the British Museum and its Bearing on the Age of Babylonian Astronomy" *PSBA* 35 41–46

Kinnier-Wilson J.V. 1972 *The Nimrud Wine Lists = CTN*1

Kirk G., J. Raven and M. Schofield 1957, 83 *The Presocratic Philosophers*

Klein J. 1997 "The Sumerian me as a Concrete Object" *AoF* 24 211–218

Koch J. 1989 *Neue Untersuchungen zur Topographie der babylonischen Fixsternhimmels*

1991 "Der Mardukstern Nēberu" *WO* 22 48–72

1991/2a "Zu einiges astronomisches 'Diaries'" *AfO* 38–9 101–9

1991/2b "Irrungen und Wirrungen einer Rezension" *AfO* 38–9 125–30

1992 "Der Sternkatalog BM 78161" *WO* 23 39–67

1993 "Das Sternbild ^{mul}maš-tab-ba-tur-tur" *Graz Vol.* 185–198

1995 See Donbaz

1996 "AO 6478, MUL.APIN und das 364 Tage Jahr" *NABU* 1996/4 97–99

1997 "Zur Bedeutung von SAG GE_6 in den "Astronomical Diaries" *ZA* 87 33–42

1999 "Die Planeten-Hypsomata in einem babylonischen Sternkatalog" *JNES* 58 19–31

Koch-Westenholz U. and Schaper et al. 1990 "Eine neue interpretation der Kudurru-Symbole" *AHES* 41/2 93–114

Koch-Westenholz U. 1993 "Mesopotamian Astrology at Hattusas" *Graz Vol.* 231–46
1995 *Mesopotamian Astrology An Introduction to Babylonian and Assyrian Celestial Divination*
1999 "The Astrological Commentary *Šumma Sîn ina tāmartīšu* Tablet 1" *Res Orientales* 12 149–64
Kramer S.N., T. Baqir and J. Levy 1948 "Fragments of a Diorite Statue of Kurigalzu in the Iraq Museum" *Sumer* 4 1–38
Kramer S.N. 1961 "New Literary Catalogue from Ur" *RA* 55 169–76
Kudlek M. 1971 and E. Mickler *AOATS* 1 *Solar and Lunar Eclipses of the Ancient Near East from 3000 BC to 0 with Maps*
Kugler F.X. 1900 *Die babylonische Mondrechnung*
1907–24 *Sternkunde und Sterndienst in Babel* I(1907), II(1909–24) = *SSB*. Ergänzungen ÌII(1913–4) = *SSB* Erg. For *SSB* Erg.III see Schaumberger (1935).
Kuhn T.S. 1962 *The Structure of Scientific Revolutions* (2nd ed. 1972)
1970 "Logic of Discovery or Psychology of Research?" and "Reflections on my Critics" *Criticism and the Growth of Knowledge* eds. Lakatos & Musgrave 1–23 and 231–78
Kuhrt A. 1982 "Assyrian and Babylonian Traditions in Classical Authors: A Critical Synthesis" *Nachbarn*
Kuyper J. 1993 "Mesopotamian Astronomy and Astrology as seen by Greek Literature: The Chaldaeans" *Graz Vol.* 135–7
Kwasman T. and S. Parpola 1991 *SAA* 6 *Legal Transactions of the Royal Court of Nineveh*
Labat R. 1933 *Commentaires assyro-babyloniens sur les présages*
1939a *Hémérologies et ménologies d'Assur*
1939b *Le charactère religieux de la royauté assyro-babylonienne*
1957 "Nouveaux textes hémérologiques d'Assur" *MIO* 5 343
1959 "Esarhaddon et la ville de Zaqqap" *RA* 53 113–8
1965 *Un Calendrier babylonien des Travaux, des Signes, et des Mois (séries Iqqur īpuš) = Labat Calendrier*
1975 "Hémérologies" *RlA* 4 317–23
Lacheman E.R. 1937 "An Omen Text from Nuzi" *RA* 34 1–8
Lakatos I. 1976 *Proofs and Refutations* eds. J. Worrall and E. Zapar
1978 *Philosophical Papers* Vol 1. = *The Methodology of Scientific Research Programmes,* eds. J. Worrall and G. Curie
Lambert W.G. 1957 "Ancestors, Authors, and Canonicity" *JCS* 11 6–14 & 112
1957/8 "Two Texts from the Early Part of Reign of Assurbanipal" K4449 II21–24, CBS733+1757 (Taf.XXV) *AfO* 18 382–7
1962 "A Catalogue of Texts and Authors" *JCS* 16 59–77
1964 See Grayson.
1966 "The Tamītu Texts" *CRRA* 14 119–23
1967 "Enmeduranki and Related Matters" *JCS* 21 126–33
1968 "Literary Style in First Millennium Mesopotamia" *JAOS* 88 123–32
1974 *ENŪMA ELIŠ The Babylonian Epic of Creation - The Cuneiform Text* (copied by S.B. Parker)
1976a "Tukulti-Ninurta I and the Assyrian King List" *Iraq* 38 85 n.2 in
1976b "Berossus and Babylonian Eschatology" *Iraq* 38 171–3
1976c "A Late Assyrian Catalogue of Literary and Scholarly Texts" *Fs.Kramer* 313–8
1983 "The God Aššur" *Iraq* 45 82–6
1987 "Babylonian Astrological Omens and Their Stars" (Review of *BPO*1 & 2) *JAOS* 107 93–6
1989 "A Late Babylonian Copy of an Expository Text" *JNES* 48 215–21
1996 "The etymology and meaning of Ṣal*ba*tānu" *NABU* 1996 n° 4 108

Landsberger B. 1915 *Der kultische Kalender der Babylonier und Assyrer*
1965 *Brief des Bischofs von Esagila an könig Asarhaddon*
Lanfranchi G.B. and S. Parpola 1990 SAA 5 *The Correspondence of Sargon II* Part II
Langdon S. 1928 See Fotheringham
1935 *Babylonian Menologies and the Semitic Calendar - The Schweich Lectures of the British Academy* (55 = Ivory Prism)
Largement R. 1957 "Contribution à l'Etude des Astres errants dans l'Astrologie chaldéenne" *ZA* 52 235–66
Laroche E. 1971 *Catalogue des textes hittites*
Larsen M.T. 1987 "The Mesopotamian Lukewarm Mind. Reflections on Science, Divination and Literacy" *Fs.Reiner* 203–25
Leach E. 1970 *Lévi-Strauss*
Leibovici M. 1956a "Un texte astrologique akkadien de Boghazköi" *RA* 50 11–21
1956b "Présages hittites traduits de l'akkadien" *Syria* 33 142–6
1956c "Sur l'astrologie médicale neo-babylonienne" *JA* 275–80
Leichty E. 1970 *The Omen Series Šumma Izbu TCS* 4
1993 "The Origins of Scholarship" *Graz Vol.* 21–9
Lévi-Strauss C. 1962 *La Pensée Sauvage* = Lévi-Strauss 1966
1966 *The Savage Mind* (transl. Weidenfeld and Nicolson Ltd.)
Lewy H. 1956 "Chaldaean Oracles and Theurgy" *Religion and Astrology*
Lieberman S.J. 1987 "A Mesopotamian Background for the so-called *Aggadic* "Measures" of Biblical Hermeneutics" *HUCA* 58 157–225
1990 "Canonical and Official Cuneiform Texts: Towards an Understanding of Assurbanipal's Personal Tablet Collection" *Fs.Moran* 305–36
Limet H. 1982 "Une science "bloquée": le cas de la Mésopotamie ancienne" *Akkadica* 26 17–34
Lindberg D.C. 1992 *The Beginnings of Western Science. The European Scientific Tradition in Philosophical, Religious and Institutional Context 600 BC to A.D. 1450*
Linssen M. 1997 See Brown
Livingstone A. 1986 *Mystical and Mythological Explanatory Works of Assyrian and Babylonian Scholars = MMEW*
1989 SAA 3 *Court Poetry and Literary Miscellanea*
1992 Review of Hunger and Pingree *Mul.Apin*, *Bib.Or.* 49 162–6
1997 "New Dimensions in the Study of Assyrian Religion" *Assyria 1995* eds. S. Parpola, R. M. Whiting 165–77
Lloyd G.E.R. 1979 *Magic, Reason and Experience*
1987 *The Revolutions of Wisdom – Studies in the Claims and Practice of Ancient Greek Science*
1990 *Demystifying Mentalities*
1991 *Methods and Problems in Greek Science*
1992 "Methods and Problems in the History of Ancient Science. The Greek Case" *Isis* 83 564–77
1996 *Adversaries and Authorities – Investigations into ancient Greek and Chinese Science*
Loretz O. 1986 See Dietrich
Luckenbill D.D. 1926&7 *Ancient Records of Assyria and Babylonia* I–II
Magee B. 1973 *Popper*
Mallowan M.E.L. 1966 *Nimrud and its Remains* I–II
Masterman M. 1970 "The Nature of a Paradigm" *Criticism and the Growth of Knowledge* eds. Lakatos & Musgrave 59–89
Mayer W. 1976 *Untersuchungen zur Formensprache der babylonischen "Gebetsbeschwörungen"*
McEwan G. J. 1981 *Priest and Temple in Hellenistic Babylonia*
Meyer L. de 1982 "Deux prières *ikribu* du temps d'Ammī-ṣaduqa" *Fs.Kraus* 271–8
Mikler E. 1971 See Kudlek

Militarev A. 1993 See Bobrova

Moesgaard K.P. 1980 "The Full Moon Serpent. A Foundation Stone of Ancient Astronomy" *Centaurus* 24 51–96

Moortgat A. 1969 *The Art of Ancient Mesopotamia*

Moren S. 1978 *The Omen Series Šumma Alu: a Preliminary Investigation*

Neugebauer O. 1945 "The History of Ancient Astronomy: Problems and Methods" *JNES* 4 1–38

1946 "The History of Ancient Astronomy: Problems and Methods" *Pub.Astron.Soc Pacific* 58 113f

1947a "The Water Clock in Ancient Astronomy" *Isis* 37 37–43

1947b "A Table of Solstices from Uruk" *JCS* 1 143–8

1948 "Solstices and Equinoxes in Babylonian Astronomy during the Seleucid Period" *JCS* 2 209–22

1950a "The Alleged Babylonian Discovery of the Precession of the Equinoxes" *JAOS* 70 1–8

1950b "The Early History of the Astrolabe" *Isis* 40 240–56

Neugebauer O. and A. Sachs 1952–3 "Dodekatemoria in Babylonian Astrology" *AfO* 16 65–6

Neugebauer O. 1955 *Astronomical Cuneiform Texts* I, II and III = *ACT*

1957 *The Exact Sciences in Antiquity* 2nd ed.= *ESA*

1963 "The Survival of Babylonian Methods in the Exact Sciences of Antiquity and the Middle Ages" *PAPS* 107 529–35

1967 "Problems and Methods in Babylonian Mathematical Astronomy" *Ast. J.*72 964–72

Neugebauer O. and A. Sachs 1967 "Some Atypical Astronomical Cuneiform Texts. I" *JCS* 21 183–218

Neugebauer O. and A. Sachs 1969 "Some Atypical Astronomical Cuneiform Texts. II" *JCS* 22 92–111

Neugebauer O. 1975 *A History of Ancient Mathematical Astronomy* I, II and III = *HAMA*

1979 Bibliography of O. Neugebauer *Centaurus* 22 257–80

1983 *Astronomy and History: Selected Essays*

1988 "A Babylonian Lunar Ephemeris from Roman Egypt" *Fs.Sachs* 301–4

1989 "From Assyriology to Renaissance Art" *PAPS* 133 392f

1991 See Aaboe

Neugebauer P.V. and E.Weidner 1915 "Ein astronomischer Beobachtungstext aus dem 37. Jahre Nebukadnezars II (-567/66)" *BSGW* 67 29–89

Neugebauer P.V. 1934 *Spezieller Kanon der Mondfinsternisse für Vorderasien und Ägypten von 3450 bis 1 v. Chr.*

Neumann V.H. and S. Parpola 1987 "Climatic Change and the Eleventh-Tenth Century Eclipse of Assyria and Babylonia" *JNES* 46 161–82

Neumann V.H. 1991–2 "Anmerkungen zu Johannes Koch (1989)" *AfO* 38–9 110–24, 131

1995 "Die Berichte der Astrologen an die assyrischen Könige, ihr astronomischer Inhalt und ihre zeitliche Einordnung" *WZKM* 85 239–64

Nissen H. ed 1993 *Archaic Bookkeeping: Writing and Techniques of Economic Administration in the Ancient Near East*

Norris C. 1987 *Derrida*

Nougayrol J. 1966 "La divination en Mésopotamie ancienne et dans les regions voisins" *CRRA* 14 2–6

Oates J. 1991 "The Fall of Assyria (635–609 BC)" *CAH 2nd ed.* III/2 162–93

O'Neil W.M. 1986 *Early Astronomy from Babylonia to Copernicus*

Oppenheim A.L. 1956 *The Interpretation of Dreams in the Ancient Near East*, *TAPS* 46/iii

1957/8 See Köcher

1959 *Analecta Biblica* 12 282–301

1960 "The City of Assur in 714 BC" *JNES* 19 133–47

1964 *Ancient Mesopotamia* Rev. ed. Reiner 1977 = *AM*

1966 "Perspectives on Mesopotamian Divination" *CRRA* 14 35–8

1969 "Divination and Celestial Observation in the last Assyrian Empire" *Centaurus* 14 97–135
1974 "A Babylonian Diviner's Manual" *JNES* 33 197–220
1978 "Man and Nature in Mesopotamian Civilization" *Dict.Sc.Biog* 15 634–66
Parker R.A. 1956 See Dubberstein
Parpola S. 1970 *Letters from Assyrian Scholars to the Kings Esarhaddon and Assurbanipal* Part I, *AOAT* 5/1 = *LAS* I
1972 "A letter from Šamaš-šumu-ukīn to Esarhaddon" *Iraq* 34 21–34
1979 *Neo-Assyrian Letters from the Kuyunjik Collection, CT* 53
1980 "The Murderer of Sennacherib" (x109) *CRRA* 24 171–82
1981 "Assyrian Royal Inscriptions and Neo-Assyrian Letters" *Assyrian Royal Inscriptions: New Horizons in Literacy, Ideological, and Historical Analysis* ed. F.M. Fales 117–42
1983a *Letters from Assyrian Scholars to the Kings Esarhaddon and Assurbanipal* Part II: *Commentary and Appendices, AOAT* 5/2 = *LAS* II
1983b "Assyrian Library Records" *JNES* 42 1–29
1983/4 See Hunger
1986 "The Royal Archives of Nineveh" *CRRA* 30 223–36
1987 See Neumann
1987a "The Forlorn Scholar" *Fs.Reiner* 257–78
1987b SAA 1 *The Correspondence of Sargon II*, Part 1
Parpola S. and K. Watanabe 1988 *Neo-Assyrian Treaties and Loyalty Oaths* SAA 2
Parpola S. 1990 See Lanfranchi
1991 See Kwasman
1993a "The Assyrian Tree of Life: Tracing the Origins of Jewish Monotheism and Greek Philosophy" *JNES* 52 161–99
1993b "Mesoptamian Astrology and Astronomy as Domains of the Mesopotamian "Wisdom" " *Graz Vol.* 47–59
1993c *Letters from Assyrian and Babylonian Scholars* SAA X
1997 *Assyrian Prophecies* SAA 9
Parrot A. 1955 *Nineveh and the Old Testament*
Pedersén O 1985&6 *Archives and Libraries in the City of Assur—A Survey of the Material from the German Excavations* = *ALCA* I & II
Pettinato G. 1966 *Die Ölwahrsagung bei den Babylonieren*
Pfeiffer R. 1935 *State Letters of Assyria*
Pinches Th.G. 1900 "Review of R. Brown "Researches into the Origins of the Primitive Constellations of the Greeks, Phoenicians and Babylonians" " *JRAS* 571–7
Pinches Th.G., A.Sachs and J.N. Strassmaier 1955 *Late Babylonian Astronomical and Related Texts* = *LBAT*
Pingree D. 1963 "Astronomy and Astrology in India and Iran" *Isis* 54 229–46
1973a "The Mesopotamian Origin of Early Indian Mathematical Astronomy" *JHA* 4 1–12
1973b "Astrology" *Dict.Hist.Ideas* I 118–26
Pingree D. and E. Reiner 1974/7 "A Neo-Babylonian Report on Seasonal Hours" *AfO* 25 50–5
Pingree D. and E. Reiner 1975a "Observational Texts Concerning the Planet Mercury" *RA* 69 175–80
Pingree D. and E. Reiner 1975b "TheVenus Tablet of Ammiṣaduqa" *Bib.Mes.* 2/1 *Babylonian Planetary Omens* = *BPO*1
Pingree D. 1978a The Yavanajtaka of Sphujidhuraja *HOS* 48 2 vols.
1978b "A History of Mathematical Astronomy in India" *Dict.Sc.Biog.* 15 533–633
Pingree D. and E. Reiner 1981 "Enūma Anu Enlil. Tablets 50–51" *Bib.Mes.* 2/2 *BPO*2
Pingree D. 1982 "Mesopotamian Astronomy and Astral Omens in other Civilisations" *Nachbarn* 613–31
1987a "Babylonian Planetary Theory in Sanskrit Omen Texts" *Fs.Aaboe* 91–9

1987b "Venus Omens in India and Babylon" *Fs.Reiner* 293–316
Pingree D. and C.B.F. Walker 1988 "A Babylonian Star-Catalogue: BM 78161" *Fs.Sachs* 313–322
Pingree D. 1989 "MUL.APIN and Vedic Astronomy" *Fs.Sjöberg* 439–45
1989 See Hunger
1992a Review of Koch (1989) *WO* 23
1992b "Mesopotamian Omens in Sanskrit" *La circulation des biens, des personnes et des idées dans la Proche Orient ancien* eds. Charpin D. and Joannes F. 375–9
1992c "Hellenophilia versus the History of Science" *Isis* 83 554–63
1993 "Venus Phenomena in EAE" *Graz Vol.* 259–73
1998 "Legacies in Astronomy and Celestial Omens" *The Legacy of Mesopotamia* ed. S. Dalley Ch. VI 125–137
1998 See Reiner
Pliny the Elder (d. 79 A.D.) *Naturalis Historia*
Porter B.N. 1993 *Images Power Politics. Figurative Aspects of Esarhaddon's Babylonian Policy*
Popper K. 1936 *Logik der Forschung* = Popper 1959
1959 *The Logic of Scientific Discovery*
1963 *Conjectures and Refutations: The Growth of Scientific Knowledge*
1970 "Normal Science and its Dangers" *Criticism and the Growth of Knowledge* eds. Lakatos & Musgrave 51–8
Porada E. 1987 "On the Origins of "Aquarius" " *Fs.Reiner* 279–91
Postgate J.N. 1972–5 "Huzirīna" *RlA* 4 535–6
Postgate J.N. and J. E. Reade 1977–80 "Kalḫu" *RlA* 5 303–23
Powell M. 1987–90 "Masse und Gewichte" *RlA* 7 457–517
Reade J.E. 1977–80 See Postgate
1986a "Archaeology and the Kuyunjik Archives" *CRRA* 30 213–22
1986b "Rassam's Babylonian Collection: the Excavations and the Archives" *Catalogue of the Babylonian Tablets in the British Museum Vol VI: Tablets from Sippar I*, ed. E. Leichty
1996 See Finkel
Reiner E. 1960a "Fortune Telling in Mesopotamia" *JNES* 19 23–35
1960b "Plague Amulets and House Blessings" *JNES* 19 148–55
1961 "The Etiological Myth of the "Seven Sages" " *Or.NS* 30 1–11
1973 "New Light on Some Historical Omens" *Fs.Güterbock* 257–61
1974–7 See Pingree
1975 See Pingree
1981 See Pingree
1982 "Babylonian Birth Prognoses" *ZA* 72 124–38
1985a *Your thwarts in pieces Your mooring rope cut. Poetry from Babylon and Assyria*
1985b "The Uses of Astrology" *JAOS* 105 589–95
1986 "Amulets and Talismans" *Monsters and Demons in the Ancient and Mediaeval World* eds. Farkas A. et.al. 27–36
1990 "Nocturnal Talk" *Fs.Moran* 421–4
1991 "First Millennium Babylonian Literature" *CAH* 2nd *ed.* III/2 293–321
1993 "Two Babylonian Precursors of Astrology" *NABU* 21–2
1995 *Astral Magic in Babylonia = TAPS* 85/4
Reiner E. and D. Pingree 1998 *Babylonian Planetary Omens Part Three = BPO* 3
Reiner E. 1998 "Celestial Omen Tablets and Fragments in the British Museum" *Fs.Borger* 215–302
Riley L. 1994 "The Lunar Velocity Function in System B. First Crescent Ephemerides" *Centaurus* 37 1–51
Roaf M. 1990 *Cultural Atlas of Mesopotamia and the Ancient Near East*

Robson E. 1999 *Mesopotamian Mathematics 2100–1600 BC: Technical Constants in Education and Bureaucracy* = *OECT* 14

Rochberg-Halton F. 1982 "Fate and Divination in Mesopotamia" *AfO* Bh.19 363–71

1983 "Stellar Distances in Early Babylonian Astronomy: A New Perspective on the Hilprecht Text (HS229)" *JNES* 42 209–217

1984a "New Evidence for the History of Astrology" *JNES* 43 115–40

1984b "Canonicity in Cuneiform Texts" *JCS* 36 127–44

1987a "The Assumed 29th *ahû* Tablet of *Enūma Anu Enlil*" *Fs.Reiner* 327–50

1987b "Mixed Traditions in Late Babylonian Astrology" *ZA* 77 207–28

1988a "Elements of the Babylonian Contribution to Hellenistic Astrology" *JAOS* 108 51–62

1988b "Benefic and Malefic Planets in Babylonian Astrology" *Fs.Sachs* 323–28

1988c "Nabû-rimanni" *Great Lives from History: Ancient and Medieval Series* ed. Magill F. 1439–43

1989a *Aspects of Babylonian Celestial Divination, AfO* Bh.22 = *ABCD*

1989b "Babylonian Horoscopes and their Sources" *Or.NS* 58 102–23

1989c "Babylonian Seasonal Hours" *Centaurus* 32 146–70

1991a "The Babylonian Astronomical Diaries" *JAOS* 111 323–32

1991b "Between Observation and Theory in Babylonian Astronomical Texts" *JNES* 50 107–20

Rochberg 1992 "The Culture of Ancient Science: Some Historical Reflections" *Isis* 83 547–53

1993 "The Cultural Locus of Astronomy in Late Babylonia" *Graz Vol.* 31–45

1996 "Personifications and Metaphors in Babylonian Celestial *Omina*" *JAOS* 116.3 475–85

1998 *Babylonian Horoscopes*

Röllig W. 1957–71 "Götterzahlen" *RlA* 3 499–500

Römer W.H. 1986 "Die Menologie des Astrolabs B" *Deutung der Zukunft in Briefen, Orakeln und Omina* eds. Dietrich, Hecker et al. 48–53

Sachs A. 1948 "A Classification of Babylonian Astronomical Tablets of the Seleucid Period" *JCS* 2 271–90

1952a "Babylonian Horoscopes" *JCS* 6 49–75

1952b "Sirius Dates in Babylonian Astronomical Texts of the Seleucid Period" *JCS* 6 105–14

1952c "A Late Babylonian Star Catalog" *JCS* 6 146–50

1952–3 See Neugebauer

1955 See Pinches

1956 See Neugebauer

1967 "La naissance de l'astrologie horoscopique en Babylonie" *Archeologia* 15 Mars-Avril 12–19

1967 See Neugebauer

1969 See Neugebauer

1969 See Aaboe

1970 "Absolute dating from Mesopotamian Records" *Phil.Tran.R.S.Lon.* A269 19–22

1974 "Babylonian Observational Astronomy" *Phil.Tran.R.S.Lon.* A276 43–50

1976 "The Latest Datable Cuneiform Texts" *Fs.Kramer* 379–98

Sachs A. and C.B.F. Walker 1984 "Kepler's View of the Star of Bethlehem and the Babylonian Almanac for 7/6 BC" *Iraq* 46 43–55

Sachs A. 1988 See Hunger

1989 See Hunger

1991 See Aaboe

Saggs H.W.F. 1955–74 "The Nimrud Letters, 1952" Parts I–IX *Iraq* 17, 18, 20, 21, 25, 27, 28 & 36

1978 *The Encounter with the Divine in Mesopotamia and Israel* (Jordan Lectures 1976)

Sayce A.H. 1874 "Astronomy and Astrology of the Babylonians" *TSBA* 3 145–339

Sallaberger W. 1993 *Der kultische Kalendar der Ur III-Zeit*

Schaumberger J. 1935 *Ergänzungsheft* III to Kugler's *SSB*

Schaumberger J. and A. Schott 1938 "Die Konjunction von Mars und Saturn im Frühjahr 669 v.Chr. nach Thompson Reports Nr.88 und anderen Texten" *ZA* 44 271f
Schaumberger J. and A. Schott 1941/2 "Vier Briefe Mâr-Ištars an Asarhaddon über Himmelserscheinungen der Jahre –670/668" *ZA* 47 89–130
Schaumberger J. 1949 "Die Mondfinsternisse der Dritten Dynastie von Ur" *ZA* 49 50–8
1952 "Die *Ziqpu*-Gestirne nach neuen Keilschrifttexten" *ZA* 50 214–29
1954–6 "Astronomische Untersuchung der "historische" Mondfinsternisse in Enûma Anu Enlil" *AfO* 17 89–92
1955 "Anaphora und Aufgangskalendar in neuen Ziqpu-Texten" *ZA* 51 237–51
Schmidt O. 1994 See Brack-Bernsen
Schoch C. 1928 See Fotheringham
Schott A. 1936 "Marduk und sein Stern" *ZA* 43 124–45
1938 "Nabû-aḫḫē-erība, der Astrologe mit den SilbenleSungen" *ZA* 44 194–200
1938 See Schaumberger
1941/2 See Schaumberger
Sivin N. 1969 "Cosmos and Computation in Early Chinese Mathematical Astronomy" *T'oung Pao* LV 1–73
Sjöberg Å.W. 1960 *Der Mondgott Nanna-Suen in der sumerischen Überlieferung*
Sjöberg Å.W. and E. Bergmann 1969 *The Collection of Sumerian temple Hymns*
Slotsky A. 1993 "The Uruk Solstice Scheme Revisited" *Graz Vol.* 359–65
Smith S. 1969 "Babylonian Time Reckoning" *Iraq* 31 74–81
Soden W. von 1936 "Schwer zugängliche russiche Veröffentlichungen altbabylonischer Texte 1/ Ein Opferschaugebet bei Nacht" *ZA* 43 305–8
1954 *Herrscher im alten Orient*
1965 "Leistung und Grenze sumerische und babylonischer Wissenschaft" *Wissenschaftliche Buchgesellschaft Darmstadt* (Orig. 1936 *Die Welt als Geschichte* 2 411f)
1966, 68 & 77 "Aramäische Wörter in neuassyrische und neu- und spätbabylonischen Texten. Ein Vorbericht I, II & III" *Or.*35, 37 & 46
Sokoloff M. 1989 See Greenfield
Soldt W.H. van 1995 *Solar Omens of Enūma Anu Enlil: Tablets 23 (24) - 29 (30)*
Solla-Price D.J. 1964 See Aaboe
Spalinger A. 1994 "Calendrical Comments" *Bib.Or.* 51 5–20
Spek R.J. van der 1993 "The Astronomical Diaries as a Source of Achaemenid and Seleucid History" *Bib.Or.* 50 91–101
Stahlman W.D. 1963 See Gingerich
Starr I. 1985 "Historical Omens of Assurbanipal and Šamaš" *AfO* 32 60–7
1986 "The place of the Historical Omens in the System of Apodoses" *Bib.Or.* 43 628–42
1989 *The Rituals of the Diviner Bib.Mes.* 12
1990 *Queries to the SunGod* SAA 4
Steele J. 1997/8 "Canon of Solar and Lunar Eclipses for Babylon: 750 BC – AD 1" *AfO* 44/45
Steele J. and F.R. Stephenson 195–209
Steinkeller P. 1995/6 Review of Englund, Nissen & Damerow *ATU* 3 *AfO* 42/43 212f
Stephenson F.R. 1974 "Late Babylonian Observation of 'lunar sixes' " *Phil.Tran.R.S.Lon.* A276 118–21
1985 See Hunger
1991 See Baolin
1994 See Fatoohi
1997/8 See Fatoohi
1997/8 See Steele
Stieglitz R.R. 1981 "The Hebrew Names of the Seven Planets" *JNES* 40 135–7

Strassmaier J.N. 1955 See Pinches
Streck M. 1916 *Assurbanipal und die letzten assyrischen Könige bis zum Untergang Nineveh's*, *VAB* 7
Swerdlow N. 1993 "Otto E. Neugebauer" *PAPS* 137 139–65
1998 *The Babylonian Theory of the Planets*
1999 ed. *Ancient Astronomy and Celestial Divination*
Šilejko V.K. 1921 "A Tablet of the Prayer to the Gods of the Night in the Lichacev Collection" *Izvestija Rossijskoj Akaemii Istorii Material'noj Kul'tury* 3 144–52 and Pl.VIII (in Russian, see Dossin 1935)
1927 *Acad.Sci.URSS* 125–8 (in Russian, see Bauer 1936)
Tadmor H. 1982 "The Aramaization of Assyria: Aspects of Western Impact" *Nachbarn* 449–70
Tigay J. 1983 "An Early Technique of Aggadic Exegesis" *History, Historiography and Interpretation: Studies in Biblical and Cuneiform Literatures* 169–89
Thureau-Dangin F. 1912 *Une Relation de la Huitième Campagne de Sargon* (2nd ed. 1973)
1913 "Distances entre étoiles fixes d'après une tablette de l'époque Séleucides" *RA* 10 215–25
1914 "L'exaltation d'Ištar" *RA* 11 141–58
1919 "Appendice" AO 2163A/B *RA* 16 171
1922 *Tablettes d'Uruk à l'usage des prêtres du temple d'Anu au temps des Seleucides* (= *TCL* 6) 11–30 and 41
1931a "Les mesures angulaires ammatu et ûbanu" *RA* 28 23–5
1931b "La tablette astronomique de Nippur" *RA* 28 85–8
1931c "Mesures de temps et mesures angulaires dans l'astronomie babylonienne" *RA* 28 111–4
1932 "La clepsydre chez les Babyloniens" *RA* 29 133–6
1936 Review of *SSB* Erg.3 *RA* 33 197–8
Toomer G.J. 1984 *Ptolemy's Almagest*
1988 "Hipparchus and Babylonian Astronomy" *Fs.Sachs* 353–62
Tuckerman B. 1964 *Planetary, Lunar and Solar positions, 601 BC to A.D. 1*
Tuman V. 1986 "Immortality etched in stone. Dating Ancient Mesopotamian Stones from Astronomical Symbols" *Griffith Observer* 50/1 10–19
1987 "Astrochronology versus Brinkman's Historical Chronology" *CRRA* 34 *Abstract* 52–3
1993 "Astronomical Dating of Observed and Recorded Events in the Astrolabe VR 46" *Graz Vol.* 199–209
Unger E. 1938 "Dûr-Šarrukên" *RlA* 2 249–252
Ungnad A. 1944 "Besprechungskunst und Astrologie in Babylonien" *AfO* 14 251–84
Veenhof K.R. (ed.) 1986 *Cuneiform Archives and Libraries* = *CRRA* 30 and in this volume 1–36 "Cuneiform Archives. An Introduction"
Virolleaud C. 1902 "Présages tirés des éclipses de soleil et de l'obscurissement du soleil ou de ciel (par les nuages)" *ZA* 16 201–39
1905–12 *L'astrologie chaldéenne, le livre intitulé "enuma (Anu) iluBêl", publié, transcrit et traduit*[sic] fascicules 1–14 = *ACh.*
1910a "De Quelques Textes Divinatoires" *Babyloniaca* 3 133–5, 137–40, 197–200, Pl.X 287ff
1910b "Fragments Astrologiques" *Babyloniaca* 3 268–86, 301–3
1911 *Babyloniaca* 4 109–16, 120–6
1912 "Etudes Astrologiques" *Babyloniaca* 6 115–28, 251–61
1922 "Les Origines de l'Astrologie" *Babyloniaca* 7 99–104
Waerden B.L. van der 1941 "Babylonian Astronomy I. The Venus Tablets of Ammiṣaduqa" *JEOL* 10 414–24
1949 "Babylonian Astronomy II. The Thirty-Six Stars" *JNES* 8 6–26
1951 "Babylonian Astronomy III. The Earliest Astronomical Computations" *JNES* 10 20–34
1952/3 "The History of the Zodiac" *AfO* 16 216–30
1957 "Babylonische Planetrechnung" *Vierteljahrsschrift d. Nat. Ges. Zürich* 102 39–60

1963 "Das Alter der babylonische Mondrechnung" *AfO* 20 97–102
1968 "The Date of the Invention of Babylonian Planetary Theory" *AHES* 5 70–8
Waerden B.L. van der and P.J. Huber 1974 *Science Awakening 2 The Birth of Astronomy = BA*
Waerden B.L. van der 1978 *Dict.Sc.Biog* 15 667–80 "Mathematics and Astronomy in Mesopotamia"
Walker C.B.F. 1977 See Hunger
1982 "Episodes in the History of Babylonian Astronomy" *BSMS* 5 10–26
1983 "The Myth of Girra and Elamatum" *AS* 33 145–52
1984 See Sachs
1984 "Notes on the Venus Tablet of Ammiṣaduqa" *JCS* 36 64–6
1985 "Halley's Comet in Babylonia" *Nature* 314 576–7
1987 "Halley's Comet in Cuneiform: The First Recorded Observation in Babylonia" *BSMS* 13 1–20
1988 See Pingree
1991 See Britton
1993 "Bibliography of Mesopotamian Astronomy and Astrology" *Graz Vol.* 407–49
1996 *Astronomy before the Telescope* (ed.) and pp 42–67 "Astronomy and Astrology in Mesopotamia" with J. Britton
1997 "Achaemenid Chronology and the Babylonian Sources" *Mesopotamia and Iran in the Persian Period: Conquest and Imperialism 539–331 BC* ed. J. Curtis
1999 "Babylonian Observations of Saturn during the reign of Kandalanu" forthcoming in Swerdlow (1999)
1999 See Brown
Wasserman N. 1996 See Horowitz
Watanabe K. 1988 See Parpola
Waterman L. 1930–6 *Royal Correspondence of the Assyrian Empire* I–IV
Weidner E. 1911a "Beiträge zur Babylonischen Astronomie" *Beiträge zur Assyriologie* 8 1f
1911b "Zur babylonischen Astronomie" *Babyloniaca* 4 162f
1912a "Zur babylonischen Astronomie" *Babyloniaca* 6 8f, 65f,134f
1912b "Der dreißigste Tag" *ZA* 27 385–8
1915 *Handbuch der babylonische Astronomie*
1919 "Babylonische Hypsomatabilder" *OLZ* 22 10–16
1923/4 "Ein babylonisches Kompendium der Himmelskunde" *AJSL* 40 186f
1925 "Ein astrologische Kommentar aus Uruk" *StOr* 1 347–58f
1928/9 "Historisches Material in der babylonischen Omina-Literatur" *MAOG* 4 231, 236
1941/4a "Die astrologische Serie Enûma Anu Enlil" *AfO* 14 172–95 and 308–18
1941/4b "Der Tag des Stadtgottes" *AfO* 14 340–2
1952/3 "Die Bibliothek Tiglathpileser I" *AfO* 16 197–215
1957/71a "Fixsterne" *RlA* III 72–82
1957/71b "Geheimschrift" *RlA* III 185–8
1954/6 "Die astrologische Serie Enûma Anu Enlil" *AfO* 17 71–98
1959–60 "Ein astrologische Sammeltext aus der Sargonidzeit" *AfO* 19 105–13
1966 "Ein Omenkommentar des Nabû-zuqup-kênu" *AfO* 21 46
1967 *Gestirn Darstellungen auf babylonische Tontafeln, Österreiche Akademie der Wissenschaften Phil-Hist Klasse Band 254 Abh.2*
1968–9 "Die astrologische Serie Enūma Anu Enlil" *AfO* 22 65–75
Weiher E. von 1971 *Der babylonische Gott Nergal*
Weissbach F.H. 1903 *Babylonische Miscellen*
Weitemeyer M. 1956 "Archive and Library Technique in Ancient Mesopotamia" *Libri* 56/3 217–38
Westenholz J.G. 1997 *The Legends of the Kings of Akkade – The Texts*
Whiting R. 1995 See Kataja

Wilhelm G. 1989 *The Hurrians* (Trans. J. Barnes)
Wiseman D.J. 1950 "The Nimrud Tablets" *Iraq* 12 184–200
1953 *The Alalakh Tablets* 114 and pls. 52–4, Nos. 451–2
1955 ND 3579 and 3557 *Iraq* 17 6f
"The Vassal Treaties of Essarhaddon" *Iraq* 20 1–99
1968 "The Nabû temple texts from Nimrud" *JNES* 27 248–50
1988 "A Note on Some Prices in Late Babylonian Astronomical Diaries" *Fs.Sachs* 363–73
1991 "Babylonia 605–539 BC" *CAH* 2nd *ed.* III/2 229–51
Wiseman D.J. and J.A. Black 1996 *Literary Texts from the Temple of Nabû CTN*4
Yau K.K.C. 1985 See Hunger
Zadok R. 1986 "Archives from Nippur in the first Millenium B.C." *CRRA* 30 278–88
Zawadzki S. 1994 Review of SAA 8 *ZA* 84 308–10

Indices

Index of Cuneiform Texts

ABL 209 p38
ABL 965 p39
ABL 1406 p256 (App.1 §22)
ABRT 1 30 p64
ACh. Ištar 4 n184
ACh. Ištar 16 pp267, 274 (App.2 Years 672, 667)
ACh. Ištar 19 p134
ACh. Ištar 39 p257 (App.1 §26)
ACh. Sîn 1 p255 (App.1 §21)
ACh. Sîn 3 pp200–3, nn51, 354
ACh. Sîn 22 p140
ACh. 1Supp.53 p267 (App.2 Year 672)
ACh. 2Supp.19 pp140–1, n304
ACh. 2Supp.57 pp137, 140
ACh. 2Supp.58 p137
ACh. 2Supp.60 p274 (App.2 Year 667)
ACh. 2Supp.62 p267 (App.2 Year 672)
ACh. 2Supp.69 p137
ACT 135 pp127, 226–7
ACT 180 n482
ACT 200 n432
ACT 310 n400
ACT 600 p8
ACT 654 n400
ACT 655 n400
AO 6478 p259 (App.1 §33), n305
AO 6769 p250 (App.1 §11)
Assur 10145 p252 (App.1 §15)
Astrolabe B pp37, 65, 221, 251–4, 257–8 (esp. App.1 §16), nn293, 318
Atraḫasīs pp234, 250 (App.1 §11)
BM 3715 p262 (App.1 §43)
BM 17175+ pp3, 128–9, 251 (App.1 §8), n3
BM 22696 see *ABCD* p20 "OB Text A"
BM 26472 p248 (App.1 §6)
BM 29371 p127, nn322, 428
BM 33552 p262 (App.1 §43)
BM 34162 p263 (App.1 §44)
BM 36722 n407, see "Text K"
BM 36731 pp98, 193, 198, 261 (App.1 §38), nn306, 422
BM 36746 p264, n389
BM 37266 see "Text F"
BM 37467 n437, see "Mercury Records"
BM 40094 p263 (App.1 §44)
BM 41004 n408, see "Text E"
BM 45728 pp176, 178, nn412, 414
BM 46083 p177
BM 76738+ p261 (App.1 §41), see "Saturn Records"
BM 77054 p259 (App.1 §30)
BM 77242 p260 (App.1 §33)
BM 78161 pp98, 260 (App.1 §34), n248
BM 78903 n47
BM 86381 see *ABCD* p20 "OB Text B"
BM 92685 n52, see DT 78
BM 97210 p248 (App.1 §6)
BM 121034 p252 (App.1 §15)
Boghazköy Star-List p257 (App.1 §26)
BOR 4, 132 pp222–4
Borger Ass.A. pp23, 25 (Table1) nn42, 355
Borger Nin.A. n42
BPO2 III:7a p138
BPO2 III:28 n318
BPO2 IV:5a p145
BPO2 VI:5–5a p145
BPO2 IX:8,13 p151
BPO2 XV:24 p67
BPO2 XVI:4 p108
BRM IV 6 nn48, 380
BRM IV 19–20 pp22, 259 (App.1 §29), 264
Cambyses 400 p262 (App.1 §43), n384
Catalogue of Texts and Authors pp35–6, 46, 108, 222, 242, n142
CBS 574 p250 (App.1 §11)
CBS 733+ p23, n42
CBS 1471 n75, see "x295"
CT 22 1 nn57, 59
CT 13 31, see K7067
CT 31 40 n320
CT 33 9 p258 (App.1 §26)
CT 33 11 p253 (App.1 §17)
CT 46 55 p254 (App.1 §20)
CT 49 144 pp223–4
CTN 2 246 n110
CTN 4 10 p255 (App.1 §21), n71

Diary -651 pp55–6, 98–101, 191, 193, 259, n249
Diary -567 pp4, 55–6, 98, 100–1, 166, 191, 193, 199–200, 262 (App.1§43), nn459, 465
Diary -418 pp 55, 100–1, 192
Diary -382 pp 100–1, 192
Diary -366 n439
Diary -346 p192
Diary -321 p103
DT 72+ pp 16, 19, 193–5, 198, 227, 229, 258 (App.1 §29)
DT 78 p194, n52, see DT 72+
EAE 1 pp235, 255 (App.1 §21)
EAE 8 p255 (App.1 §21)
EAE 14 pp98, 109, 114–5, 125, 128–9, 149, 158, 164, 166, 188, 217, 249 (App.1 §8), 252 (App.1 §16), 254 (App.1 §21), nn5, 71, 289–90, 387, 428, 457
EAE 15 pp252, 279
EAE 16 pp140–1, 279
EAE 17 p279, n339
EAE 18 p279, n339
EAE 19 pp 142, 279, nn339, 389
EAE 20 pp133, 140–1, 254 (App.1 §21), 279–80, n389
EAE 21 p280, n339
EAE 22 pp235, 252 (App.1 §14), 254 (App.1 §21), n339
EAE 23 p280
EAE 24 p280
EAE 25 pp280–1
EAE 26 p281
EAE 27 p281
EAE 28 p282
EAE 29 pp282–3
EAE 50 pp108, 115, 125, 138
EAE 51 pp67, 115, 150, 252 (App.1§16), 254 (App.1 §21), n228
EAE 55 p254 (App.1 §21)
EAE 56 pp118, 125, 128, 150, 254 (App.1 §21), nn234–5, n301
EAE 59–60 p283
EAE 61 pp66, 151, 256 (App.1 §21), 283, n190
EAE 62 p283
EAE 63 = *Venus Tablet of Ammiṣaduqa* pp114–5, 118, 123, 125, 150, 154, 249–50 (App.1 §9), 254-6 (App.1 §21), 284–5, nn235, 545
EAE 64 pp118, 255 (App.1 §21), 283–4, n300
Proto-EAE pp107, 126–7, 248 (App.1 §7), 254
Enki and the World Order pp236–7
Enūma Eliš pp64, 124–5, 128, 138, 235–6, 253–4 (esp. App.1 §19), nn14, 179
Erra and Išum pp13, 23, 256 (App.1 §24), n144
*Etana 250 (*App.1 §11)
Gilgameš p250 (App.1 §11), n511
Gu Text see BM 78161
Gudea Cylinder A pp123, 247 (App.1 §3), n514
GSL pp59–60, 62,66, 140–1, 144, 257–8, nn170, 191
HS 229 = HS 245
HS 245 pp124, 129, 253–4 (esp. App.1 §17), n28
HS 1897 p251 (App.1 §13)
HSM 1490 pp259, 261 (App.1 §§32, 41), see "Mars Records"
IM 62257 p140
IM 80213–4 p250 (App.1 §11)
i.NAM.giš.ḫur.an.ki.a pp107, 109, 114, 120, 125, 127–9, 164, 166, 227, 236, 260 (Ap. 1§35), nn30, 306
Inana and Enki n521
Iqqur īpuš pp107, 109, 114, 125, 254, 260 (App.1 §35), nn232, 545
K35 pp142, 283
K148 pp151–2, 283, n190
K250+ p258, nn28, 170, see "*GSL*"
K2164+ p114, see *i.NAM.giš.ḫur.an.ki.a*
K2226 pp143, 155, 283, n251
K2248 pp35, 38, 46
K2313 p254 (App.1 §20)
K2346+ p66,143, 145, n300
K2486+ p46, n145
K3111+ p283, n338
K3357+ n319
K3601 p284, n338
K4364+ = *BBR* 24 p127, n319
K5981 255 (App.1 §21)
K6153 n437, see "Mercury Records"
K7067 = *CT* 13 31 p254 (App.1 §20)
K9794 p259 (App.1 §33)
K10817+ 11118 p254 (App.1 §20)
K11151 p259 (App.1 §29)
K11867 p255 (App.1 §21)
KAR 44 p34, n69
KAR 307 (=SAA3039) pp112, 236, 257 (App.1 §24), n179
KAR 366 p252 (App.1 §15)
KAR 421 n43
KAV 218 pp37, 252 (App.1 §16), see "*Astrolabe B*"
Keš Temple Hymn pp246–7 (App.1 §3f)

King of Battle n41
KUB 4 47 pp250–1 (App.1 §§11, 13)
KUB 4 63 pp248, 251 (App.1 §§5, 13)
KUB 34 12 p251 (App.1 §13)
LBAT 1197–1201 p236 (App.1 §46)
LBAT 1285 p5
LBAT 1366 p190
LBAT 1386 pp261–2
LBAT 1393 p262
LBAT 1411 p262
LBAT 1413 p9, 98, 175, 190, 205, 259, Table 1
LBAT 1414 pp190, 201, 205, 259, nn384, 425
LBAT 1415+ pp190, 201, 259, n425
LBAT 1416 pp190, 259
LBAT 1417 pp5, 190, 259, n7
LBAT 1419 pp190, 259, n425
LBAT 1420 p190
LBAT 1436 p190
LBAT 1499 p257
LBAT 1521–77 p255
LBAT 1578–80 p264
LBAT 1586–7 p264
LBAT 1596 p264
LBAT 1604–5 p264
LKA 29 n365
Mars Records pp97, 102, 262 (App.1 §41)
Mercury Records pp97–100, 102, 163, 191–2, 259, 262 (App.1 §§32, 41)
MLC 1885, see BRM IV 19
MLC 2190 n401
MLC 2195 n392
Mul.Apin pp20, 22, 30, 57, 6–62, 65, 71 , 92, 94, 96, 98, 103, 107, 109, 114–21, 123–5, 128–9, 137, 143, 149–50, 156–8, 162, 164, 166, 169, 195, 213, 217, 247, 249, 253–4, 257–8, 259 (App.1 §30), 260, nn3, 14, 28, 30, 168, 173, 185, 190, 235, 272, 290, 294, 306, 367, 445, 449, 457
Myth of Girra and Elamatum p250 (App.1 §11)
ND 5427 p38
Ni 1856 p252 (App.1 §15)
Nimrūd or *Sargonid Ivory Prism* p259 (App.1 §31), n457
Ninurta and the Turtle p236
Nippur Star List p250 (App.1 §10)
NV 10 p258 (App.1 §26)
OB Coefficient Lists pp114, 249 (App.1 §8), 251, 254
OB Text A p142
OB Text B p142
OB Water Clock Texts p251 (App.1 §12)
OECT 4 161 p122
OECT 6 74–75 pl.XII p250 (App.1 §11)
OIP 2 94 pp123, 236
PBS II/2, 123 p 252 (App.1 §15)
Prophecy Texts A-D n43
Rm 150 n321
Rm 2,38 p140
Rm 2,303 n437, see "Mercury Records"
Rm 2,361 n437, see "Mercury Records"
SAA2006 p30
2249 p98
SAA3001 p112, 257 (App.1 §24), n126
3002 pp55, 65, 257 (App.1 §24)
3025 p257 (App.1 §24)
3032 p57, n192
3034–5 nn49, 159
3039 pp23, 38, 48, 57, 112, 257 (App.1 §24), n190, see *KAR* 307
SAA4326 n154
SAA7001 p35, nn97, 154
7049–56 p35
*SAA8001 n86
8004 pp50, 60, 89–90, 133–4, 142, 145, 148, 202, n229, 465
8005 pp50, 61, 135
8006 pp89
8007 pp86, 148
8011 pp86, 147
8014 pp86, 147
8015 pp88, 147–8
8019 pp30, 258
8020 p135
8022–3 p34
8025 pp78, 88
8027 pp87, 149
8029 pp66, 138
8039 pp57, 61, 93
8040 pp62, 89, n228
8041 pp62, 92
8042 p201
8046 pp88, 201
8048 pp63, 91
8049 pp59, 62, 91–2, 133, 145, n200

* Reports and Letters discussed in App.2 are not listed here, since the dating provided for each text in those volumes provides sufficient information to locate my comments in that appendix.

8050 pp63, 86, 94
8051 pp55, 57, 92, 145, nn226, 229
8052 p197
8055 pp60, 62–3, 67
8056 p87
8060 p198
8063 p147
8064 pp58, 62, 90, 92
8067 p201
8069 p58
8072 pp59, 63, 150
8073 pp60–1, 71, 90, 93, 134, n173
8074 p62
8080 p92
8081 p93, n200
8082 pp57, 59, 61, 71, 88, 91, 96, 148, n436
8083 p147, n534
8084 p92
8086 p147–8
8087 p201
8088 pp88, 148, n236
8091 pp88, 148
8093 pp57–8, 86, 102, 149, 187
8095 pp57–8, 144
8098 pp150–1, 197, n452
8100 pp88, 144
8101 pp58, 88, 91
8102 pp57, 70, 91
8103 pp2, 89–90, 135
8104 pp90, 94
8107 p133
8110 pp57, 88, 147
8112 pp39, 135
8113 p57, n228
8114 pp56, 58, 62, 79, 86, 91, 142–4
8115 pp60–1, 86, 92, 133, n229
8118 p89
8121 p6
8125 pp58, 145
8136 n156
8140–2 pp102, 176, 197
8145 pp87, 149, 197
8146 n228
8147 pp57–8, 133, n206
8154 p57
8157 pp57, 86, 93, n223
8158 pp57, 59, 133, 144, 158
8160–3 p34
8166 pp57, 88, 144
8167 pp87, 149
8168 pp57, 89, 133
8170 pp60–2, 86, 92, n229
8173 p88, n236
8174 p141
8175 pp62, 86, 92, 96, n209
8177 pp88, 141, 148
8180 pp57, 91
8181 pp34, 133
8184 p86, nn184, 223, 229
8185 n229
8188 p89
8189 p89
8191 p86
8192 p86
8205 p62
8207 n450
8210 p89
8211 p60
8212 pp58, 91, 145
8213 p40
8214 pp58, 145
8218 n229
8219 pp59, 105
8231–6 n272
8232 n86
8237 n270
8244 pp57, 91, 145
8245 pp80, 93, n228
8246 p197
8247 pp62, 87, 150, n209
8250 pp201, 203
8251 pp149, 201
8252 p60
8253 p93
8254 pp55, 58, 86, 90, 137, n229
8257 p198
8259 p89
8266 pp39, 198, n156
8271 p198
8274 pp60, 86, n229
8275 p141
8278 p135
8279 p202
8281 pp60, 87, 150
8284 pp56, 63
8288 pp56–60, 91, 142, n200
8288–91 p40
8290 pp86, 146
8293 pp88, 147, 198
8294 pp89

8295 pp88, 148, n236
8296 p92
8297 pp57–8
8300 pp90, 140–1, 168
8301 p57
8302 pp60, 144
8304 pp86, 133
8306 pp88, 147
8308 p91
8309 p40
8311 pp57, 59, 89, 144
8316 pp29–30, 89–90, 141
8317 pp57, 89
8320 pp6, 78, 197–9, 201
8321 p201
8323 pp39, 58
8324 pp88, 92
8325 pp39, 59–60, 93, n172
8326 pp57, 60, 91–2
8329 pp87, 150
8334 pp40, 59
8336 pp90, 142
8337 pp 80, 93
8338 pp50–1, 122
8339 pp59, 87, 150
8341 pp57–8, 93
8344 p201
8346 pp87, 149, 166, 201, n460
8350 pp57, 89
8351 p92
8356–8 p40
8357 pp59, 62, 86, 92, n223
8363 p93, 202
8365 p96
8369 p93
8370 n229
8371 p93, n228
8376 p63
8377 p89
8378 pp89, 136–7
8380 p92
8382 p201, n460
8383 pp57, 60, 71, 80, 89, 144
8384 pp28, 41, 90
8385 p96
8386 p88
8387 pp28, 41, 88, 94, 191
8388 pp204, 240
8391 p102, 147, 149
8398 p58
8400 p92
8403 p87
8403–7 p39
8408 p92
8410 p198
8411 p88
8412 pp60, 89, 93, 144
8413 p89
8414 p61, nn119, 209
8415 pp63, 92
8416 p57
8416–7 p40
8418 pp28, 41, n236
8419 p60
8426 p58
8430 p93
8431 pp89
8437 n228
8438 pp58, 88, 144
8443 p92, 150
8445 p39
8447 pp39, 201
8448 pp39, 91
8449–53 p40
8452 pp58–9, 92
8454 pp57, 92
8455 pp39, 92
8456 pp39, 59, 86–7, 150, n171
8457 pp86, 147–9
8458 pp88, 147, n236
8461 p55
8463 p40
8466 p93
8469 p199
8472 p39
8474 p148
8481 pp40, 88
8486 pp57, 86
8487 pp28, 41, 90
8489 pp62, 90, 92, n436
8491 pp40, 56, 60, 63, 79, 86, 90, 144
8499 pp33, 40
8500 pp92, 94, 96, n436
8501 pp28, 58, n62
8502 pp45, 56, 59, 87, 93, 145, 205–6, n228
8503 pp57–8
8504 n228
8506 pp86–7
8507 pp62–3
8513 n226

8516 p199
8517 p40
8521 p58
8536 pp62, 67
8537 p59
8538 pp62, 67, 143
8541 pp56, 91–2, 145, n229
8544 p61
8545 n228
8546 pp60, 71, n62
8547 pp61, 69, n229
SAA9002 n45
9008 n46
SAAX001 n108
x006 n272
x008 pp87–8, 94, 145, 197
x010 p34
x013 p39
x020 p30
x023 p197, n534
x024 p46
x026 pp201, 203
x030 n343
x031 p197
x033 p47, n62
x038 pp47, 242
x039 p49, n103
x042 p196
x043 pp58, 65
x044 nn314, 368
x045 pp203, 241
x046 p201
x047 pp91, 96, n436
x048 p90, nn208, 314, 454, 458
x050 p197, n534
x051 pp56, 58, 197, n534
x052 pp56, 241, n229
x053 n368
x056 p46, 47, n471
x057 p89
x059 n226
x060 p157
x062 p30, n272
x063 pp63, 92
x064 pp56, 92
x067 p91, n228
x071 pp200, 203
x072 pp87, 197, 240, nn223, 534
x073 p56
x074 pp56, 90, 144, 197
x075 p89
x076 p19
x078 pp203–4
x079 pp86, 91
x084 n436
x088 p62
x090 p89
x093 p39
x094 pp88, 148
x095 p41
x096 n151
x096–8 p41
x097 n151
x099 p41
x100 pp79, 81, 86–7, 91, 102, 143, 149–50, 187, 191–2, n310
x102 p48
x104 pp88, 91, 135
x105 pp88, 148
x107 p41
x109 pp19, 28, 30, 44, 92, 102, 136, n119
x110 p41
x111 pp46, 134, n118
x112 pp34, 157–8, n118
x113 pp45, 57, 89
x114 pp22, 40, 201
x118 p40, nn118, 147
x132 p200
x133 pp200, 259
x134 pp39, 98, 259
x135 p200
x136–42 p39
x137 p19
x143 pp39, 48
x147 p202
x148 pp90, n253
x149 pp27, 90, 98, 190, 259, nn65, 253
x154 n154
x155 p40
x159 p200
x160 pp34–5, 47, 49, 61–2, 93, 97, 225, nn147, 154
x163 pp40, 45, 47
x164 p45
x166 p47
x167 p39
x168–9 p39
x169 p40, n118
x170 pp201, 241
x171 p48

x172 pp41, 61, 93–4, 96, 191
x173 p47
x176 pp41, 45, n154
x177 n154
x182 nn114, 117
x192 p47
x199 p46
x205 n154
x206 p46
x216 p201
x220 p200
x221 n114
x224 pp37, 200, 204
x225 pp102, 149
x226 pp37, 38, n146
x227 p37, n146
x232 n108
x240 p204
x250 n62
x251 p38
x253 p151, n452
x257 p38
x270 p45
x279 p39
x291 p38
x294 pp37–9, 45, 47–8, n104
x295 (= CBS 1471) p40, n75
x297 n108
x347 pp19, 22, 203, n48
x349 p41
x351 pp202, 204, 242
x353 p41
x354 p41
x358 pp204–5
x359 p41
x362 pp58, 86–7, 92, 94, 150, 191, 197
x363 pp200, 204, n367
x364 p102
x371 pp39–40
x377 pp45, 201
SAAXI124 n98
xi156 n102
Sargon's 8th Campaign pp14, 20, 23, 56, 158, 256 (App.1 §23), nn41–2, 95
Saros Canon pp190, 201–2, 259, 261 (App.1 §§32, 39), n425,
Saturn Records pp29, 97–100, 102, 163, 191–2, 259, 262 (App.1 §§32, 41)
SH 81-7-6,135, see BM 45728, p262 (App.1 §41), n412
Sm.2189 p254 (App.1 §21)
SpTU 94 p180
SpTU 4 159 p264
STC II 49 p255 (App.1 §21)
STT 73 pp21–22
STT 300 p22, 259 (App.1 §29)
STT 329–339 p22
STT 340 p260 (App.1 §33)
Šamaš Hymn p250 (App.1 §11)
Šumma ālu pp34, 254 (App.1 §21), n270
Šumma Izbu pp34, 157, 254 (App.1 §21), n30
Šumma Sîn n51, see *ACh.* Sîn3
"Text A" pp186, 262 (App.1 §43)
"Text C" p262 (App.1 §43)
"Text E" pp176, 178, 194, 199, 227, nn408, 412–4, 420, 445
"Text F" p262 (App.1 §43), n403
"Text G" p262 (App.1 §43)
"Text H" p262 (App.1 §43)
"Text K" pp200, 262 (App.1 §43), n407
"Text L" pp186, 262 (App.1 §43)
"Text M" p262 (App.1 §43)
"Text S" pp186, 261–2 (App.1 §§39, 43), n431
The Babylonian Diviner's Manual pp107–8, 120–2, 125, 151, 195, 260 (App.1 §36), n310
The Exaltation of Ištar pp236, 254 (App.1 §20), see TCL 6 51
The Farmer's Almanac p246 (App.1 §1)
The Prayer to the Gods of the Night pp35, 124, 250–1, 258 (App.1 §§11, 13, 26), n47
The 12 Names of the Marduk Planet pp59, 61–2, 65
TCL 6 9/12–14/19/20 p264
TCL 6 21, see AO 6478
TCL 6 51 p254 (App.1 §20)
TU 11 pp174, 199, 216, 262 (App.1 §43), 264
TU 14 p261 (App.1 §39)
TU 17 p255 (App.1 §21)
UET 1 300 p247 (App.1 §3)
Ur_5-ra XX-XXIV forerunner p250 (App.1 §10)
VAB 4 270 p263 (App.1 §47)
VAS 24 120 p252 (App.1 §16)
VAT 7814 p255 (App.1 §21)
VAT 7815/6 p264
VAT 7827 p255 (App.1 §21)
VAT 7847/51 p264
VAT 9740+11670 pp 252, 254 (App.1 §§15, 21)
VAT 9803 p252 (App.1 §15)
VAT 10218 pp144–5, 283, n386
VR46 pp62, 64, n178

VR64 p66
YOS 1 39 p264 (App.1 §47)
81–6-25,136, see DT 72+
81–7-27,81 = *ACh. Išt.* 39
83–1-18,2434 p264 (App.1 §47)

Index of Akkadian and Sumerian Words

adannu / *edānu* pp87, 142, 150–1, 197, 276, nn367, 542
adāru pp92, 256, nn229, 364
adê pp13, 268
aga – see *agû*
agû = aga, p89, n226
aḫû pp158, 258, nn30, 373
ana igi gub p91, n237
an.dùl – see *ṣalmu*
an.ta-*nu* – see *šaplānu*
gišan.ti.bal – see *ṣaddum*
ár p99, n249
arki p99
a.rá – see *arû*
arû = a.rá, pp46, 127–9, 138, 26–7, 237, nn145, 323–4
ašar / *bīt niṣirti* pp68, 145, n193
baʾālu pp80, 100, n229
bēru pp98, 250, 257, 259–60, n365
bibbu = udu.idim pp57, 60 , 68, 256, nn191–2
biblu / *biblāni* pp121, 260, n272
bīt ṭuppāti n53
darû p151& n360
dib – see *etēqu* / *etāqu*
dibbū n514
didabû p48, n151
dim_4 – see *sanāqu*
e p99
egirtu p12, n33
elēnu = ki.ta-*nu*, p99
elû = nim(-a), pp101, 192
lúengar – see *ikkāru*
epāšu / *epēšu* n86
ér n7
erēbu p92
eṣēru p112, 125, nn282, 315
ešēru p88, n236
etēqu / *etāqu* = dib, pp86, 91, 92, 99, 201, n237
ezi p88, n236
gerginakku = im.gú.la, n53
gi_6-*ma* – see *ṣalimma*
gi_6(.zal) n387
gi.na – see *kânu*
giskim – see *ittu*
gub – see *uzuzzu*
gùb pp99, 192
gummuru pp49, 157, 283
g̃arza - see *parṣū* n520
gišgal – see *manzāzu*
g̃iš.ḫur(.meš) = *uṣurtu* / *gišḫurru*, pp235–7, 246, nn518, 522, 525–6
ikkāru = lúengar p267
im.gú.la – see *gerginakku*
ina dal.ban p99
ina egir p99
ina igi p99, n249
ina šēreti ikūn n223
iškaru pp75, 158, 258, 279 n30
ittananbiṭu – see *nabāṭu*
itti alāku p91, n237
ittu = giskim, pp235, 247, nn279, 514
kaiamānu p56, 68–9, 77 n444
kakkabū minâti = mulšid.meš, nn250–1
kânu / *kīnāta* = gi.na (-*ta*) pp69, 86, nn220, 223, 367, 444
kaqquru n462
kašādu = kur, pp59, 91, n237
ki – see *qaqqaru*
kibrāt arbāʾi, kibrāt erbetti = ub.da.límmu.ba n345
ki.gub - see *manzāzu*
ki.ta(-*nu*) – see *elēnu*
kittu pp69, 77
kunsaggû p253
kur - see *napāḫu* or *kašādu*
kùš (= *ammatu*) pp5, 92, 96, 98, 217
la p30
lā n86
leʾu p30
leʾû p49
liginnu p258
liqtu p258
lúengar- see *ikkāru*
lumāšu n284
manzāzu / *nanzazu* = ki.gub / gišgal pp55, 65, 69, 89, 149, 235, nn41, 146, 179, 194, 223, 386, 515
mašāḫu pp100, 191–2, 223, 256
mašʾaltu p258
mātu n12

me (gal.gal.la) (≈ *parṣū*) pp235–7, 239, 247, 255, nn523, 525
minâtu pp86, 148, 200, 202, nn367, 465, 542
mitḫuru p91, n237
mí.urì – see *niṣirtu*
^{mul}šid.meš – see *kakkabū minâti*
mišiḫiti – see *mašāḫu*
mukallimtu pp255, 257–8, nn51, 85
murruru p49
muṣlālu p278
muššuḫ - see *mašāḫu*
na for NA n386
nabāṭu pp92–3, n229
nam (≈ *šimtu*) n526
namāru n229
napāḫu = kur, pp58, 86
nēḫu p88
nīdu pp86, 150, 282, n221
nim(-a) – see *elû*
niṣirtu = mí.urì p227
parṣū = g̃arza, n520
pišru n514
qaqqaru = ki, n65
qerēbu p92, n237
rēḫtu n139
rēš našû n147
rikis girri p258
saḫāru p87, n237
sanāqu = dim$_4$, pp92, 99
simanu pp88, 148, 259, n367
ṣaddum = ^{giš}an.ti.bal, nn279, 514
ṣalimma n228
ṣalmu = an.dùl, pp69–70, n363
ṣâtu / *ṣâti* pp46, 127–8, 134, 138, 157, 255, 258, nn145, 319–21
ṣubbû pp49, 216, n243
šamallû = (^{lú}šamán.(mál).lá) p48
^{lú}*šaniu* p48
ša pî ummâni p89, 158–9, 258, nn30, 373
šaplānu / *šap* p99
šaqû pp102, 192
šašû p49
šaṭāru p48
še.er.zi (= *šarūru*) pp61, 66, 79, 90, n229
šiknu nn343, 526
šitqulu pp88, 95, 119, 121, 197, n304
šiṭir burūmê p112, nn283, 315
šiṭir šamê p112, n284
šumu n514
šu.si / si (= *ubānu*) pp5, 90, 96, 98, n247
šutātû pp88, 95
tarbāṣu = tùr, p89, n225
te – see *ṭeḫû*
tùr – see *tarbāṣu*
ṭeḫû / *ṭaḫû* = te, pp59, 91, 99, 256, nn237, 375
ṭuppu n53
ṭupšarrūtu pp48, 51
ub.da.límmu.ba – see *kibrāt arbā'i*
ud.da.zal(.lá)-*e* (= *uddazallê*) p121, n306
udu idim – see *bibbu*
u'iltu p12
^{lú}*ummâni dannuti* p48, n104
ummânu (= um.me.a) pp20, 33–6, 40, 43, 45–6, 48–9, 157, 256, nn35, 103, 136, 151
uṣurtu – see g̃iš.ḫur
UŠ pp4–5, 98, 101–2, 117, 129, 177, 181–4, 188, 190–2, 199, 201–2, 205, 211, 216–7, 239, 259–60, nn7, 238, 256, 289, 305, 415, 429, 439, 459
ūṭu / *rūṭu* pp93, 191
uzuzzu / *izuzzu* = gub, pp58, 69, 91, 277, nn146, 223, 237
zamāru p49

Subject Index

Aaboe pp5, 163, 171, 186–7, 230, nn216, 265, 488
Abstraction / Abstract pp13, 139, 213–4, 230, 236, 242, nn24, 27
Abu-Ṡalabikh p247
Accuracy pp94–103, 114–7, 234–4, 154–6, 156, 159, 161, 168, 171–3, 184, 187–9, 190–3, 210–8, 231, 240 nn231, 278, 295, 304, 394, 457
Achaemenid p162
Acronychal/Evening Rising pp4–5, 82, 95–6, 101, 117, 148, 164, 179, 186–7, 210, 266, nn236, 390–1
Active use of EAE pp78–81, 158–60
Adad pp62, 67, 91, 124, 147, 250, 252, 255, 283
Adad-apla-iddina pp252, 254
Adad-šumu-uṣur pp37–8, 41, 149, 201, 271, 273, n103
Administrative Records p13, n36
Agenda pp75, 80–1, 126, 159, n541
Akkad pp7, 22, 28, 41, 60, 66, 78–9, 106, 134–5, 137–41, 144, 146–7, 150–2, 157, 205, 212, 247, 257, n165

Akkullanu pp39, 41, 48, 50, 79, 102, 149, 276–8, n310
Alalakh p252
Almagest nn22, 528–9
Almanac pp163–70, 172, 176–7, 210–1, 220, 223, 229, 263–4, n392
  Normal Star Almanac pp165, 218, 263
Altitude pp102, 179, 192, 277
Ammiṣaduqa pp114–5, 249
An (=*Anu*) pp70, 235
Anomaly Producing pp153, 156, 209, 212–3, 215
Antiquarianism pp108, 162, n29
Anu p70
Apodosis/Apodoses (defined) p108
  Variant Apodoses pp77–8
Apotropaic Ritual (namburbi) pp6, 45–7, 121–2, 151, 168, 178, 231, 241
Apprentice pp48–9, 51, 223, 225, n151
Appropriate (model) pp182, 184, 238, n415
Apsidal Lines n426
Arabs pp23, 44, 67, n101
Aramaeans pp31, 33, 44, nn98, 537
Aramaic pp30–31
Arba'il p39, n12
Archive pp17–22, 28–9, 35, 42–3, 75, 219–20, 222, 226, 258, nn53, 64, 377
Arguments (of a function) p182
Aristotle pp189, 238–9, n531
Artefact of the PCP Paradigm pp168–9, 210, 217, 225, 232
Artefact Paradigm p126
Associations pp54, 63–81, nn190, 201, 203
  Basic Associations pp63, 65, 67, 70–2, 213
  Learned Associations pp64–72, 75–81, 132, n201
  Observational Asociation pp64, 66–70, 72, n201
  Symbolic Association pp64, 69–70, 72
  Theological Association pp63–6, 68, 70, 72
Assur pp20–23, 34–42, 50, 69, 221, 252, 254–6, nn12, 42, 110
Assurbanipal's Library pp18, 23, 30, 227, n30
Assyria (defined) p7, n12
Assyrianisms p30, nn85–6
Astral Magic p264
Astrolabe pp37, 65, 107, 115–7, 121, 123–5, 130, 150, 157, 2221, 246, 251–4, 257–8, nn28, 179, 292–3, 318
Astrologer(s) p33, n536
Astrologia p7, n15
Astrology (Babylonian) p264
Astrology-Astronomy (defined) p7
Astronomia n15, n294
Astronomical Archive p22, 42, 219–20, n377
Astronomical Bureau pp224–5
Astronomical Dating pp23–29, App.2, nn28, 65
Astronomical Terms (modern, described) pp81–5
Asû (= physician) pp33, 35, 47
Ašaredu pp39–40, 50, 122, n154
Āšipu (= exorcist/healer-seer) pp33–5
Aššur (god) pp42, 70, 242, nn30, 122, 126
Atmospheric Effects pp96, 134, 143, 252, 281, nn228–9
Athtar p67
Auxiliary Hypotheses pp189, 229–30
Auxiliary Texts pp165, 262, n382

Babylon pp3, 5, 7, 9, 16–17, 33–4, 40, 41–2, 50, 52, 57, 81, 97, 129, 134, 146, 162, 219, 221–3, 225, 246, 254, 256, nn5, 21, 135, 326, 343, 377, 405
Babylonia (defined) p7
Babyloniaca n23
Babylonianisms p30
Babylonian Question p8, n21
Balasî p47, 49, 96, 150–1, 157, 196, 240, 269, nn103, 155, 368
Bārû (= haruspex/extispicer/diviner) pp13, 21, 35, 40, 47, nn72, 96
Base 60 / Sexagesimal - pp114, 128, 236
Beards p67, n151
Bearing Radiance = Being Bright pp67, 79, 90, 100, 256, nn184, 223, 229
Bel-ušezib pp27, 41, 40, 43, 157, 206, 265–6, nn118–9, 154
Benefic pp137, 143–4, 150–1, 155, 212
Berossus pp9, 46, 190, n23
Bīt-Ibâ pp18, 22
Bīt-Rimki n48
Boghazköy pp248, 250–1, 254, 257
Bottéro pp125, 228, 234, 246, 252, nn27, 69, 93, 269, 275, 343, 487
Brack-Bernsen pp63, 173, 185, 188, 262, nn8, 215, 254, 265, 407, 430, 433
Bricoleur p232, n538
Britton pp171, 185–6, 225, 259, 261–3, nn265, 377, 405, 411, 427, 430, 441, 465, 488

Burnaburiaš pp252, 254
By-Product of the PCP Paradigm pp169, 172–3

Calendar
Ideal pp107, 249, 254, n290
Lunar pp83, 115, 118–20, 175, 178, 195, 246, 248
Luni-Solar pp169, 174, 176, 188, 195–6, 216, n304
Metonic pp85, 175–6, 211, n411
Nippur/Sumerian p247
Seasonal/Solar p120
Seleucid p4
Semitic p247
Sidereal p115
Well-Regulated pp83, 168, 170, 176–7, 189, 192, 195–7, 210, 216, 218, nn153, 156
Canon/Canonical/Canonisation pp11, 16, 23, 31, 75, 78, 106, 156–60, 204, 224, 248–9, 252, 254–5, 258, nn30, 53
Categorisation pp106, 111–2, 124, 126, 139, 153–6, 212–3, 233
Causality pp15, 79, 109–12, 233–4, nn277–8
Chaldaeans pp1, 34, 41, 44, nn23, 118, 377, 537
Characteristic Periods pp4, 83, 173, 176, 193–5, 207, 216, 218–9, 222, 225, 227, 229, 232, nn445, 469
China/Chinese pp10, 224–5, nn393, 480
Chronographic Texts pp13, 33, n35
Circle of the Heavens (*kippat šamê*) p254
Cicero pp110–1, 264
Citizens (*mārē āli*) pp44, 219, n146
City Names pp22, 39–42
Code pp106, 111–3, 126, 134, 136–7, 139–53, 157, 159, 172, 187, 191, 209, 213–5, 229–30, 242, 280–284, nn170, 332, 348, 361, 365
Cognitive/Cognition pp3, 10, n343
Column/Function Ø pp185–7, 200, 205, nn254, 432, 465
Column F n432
Colophons pp13, 20, 22, 37, 48, 128, 193, 224, 26–7, 253, nn38, 53, 84, 403, 481–2
Colours pp143, 153, n363
Comet pp59, 61, 212, 251, n171
Commentary Texts pp16, 18, 29, 124, 151, 200, 202, 250, 253, 255, 258, 283, nn51, 190
Common People n128
Conjunction Terms pp91–2, 96, 99, n237
Constellations n312
Continuous Record pp97, 102, 164–5, 167, 188, 193, 210, 216–8, 222, 225, 229, 232, 237
Core Hypotheses pp159, 161, 163, 170, 172, 187, 189, 207, 211–7, 225–6, 229–31, 234, 236, nn396, 497
Correspondence pp12, 17, nn31, 64
Corruption (of texts) pp54, 116, nn161, 295, 375
Cosmogony pp12, 153, n14
Cosmology pp12, 233, nn14, 525
Cubit (= kùš) pp5, 92, 96, 98, 120, 190–1, 210, 217, n415
Culmination – see *ziqpu*
Cutha pp22, 40, 45
Cycles of Lunar Latitude pp181, 195
Cycles of Lunar Velocity pp179–80, 183, 185, 195

Daily/Day-To-Day Motion pp164, 170, 184, 211, 218, 229, n400
Daniel pp33–4, 45, 162, n136
Daylight Scheme pp2, 128–9, 249
Declination p259
Decoding (as writing) pp138–9, 242
De Divinatione pp110, 264
Deductive Divination pp234, 246, n511
Der pp22, 42, 146
Derrida p234, nn93, 286
Design/Designed pp112, 123–6, 153, 213–4, 235–7, 239, 242, 246, 253–5, 260, nn282, 315, 518, 522, 525-7
Dialects pp29–32, nn30, 83
Diaries (described) pp97–102, 263 - see the text index above for specific Diaries
Directly Ominous pp101, 168, 173, 187–8, 190, 209, 210, 215
Distancing of the Gods p234
Divine Powers (= me) pp235–7, 239, 255, n525
Dodekatamoria pp22, 259, 264
Dossier p29
Dūr-Šarkēn pp21–2, 36

Ea pp46, 65, nn182, 511
Ease of Use pp188, 231
Ebla pp246–7
Eclipse – Modern Description pp84–5, 180–1
Eclipse - Ominous Aspects pp89–90, 95, 111, 132, 140–2, 145–6, 168, 279–80, nn253, 273, 373
Eclipse Periods pp148, 173, 199, n354, see also Saros

Eclipse Possibilities pp4, 179–80, 182, 261–2, n425, see also Saros Canon
Eclipse Prediction pp4, 6, 84, 164–5, 171, 178, 195, 200–207
Eclipse Records pp4, 7, 9, 25, 29, 81, 97–9, 100, 102, 163, 165–6, 168, 188, 190, 201–2, 205–7, 216–7, 222, 229, 259, 261 – see also *LBAT* 1413f, in the text index above
Eclipse – Recorded Aspects in NMAATs and MAATs pp98–101, 164–170
Eclipse Ritual p162, nn7, 47–8, 376, 380
Eclipse - Solar pp90, 95, 164–5, 202–3, 248, 252, 255–6, 261, 288, nn253, 385
Ecliptic pp59–61, 66, 82–5, 96, 100, 116–7, 134, 136, 154–5, 167–8, 173, 178–82, 199, 211, 250, 260, nn185, 189, 215, 250
Ecstatics p13
Education of Scholars pp48–9
Egibi pp39–40, 222
Egypt pp33–4, 107, 220, 269, n266
Einstein/Einsteinian n396
Ekur-zākir p222
Elaborations pp16, 78, 114, 120, 126–7, 139, 155–7, 159, 188, 213, 215, 237, 239, 246, 285, nn203, 323
Elamite p252
Ellil n129
Elongation pp83–4, n434
Emar pp251, 254
Eme-sal p161, n376
Empedocles n531
Empiricist Models pp109–112
Encoding p111–2, 132–3, 146, 152, 155–6, 166, 180, 209, 212–3, 215, 279, 282–3
Enki (=*Ea*) pp234–6, n521
Enlil (=*Ellil*) pp235, 247
Entourage – "standing before the king" pp36, 41, 45–7, 49–51, 215, 220, 223, 226, 240, nn146, 151
Epact p261, n306
Ephemerides (defined) pp163–4, n382
  Lunar p164
  Planetary p164
Epoch nn19, 22
Equinox pp85, 102, 113–5, 120–1, 164–5, 169, 176, 184, 193, 197, 257, 259, nn65, 451
  Autumnal Equinox p124
  Vernal Equinox pp115, 158, 195, 197, 249, 257, nn65, 304, 414
Eridu pp22, 38, 146
Erra pp13, 23, 56, 256, n144
Error pp117, 171, nn295, 423
Error in Periods pp175–80, 185, 210, 216, 218, 227, nn408, 412–5, 446
Ešnunna pp247, 282
Eššešu = éš.éš p113, 248
Eudoxus p264
Eunuch pp43, 45, n151
Europe/European pp10, 215, n506
Event Number p182
Exeligmos pp85, 261
Exorcist/Exorcism pp20, 22, 28, 33–6, 48–50
Exorcist's Manual – see *KAR* 44
Explanatory Works pp16–17, 66, 107, 127, 258–60, nn50, 497
Extispicer pp33, 35, 46, 256, nn112, 368
Extispicy pp112, 247, 250, 264, nn30, 105, 285
Extispicy Reports p13, n37

Falsifiability pp221, 231, n502
Family Relationships pp36–9
Fara p247
Finger (= šu.si) pp5, 90–2, 96–9, 171, 175, 185, 188, 191, 210, 217, nn247, 415, 429
Folklore pp111, 213, 246
Foucault n161
Functionalist pp9, 155

Gaining Radiance / Scintillating (= *ittananbiṭu*) pp93, 100, n229
Garagaṣamhitā p255
Gasur p247
Gatamdug p247
Gematria pp127–8, 130, 138, nn205, 325
Gnomon pp126, 259, n451
Godelier pp232–3
Goody pp9, 13, 76, nn24, 195, 203, 343, 509
Grammatical Texts pp230–1, n500
Graphic Play pp70, 76–7, 128, 132, 138, nn93, 203
Gravity p234, n277
Greece / Greek pp3, 10, 29, 44, 107, 124, 162, 189, 219–20, 224, 226–7, 238–40, 258, 264, nn27, 193, 205, 264, 266, 379, 480, 527–8, 533
Guild pp218, 222–7, 231, 240–1, n481
Guti pp67, 141–2, 151, 246, 256
GYTs (defined) p5, 163, 165–8, 176

Haireseis pp224–6
Hammurapi p7
Hand Lifting Prayers (= šu.íl.la) p273, n47
Haruspex pp21, 33, 37, nn117, 154
Hebrew p107, nn204–5, 264
Heliacal Phenomena (defined) pp81–5
Heliacal Phenomena (cuneiform terms) pp86–90
Hemerologies p34, 121, nn272, 368
Hermeneutic pp45, 109, 138–9, 159, 231, n205
Hesiod p246
Hierarchy pp35–6
Historical Omens pp106, 110, 133, 246, nn35, 281, 337
Historiette n337
Historiography / Historicising pp46, 106, nn19, 35, 176
Hittite pp250–1, 254
Homophony pp70, 125, 128
Horoscopes pp161–2, 169–70, 177, 211, 215, 217, 259, 263–4, n418
House of Succession (= *bīt redûti*) p43, n54
Hunger (on associations and omens) p109, nn160–1
Hurrian p251
Huzirīna = Sultantepe pp21–2, n53
Hymns p15, n47
Hypothetico-Deductive Method pp9, 114, 137, 230–1, n502
Hypsomata p215, n467

Iatromathematics p264
Ideal Periods (defined) pp155, 212
Ideal Period Schemes (defined) pp106, 113–22, 125–6, 146–51, 155–6, 209, 213–4
  Ideal Acronychal Rising Date Scheme pp116–7
  Ideal Astrolabe Scheme pp115–6, 130, 150, 208, 252
  Ideal Calendar pp107, 249
  Ideal Intercalation Scheme pp119, 125, 130, 150, 195, 209, 212
  Ideal Lunar Visibility/Invisibility Scheme pp114, 120, 125, 130, 148–9, 166, 249, 260, n323
  Ideal Month pp113, 120, 125–7, 129–30, 147, 155, 166, 209, 217, 236, 260, n220
  Ideal Planetary Visibility/Invisibility Scheme pp118, 155
  Ideal Seasonal Hour pp115, 130, n457
  Ideal Solar Movement pp117–8, 130
  Ideal Venus Scheme pp115, 16, 249
Ideal Year (defined) pp113, 129, 184, 209, 249–50, 252, n290
Ideograms pp7, 138–9, n343
Ill Health p28
Impossible Protases pp109, 136–7, 203, 214, 230, nn273, 338, 375
Inana (= *Ištar*) pp67, 246, n521
Incantations p15, n47
Incommensurability p228, n26
India(n) pp3, 10, 107–8, 162, 220, 255, nn264, 266, 268
Indirectly Ominous pp173, 186–8, 209–10, 216–7
Infancy pp6, 16, 190, 192, 200, 205, 218–9, 221, 230
Intellectuals pp13, 220
Intellectual Interest pp161, 220, 243, nn381, 402
Intercalation pp85, 118, 12–2, 151, 154, 176, 196, 247, 261, nn287, 367 304, 411, 414, 425
The "Interval" p8
Intuitive Divination p234, n511
Invented Omens (discussed) pp130–2
Irreducible Numbers p129
Isidore of Seville n16
Islam(ic) p10
Issār pp51, 55, 241
Issar-šumu-ereš pp30, 38, 45, 50, 78, 93, 133, 240, 242, nn46, 86, 103, 154
Ištar pp51, 67, nn158, 164, 195, 205
Išum pp13, 256

Jia pp224–5
Jing p224

Kalḫu = Nimrūd pp21–3, 27, 31, 36–8, 41–3, 255, nn12, 21, 56, 71–2, 97, 103, 151
Kalû / Nargallu (= lamentation-chanter) p20, 33–5, 42, 161, nn376, 380
Kassite pp44, 65, 252–4
Khorsabad – see Dūr-Šarkēn
Kirk nn504, 531
Kiṣir-Aššur pp23, 36, 38, 48
Kiš pp22, 146
Kudurru pp35, 38, 40, 274, 277
Kudurrus pp124, 253, n284
Kudur-Mabug p247

Kuhn pp9, 126, 159, 161, 229, nn4, 10, 475, 502, 505
Kuyunjik p18, n60

Lamentation-chanter pp33, 102, n376
Lakatos pp157, 229, nn466, 475, 497, 505
Larsa pp22, 146
Larsen pp111–2, 233, nn24, 27, 275, 286, 507
Latitude pp82–4, 95–6, 171, 179–81, 185, 188, 195, 199, 202, 212, 214, 228, 237, nn403, 418, 420
Leach p139, nn329, 333, 344
Learned Techniques pp64, 75–81, 132, n445 – see also Elaborations, *arû*, *ṣâtu*
Legitimation p44, 51, 220, 222, 225–6, 241
Letter (defined) pp6, 7, 12
Lévi-Strauss pp130, 139, 232–3, nn329–331, 333, 344, 509–510
Lexical Material pp48, 69, 122, 157, 231, 233, nn203, 363
Library (defined) pp17–23, nn30, 53
Library Records pp18, 22, 30, 35, 108, nn58–9
Listenwissenschaft pp76, 132, 231, n203
Literary Texts/Literature pp13, 17, 22, 23, 42, 64, 75, 107, 110–2, 124–5, 239, 246–8, 250, 253–4, 256–7, nn128–9, 176, 466
Literate (knowledge, achievement) pp3, 30, 76, 109, 111–2, 125–6, 138–9, 213–5, 231, n203
Liver Models/Omens pp111, 246, n286
Lloyd pp9,37, 44, 224, 226, 240, 246, nn17, 27, 480, 489, 493, 504, 527–30, 533
Logocentrism n286
Logographic n30
Longest:Shortest Night Ratio pp113, 115, 120, 189, 239, 249, 261
Longitude pp85, 164–3, 168, 175–7, 182–4, 186, 196, 200, 211–2, 228, 262, nn407, 420
Long Periods pp178–9, 181–5, 193–4, 210, 216
Loyalty Oaths pp13–4, 23
Lunar Four pp165, 185–6
Lunar Six pp101, 164–6, 173–4 176, 180, 185–6, 188, 191–2, 199, 210–11, 216, 237, nn254, 465, 501
  NA* pp101, 166, 174, 192, 200, nn386, 407
  šú pp5, 165, 173–4, 176, 181, 185, 199, 261
  me pp5, 165, 173–4, 176, 181, 185, 199, 201
  NA pp5, 165–6, 173–4, 176, 181, 185, 199
  gi_6 pp5, 101, 165–6, 173–4, 176, 181, 185, 199, n387
  kur pp101, 168, 174, 199–200, nn407, 459

MAATs (defined) pp3–5, 163–4, 170–3, nn382, 488
Magicians pp33–4
Malefic p143
Maništušu p246
Marduk pp56–7, 63–6, 125, 235, 253, nn129, 182
Marduk Ordeal p16, nn49, 159
Marduk-šapik-zeri pp35, 49, 225, n154
Marduk Temple pp1, 5, 16, 22, 42, 97, 102, 223, nn119, 135, 377
Mari pp221, 247–8, n286
Mar-Issar pp22, 41, 58, 276, n48
Masterman p126, n4
Mathematical pp4, 10, 118, 120, 124, 128–9, 160–1, 163, 165–6, 182, 184, 186, 188, 209, 211, 214, 218, 230–1, 233, 238, 250–1, 253, n509
Mathematical Elegance p170
Mathematical Table – see *arû*
Mathematician p251
Mathematisation pp188–9, 218, 230, 239
Mean Interval pp181–2, 186–8, 198, 210–1, 216
Mechanistic p10
Mesopotamia (defined) p7
Metaphor pp70, 112–3, 131, nn123, 181, 195, 286
Metaphoric Relationships pp130–2, 137, 142, 209, 213, 279–84, n333, 348, 541
Metaphysical Paradigm p126
Metathesis pp128–9, 155, n326
Meteor pp2, 59, 61, 65, 134, 158, 212, 251, n171
Metonic Cycle pp85, 176, 179, 181, 184, 193, 195–6, n219, see Calendar (Metonic)
Mina (= mana / *manû*) pp45, 117
Minds / Mind-Set / Savage Mind pp10, 46, 52, 54, 139, 172, 218, 228, nn24, 161, 203, 286, 506, 509
Month = Lunation = Synodic Month (defined) p84
  Anomalistic p85
  Draconitic/Nodal p84
  Ideal – see Ideal Month
  Sidereal p84

Morning/Cosmical Setting pp2, 82, 95–6, 101, 166, n236
Mukallimtu pp255, 257–8, nn51, 85

Nabû pp56, 67, 77, n41, 129, 190
Nabû'a pp50, 197, 268
Nabû-aḫḫe-eriba pp67, 240, 269, 275
Nabû-šallim-šunu p20, 36, 38, n110
Nabû-zeru-iddina pp38, 42
Nabû-zuqup-kēna pp21, 36–8, nn56, 84, 103, 323
Nanna pp57, 72, 247
Narām-Suen p246
Naturalis Historia nn5, 9
Nature of the Universe pp173, 189, 237–9, n526
Nēbiru pp54, 58, 60, 64, 68, 73, 87, 116, 150, 253–4, nn168, 179, 202
Neugebauer pp120, 143, 163, 171, nn5, 10, 153, 217, 264–6, 307, 381, 402, 410, 441, 481, 488
Newton/Newtonian pp126, 189, n396
Nimrūd – see Kalḫu
Ninda = *nindānu* p117
Nineveh pp1, 17–20, n12
Ninĝirsu pp57, 123, 247
Ninĝišzida p247
Nippur pp22, 40, 129, 146, 158, 247, 249–50, 252–3, nn5, 47, 289
Nisaba pp123, 247
NMAATs (defined) pp4–5, 97–103, 164–73, n488
Nodes p84, n420
Normal Science pp54, 69, 78–9, 100, 126, 159, 160, 189, 214–5, 221, 226, 229–30, 240, n502
Normal Star Almanacs – see Almanac
Normal Stars pp99–100, 155, 165, 168–9, 176–7, 179, 188, 191, 210, 217–8, nn250, 251, 420
North Palace pp17–9, 28, nn54, 75
Notarikon p138, nn204–5
Number Play pp106, 128–30, 137–8, 155, 209, 213, 227, 230, 237
Nuzi p252
Nychthemeron pp250, 259

Observation (Checking of) pp171–2, nn394, 427
Observation (Role of) pp3, 15, 28–9, 49, 109–12 n332
Observations of Times pp186–7
Occultation pp88, 88, 136, 151.155, 283
Oil Omina pp108, 248, n270
Omens (defined) pp108–9
Onomancy n205
Oppenheim pp33–4, 50, 120–2, 155, 238, nn30, 42, 105, 115, 120, 137, 140, 153, 155, 202, 263, 269, 311, 352, 370, 466, 473–4, 508
Opposition
  Lunar pp88, 95, 119, 148, 158, 166–7, 173–4, 191, 198–200, 202, 204, 210, 235–6, 260, 282, nn215, 236, 390, 501
  Planetary pp82, 88, 95–6, 99–100, 164–5, 167, 182, 210, 217, n391
Orality/Oral Wisdom pp76, 112, 137–9, 152, 157–8, 229–30, 234, 237, 258, nn286, 497
Origin pp78, 112–3, 139, 152–3, 220, 246 nn144, 286
Original State of the Universe pp125, 235–9

Pāli Dīghanikāya p255
Paradigm (defined) pp3, 126, 229–32, nn4, 10
Parpola (on Scholars and Wisdom) pp33–6, 45, 49–50, nn141, 152
pars familiaris pp142, 152, 212, 282, n348
pars hostilis pp142, 152, 212, 282, n348
Persia(n) pp3, 22, 31, 162, 220, 259
Phaedrus n93, n286
Pharmakon n93
Physician pp28, 33–5, 38, 50, n146
Physics/Physicists pp189, 234, 238, nn396, 528
Physis pp185, 235
Pictogram p61
Pinches-type Astrolabe pp253, 257
Pingree (on text corruption and science) pp10, 116–7, 228, nn295–6
Planet (defined) p7
Planet Names pp55f
Planet Order p143, n347
Planet(ary) Records pp81, 98, 164–5, 187–8, 207, 216–7, 220, 262 – see also Mars, Mercury, Saturn Records
Plato pp234, 238, nn93, 286, 527
Play on Sounds p138
Play on Words – see Word Play
Pleiaden-Schaltregel p96, 118–9, 125, 129–30, 150, nn302, 304
Pliny p5, nn5, 9
Popper p110, nn276, 502, 504
Positivistic Stance p64, nn176, 524

Post hoc ergo propter hoc p109
Prayers pp15, 20, 204, nn47–8, 149, 365 – see also Hand Lifting Prayers and *The Prayer to the Gods of the Night*
Precession p246, n290
Predictive Astronomy pp113, 155–6, 180, 193–4, 207, 216, 221, 230, 242, n82 – see also *Astronomia*
Premises pp16–17, 81, 105–6, 112, 115, 126, 156–62, 213–5, 230, nn502, 527 – see also Core Hypotheses
Priests nn64, 151, 376
Primitive Astronomy pp106, 118, 213, 229, nn267, 488
Procedure Texts pp163, 222, 263, nn382, 432
Prophets/Prophecy pp13–15, 33, nn43, 46
Propitious Timing p123, n314
Prosperity p30
Protasis/Protases (defined) p108
Protective Belt pp157, 159, 215, 225, 229
Provoked/Unprovoked Divination n105
Ptolemy pp9, 238–9, nn17, 22, 528

Qatna p252
Quadrants p141

Rational pp63, 157, 227, nn161, 168, 229, 487
Reflexive/Reflexivity pp10, 46
Regnal Year n18
Relative Orientations pp99, 140, 154, 282, n249
Relativity/Relativistic pp10, 228, nn26, 490
Religion/Religious pp23, 42–3, 47, 51–2, 64, 107, nn123, 129, 159, 176, 262
Remuneration p45
Report (defined) pp6, 12
Retrocalculation pp164, 186, 202, nn384, 391, 401
Retrograding pp82–3, 87–88, 94–5, 100, 197, 269, 273
Retrophony p138
Revolution/Revolutionary pp2, 6, 9, 10, 31, 126, 207, 209–10, 216–9, 221, 223, 226, 230, 232, 234, 240, 242, nn10, 502, 505, 537
Right Ascension pp98, 259–60, n438
Rituals pp15, 20, 34, 46, 55, 271, 273–4, nn48, 96 – see also Apotropaic and Eclipse Ritual
Rochberg pp124, 139, 141, 148, 163, 200, 248, 252–5, 258–60, 262, 264, 279–80, nn26, 30, 123, 149, 161,
212, 273, 340, 354, 373, 503
Roi-Soleil p72
Rôle of King pp42–7
Rome/Roman p3, n130
Round Numbers pp113–4, 118, 125, 129, 154, 156, 183, 185, 188, 212, 236, nn304, 415
Royal Death pp106, 132, 145–6, 152, 285
Royal Inscriptions pp14, 23, 30, 33, 42, 51, nn40–2, 103
Rules pp78, 105–6, 111–3, 126–39, 144, 152–3, 156–9, 172, 209, 213–4, 229–30, 280–3, nn343, 356

Sachs pp163, 169, 177, 262–4, nn5, 265, 397
Sages = I(*apkallū*) pp46, 49, 242, nn151, 343
Sanskrit pp181, 220
Saros Period (= 223 months) pp5, 85, 199–202, 204–7, 210, 261, n465 - see also *Saros Canon*
Saving the Phenomena pp219, 238–9, n530
Schema/Schemata (defined) p139
Scholars (defined) pp7, 33–6
School pp20–1, 41, 47, 52, 80, 159, 214, 218, 224, 226, 255, nn5, 53, 210
Science (discussed) pp5, 7, 9–10, 110–1, 126, 157, 163, 218–9, 224, 226–34, 237–9, 242–3, nn4, 26, 30, 277–8, 488–9, 490, 496, 499, 502–3, 506, 509, 533
Pre-science nn488, 499
Proto-science p10, nn26, 499
Science du concret pp232–3, nn506, 509
Scintillate pp93, 100, n229
Scribe (of EAE) (defined) pp20, 23, 31, 33–6, 38, 48, 222, nn56, 103, 108
Secularisation pp23, 226
Seleucid Era pp4–5, 175, 183
750 BC (discussed) pp8–9, nn19, 22–3
Shadow Clock pp120, 129
Shooting stars pp20, 22
Simple Code pp112, 151–3, 156, 159, 209, 213, 215, 230, 282
Sîn pp57, 65, 72, nn149, 164
Sîn-lēqe-unnini p222
Sippar pp3, 22, 40, 146, 249, 264, nn5, 75
Sirius Phenomena pp165, 169, 176, 193, 216, 261, nn397
612 BC (discussed) p8
Sociological Paradigm p126
Solstice pp102, 14, 117, 120, 154, 164–5, 169, 176, 184, 193, nn299, 450
South West Palace pp17–19, 28, nn55, 64

Standard Babylonian pp29, 64
Star Path pp94, 116, 118, 123, 145, 154, 234, 251, 253, nn233, 291
Star Tablet (= dub.mul.an) pp123, 247, n316
Stars of Elam/Akkad/ Amurru p257
Stationary pp4, 69, 82–3, 87–8, 94, 99–100, 170, 180, 191, 269, 277, n223
Step Functions pp4, 128–9, 182–4, 188–9
Step-Wise Linear pp211, 217
Stock Apodoses pp131–2, 285, n334
Stoic Divination pp110–1
Stream of Tradition p75, nn30, 202
Strings (= gu) p260 – see also BM 78161
Structuralist pp106, 130–2, 139, nn329–30, 333, 344
Student Terms p48
Subartu pp6, 56, 70, 78–9, 137, 140–1, 144, 148, 151, 157, 221, 275, 282, n165
Suen (=*Sîn*) pp72, 235, 255
Sultantepe – see Huzirīna
Sun-Distance Principle (= *Sonnenabstandprinzip*) nn256, 434
Sumer p7, n165
Sumerian pp2, 29, 31, 152, 246–8, n13
Sumerograms pp31, 125, 128, 247, n173
Susa p252
Swerdlow pp163, 186–7, nn82, 256, 381, 391, 434
Syncretism pp65, 69, n182
Synodic Arc p187, n256
Synodic Time p187
Syntagmatic Relationships pp130–2, 134–6, 138, 151, 209, 213–4, 230, 280–3, nn333, 343, 541
System A pp171, 183–6, 225, 262–3, nn254, 405
System B pp182, 184, 225, 263, n405
Syzygy (defined) p84

Šamaš pp35, 56, 69–70, 72, 124, 148, 250, n149
Šangû pp42, 242, nn30, 380
Šulpae pp58, 65, 68, 213, n136

Tarbīṣu p22
Technology of Communication pp9, 222, 242, n24
Technology of Listing pp76, 132, 213, 230, n203
Teleology p10
Tetrabiblos n17
Textu(al) Play pp106, 127, 130, 137–8, 159, 209, 213–4, 230, 279–80, 283
Theology pp45, 65–6, 68, 70, 72, 76–7, 111, 152, n30
Theon of Smyrna n527
Theoretical Astronomy p5
Tithis pp4, 172, 181, 183–4, 188, 193, 198, 211, 216, 225, 230, 239, 261, n422
Transmissions of Texts pp106–7, 159, 215, 254–5, nn266, 268, 468
Treaties pp13–14, 21, 23, n41
Trine p264

Ṭupšarru (=scribe) of EAE pp20, 33, 36, n103

Ugarit p252, n92
Ur pp22, 40, 133, 146, 203, 221, 246, 248
Urad-Ea pp34, 37–8, 42, 273
Urad-Gula pp37–8, 40, 45, 48, n75
Uruk pp3, 22, 40–2, 146, 162, 219, 222–3, 226, 255, 264, nn5, 43, 53, 191, 255, 380, 392, 405
UŠ – see index above
Utu (=*Šamaš*) pp56, 69, 72, 77

Variable Reducing pp153–6, 209, 212–3, 215, 232–3
Verifiability n394

Walker pp9, 190, 195, 248–51, 257, 259–63, nn18, 265
Watch of the King pp46–8, 50, n148
Watches pp3, 89, 98–9, 106, 115, 121, 133, 136–7, 140–2, 145–6, 148, 151–2, 154, 168, 190, 203–4, 206, 209-10, 249–50, 254, nn244, 273, 338, 361, 365, 373
Wax Writing Boards p30
Weather pp29, 82–4, 91, 96–7, 99–100, 102, 147, 156, 165, 168, 192, 216, 234, 241, nn82, 257
Western/The West pp106, 153, 207, 215, 243
Winds pp141, 145, 151, 279–81, n373
"Wisdom" pp45–6, 51, 128, 215, 221–3, 227, 237, n295
Word Play pp16, 69, 105–6, 125, 127, 132, 138, 230–1, 280–1, n360 – see Syntagmatic Relationships
Works and Days p246
Writing on the Sky = *šiṭir šamê/burūmê* pp112, 123, 138–9, 231, 257, nn283–4, 315

Year
Administrative pp113, 248, nn287, 399
Ideal – see Ideal Year
Lunar/Calendar/Cultic pp4, 83, 85, 117, 170, 176–8, 194, 196, 246–7, nn287, 413, 446
Seasonal/Solar/Equinoctial pp83, 85, 117, 120, 175–7, 193, 197, 261, nn287, 290, 299, 410
Sidereal/Stellar pp83, 85, 117, 119, 170, 174–9, 194, 196–7, nn290, 299, 410, 412–4, 446

Zigzag Functions pp4, 128, 182–5, 188, 198, n457
Zimrilim p248
Ziqpu (culminating) Stars (Texts) pp90, 98, 100, 107, 117, 124, 154, 190, 210, 254, 259–60, nn297, 305
Zodiac pp4, 83, 97–8, 155, 164–7, 169, 172–88, 199–200, 211, 215–6, 218, 220, 225, 230, 239, 243, 260, 262, 264, nn265, 415, 446
Zodiacal/Solar Anomaly pp85, 185–6, 188, 262, n421
Zodiacal Astrology pp161, 167, 173, 180, 187–8, 221, 264, n418
Zodiacal Constellations pp82, 99
Zodiacal Signs pp99, 165, 177, 180, 186–7, 215, 258, 264, nn388–9
Zones pp183, 185–7
Zwölfmaldrei pp251, 257–8